ENVIRONMENTAL EFFECTS OF NORTH SEA OIL AND GAS DEVELOPMENTS

ENVIRONMENTAL EFFECTS OF NORTH SEA OIL AND GAS DEVELOPMENTS

PROCEEDINGS OF
A ROYAL SOCIETY DISCUSSION MEETING
HELD ON 19 AND 20 FEBRUARY 1986

ORGANIZED BY J. I. G. CADOGAN, F.R.S.,
R. B. CLARK AND J. P. HARTLEY
AND EDITED BY J. P. HARTLEY AND R. B. CLARK

LONDON
THE ROYAL SOCIETY
1987

Printed in Great Britain for the Royal Society
by the
University Press, Cambridge

ISBN 0 85403 332 7

First published in *Philosophical Transactions of the Royal Society of London*,
series B, volume 319 (no. 1181), pages 459–677

Published by the Royal Society
6 Carlton House Terrace, London SW1Y 5AG

PREFACE

The North Sea oil and gas industry is massive, with a history of offshore exploration and production extending back some 25 years. Today there are over 70 oil- and gasfields in the North Sea, with many export pipelines and land-based terminals. Production from these fields is expected to continue well into the next century and is currently an important factor in the economies of Norway and the U.K. With hydrocarbon discoveries being made in several other areas of the European continental shelf, it was felt opportune to review the experience of 25 years of North Sea operations. This volume is the outcome of the Royal Society Discussion Meeting held on 19 and 20 February 1986, which was designed to fulfill such a review and to highlight remaining areas of uncertainty. The meeting was the third to be held under the auspices of the Royal Society that has considered marine pollution, although it differed from its predecessors in covering a single industry in a limited geographical area.

An additional stimulus was the first conference of European Ministers on the protection of the North Sea, held in Bremen in October 1984. This ministerial conference was preceded by a period of consultation with non-governmental organizations, and it became clear that the health of the North Sea was still the subject of much debate. A second ministerial conference will be hosted by the U.K. in November 1987 and it is to be hoped that this volume will provide a sound scientific base for some of their deliberations.

Since the early 1970s the oil industry has funded many environmental monitoring and research programmes into the effects of its North Sea operations. Much of this environmental work has been done by contractors and consultants. Regrettably, a fair proportion of the information has remained unpublished, either through confidentiality restrictions or simply through pressure of contract work. One achievement of this meeting has been to make some of this information more broadly available, so it can be used to build a more complete picture of effects.

Our understanding of oil pollution has been advanced recently through research using mesocosms such as those as Solbergstrand, Loch Ewe, Texel and Bremerhaven. Mesocosm studies can help to bridge the difficult gap between laboratory toxicity testing and what actually happens in the field.

It should be noted that oil industry practices and management are not static. Over the years many oil companies have developed formal policies on environmental protection and detailed systems for the assessment of environmental impacts, both of new developments and existing operations. One of the main areas of oil-related research over the past few years has been the fate and effects of oil-based drilling muds adhering to rock chippings that are discharged to sea. These oil-based drilling muds were introduced just as there were calls for attention to be diverted from offshore to coastal discharges.

This Discussion Meeting, like its predecessors on the effects of pollution, has been successful in providing a modern synthesis of our understanding of an important topic. The volume is concluded with an assessment of the problem and a discussion of those areas requiring further work.

July 1987 J. P. Hartley

CONTENTS

Phil. Trans. R. Soc. Lond. B **316**, 461–485 (1987)
Printed in Great Britain

The North Sea: an overview

By D. Eisma

Netherlands Institute for Sea Research, Postbox 59, 1790 *AB Den Burg, Texel, The Netherlands*

An overview is given of the natural systems of the North Sea: water-circulation, topography and geology of the sea floor, sediment transport, influx of trace constituents (nutrients, trace metals, organic compounds), biological systems and their interrelations. The effects of pollution and other human activities are discussed as well as the difficulties in assessing them where they are obscured by natural changes.

1. Introduction

The North Sea, with a surface area of approximately 575000 km^2 and a total volume of seawater of *ca.* 54000 km^3, has a long geological history dating from the Permian *ca.* 275 ma ago. At that time the principal basins forming the North Sea came into existence at the same time as the original continent combining present-day Europe, Greenland and North America started to break up (Pegrum *et al.* 1975). The subsequent geological history has shaped the present North Sea and also, during the last stages, has determined its bottom sediment characteristics to a large extent. During most of this history the North Sea was land or a shallow sea with the sea floor slowly subsiding. At present the sedimentary infill reaches a maximum thickness of *ca.* 7 km. The last time the North Sea fell dry was during the Pleistocene period, which altogether lasted approximately two million years (Oele *et al.* 1979). Cold periods with a sea level approximately 100 m below present level (at least during the last three periods) alternated with warmer periods like the present one with a higher sea level. During the cold periods the North Sea floor was partly covered by ice. After the last glacial period (the Weichsel or Würm period) ended (roughly 11000 years ago), the North Sea was flooded again, reaching the present level *ca.* 6000 years ago. Since then the sea level has continued to rise relative to the land in the southern parts of the North Sea because of subsidence of the sea floor, relative sea level now being 5–6 m higher than 6000 years ago. In Scandinavia and Scotland, however, relative sea level began to fall because of uplift of the land after the heavy ice cover had disappeared. Here, the old coastline is found inland at a maximum height of *ca.* 220 m above present level. The hinge lines between subsidence and uplift are located in northern Denmark and southern Scotland. During the period the North Sea had approximately its present extension, i.e. during most of the past 6000 years, only minor changes occurred; at the beginning the climate was slightly warmer than now, in the shallow areas the tidal currents were probably stronger at first, but the main changes occurred along the eastern side of the southern North Sea, where coastal land was built up (tidal flats, beachridges, beaches, dunes) which is still going on.

The present North Sea shows an intricate relation between physical conditions (waves, tides, currents, seafloor topography), water chemistry, sediments (in suspension and on the bottom), living organisms and human activities. It is the object of this paper to point out some of these relations, how they function and what we know about their historical background. This is done, however, in the realization that many of these relations are not clear and that relations may exist that we are as yet unaware of. As to the geographical limits of the North Sea I follow here the conventional ones, i.e. the Straits of Dover in the south, the gaps between Scotland, Orkney and Shetland and the 62° N parallel in the north, but including the Skagerrak, which forms an essential part of the natural system of the North Sea. The limit there is a line from Cape Skagen due east to the Swedish coast, separating the Skagerrak from the Kattegat (figure 1).

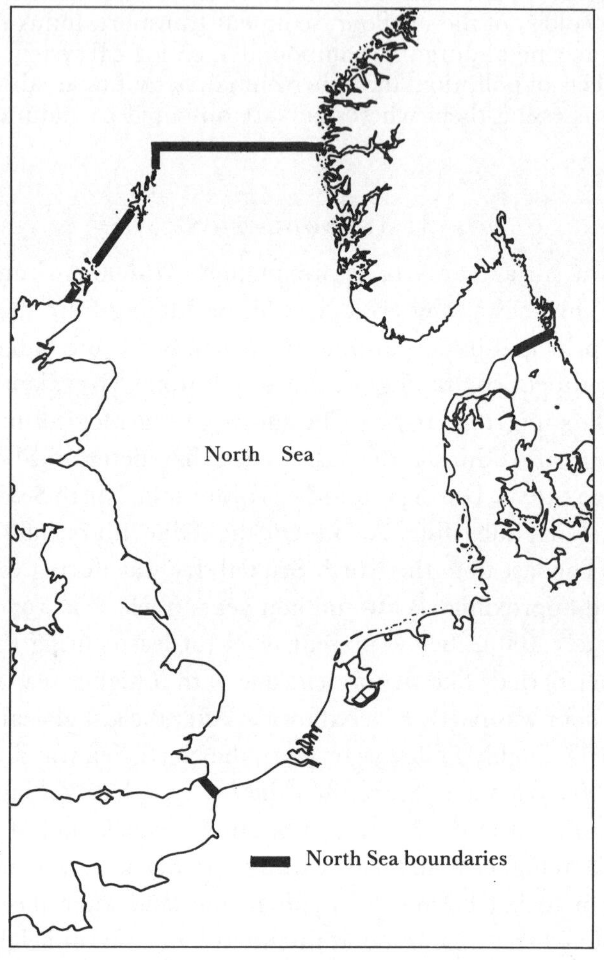

FIGURE 1. Limits of the North Sea as determined by the International Hydrographic Bureau, except between the Skagerrak and the Kattegat.

2. THE PHYSICAL ENVIRONMENT

(a) *The sea floor*

The general shape of the North Sea basin is determined by the NW–SE direction of the geological fault structures which partly reflect structural trends that are much older than the Permian. The topography of the present sea floor, however, is of much younger age and was formed primarily during the last part of the Pleistocene (roughly during the past 300000 years). Huge glaciers excavated the Norwegian Trough and the Skagerrak where an older river valley probably existed before. Here the greatest water depths are reached: 700 m in the centre of the Skagerrak with a minimum depth of 225 m in the Norwegian Trough (figure 2). The remainder of the North Sea is less than 200 m deep, except for a few holes in the central North Sea that were also excavated by ice. The glaciers that came from Scandinavia and Scotland brought large quantities of sand and gravel to the North Sea floor. Huge deposits like Dogger

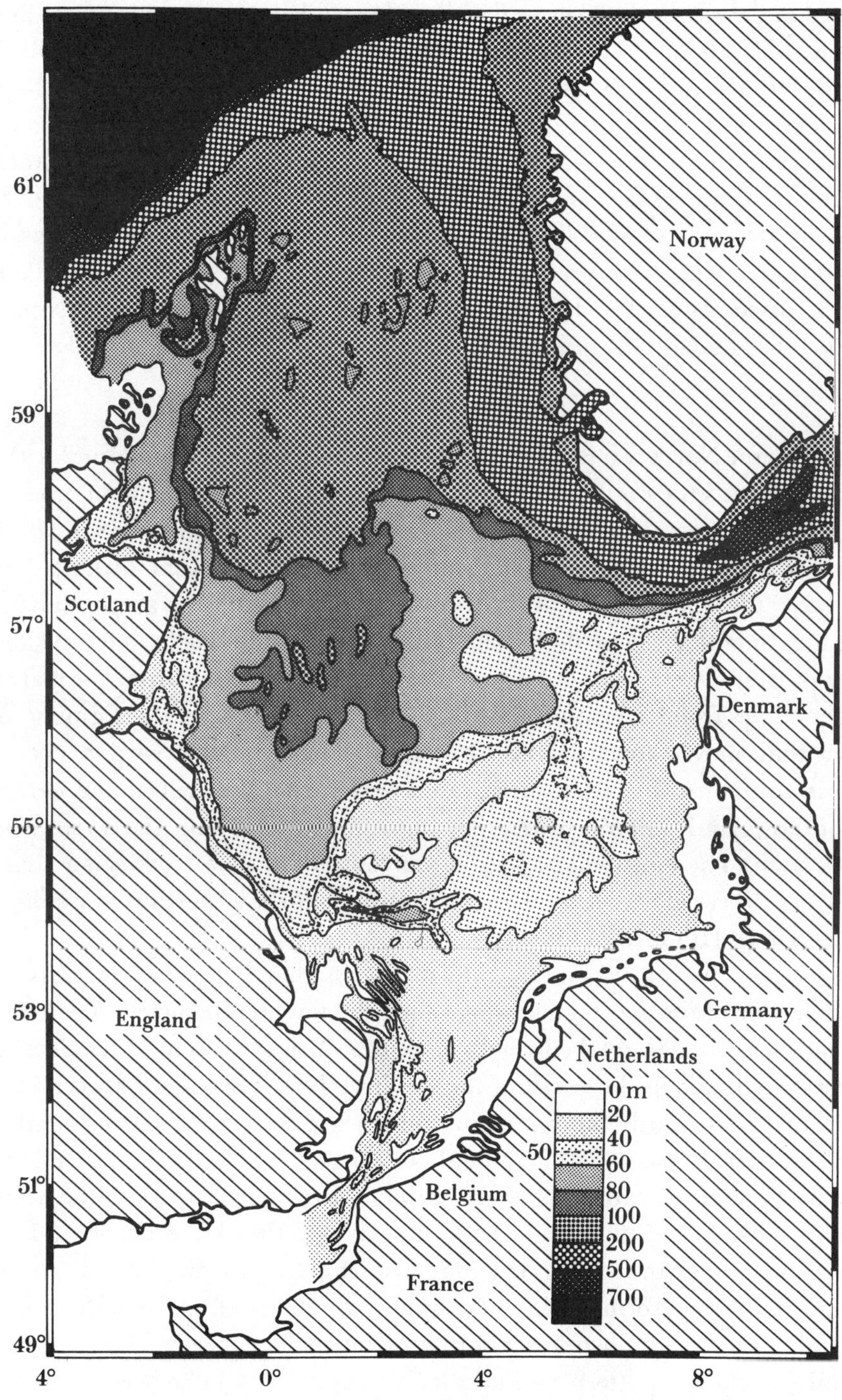

FIGURE 2. Water depth in the North Sea, in metres (after Eisma 1973).

Bank were built up, and many smaller ones which were subsequently modified somewhat when the area was flooded again by the sea. This took place mainly during the last glacial period but one (the Saale or Riss glaciation). During the last glaciation the North Sea was land again but at that time large parts were not covered by ice. On the ice-free parts large outwash plains and river valleys were formed. Their remnants can still be seen in the present bottom topography in some cases (the Deep Channel and the Elbe Rinne in the southern North Sea) or can be traced in the sub-bottom where they have been filled up with younger sediment. When the area was flooded about 7000 years ago, other sand banks such as the Norfolk Banks, the Flemish, North Hinder and Zeeland Banks were formed, possibly during an early stage when (relative) sea level was lower, water depth was less and tidal currents were stronger. The most recent sedimentary formations are found near shore or along the coast: shallow sandbanks and beaches where surface waves have a strong effect, ripple systems up to a few metres high where tidal currents are strong, and mud deposits in sheltered or deeper areas (Waddensea, German Bight, Skagerrak/Kattegat/Norwegian Trough; Oele *et al.* (1979); Eisma & Kalf (1987)). Because of this varied history the distribution of gravel, sand and mud on the North Sea floor shows a mosaic of sediment types (figure 3).

(*b*) *Hydrography*

Water flows into the North Sea through the connections with the Channel and with the North Atlantic ocean, whereas the outflow of North Sea water is concentrated along the eastern side of the gap between Norway and Shetland. Ocean water is forced into the North Sea by the predominantly westerly winds pushing the water in the North Atlantic towards NW Europe, by the tidal wave that enters the North Sea from the north and the south, moving through the North Sea in an anticlockwise direction, and by density differences that are chiefly caused by the inflow of fresh water from the coasts and low-salinity water from the Baltic. Most of the fresh water comes from rivers that flow into the southern North Sea, forming low-salinity coastal water that moves towards the Skagerrak. There the Norwegian coastal current originates. It is a mixture of coastal water, saline water from the central and northern North Sea, Baltic outflow, and Atlantic water flowing southwards along the western side of the Norwegian Trough. The Norwegian Coastal Current follows the Norwegian coast from the Skagerrak to the Norwegian Sea and there continues to flow northwards along the coast (figure 4) (Lee 1970; Hill & Dickson 1978).

The tidal currents are strongest and the tidal range is highest in the shallow parts of the North Sea, off capes and in straits and bights, where the tidal flow is obstructed. The highest ranges (up to 6 m) and the strongest tidal currents (more than 1 m s^{-1}) occur in the Straits of Dover, in the Southern Bight, along the coasts of the southern North Sea and between the islands in the north, decreasing towards the Skagerrak where the water is deepest and tidal current velocities are less than 10 cm s^{-1}. Where tidal currents are weak, the influence of the wind is more pronounced and flow tends to be more variable. In some areas, as off the northern coast of Denmark, wind-induced flow can be quite strong (in the order of 30–40 cm s^{-1}). Wind-waves, generated at the sea surface, may disturb bottom sediment at more than 100 m water depth during heavy storms. Where the tidal currents are strong too, the combination of waves and tidal currents can rework the bottom sediment intensively.

The general circulation pattern as indicated in figure 4 is influenced by more regional phenomena, as listed next.

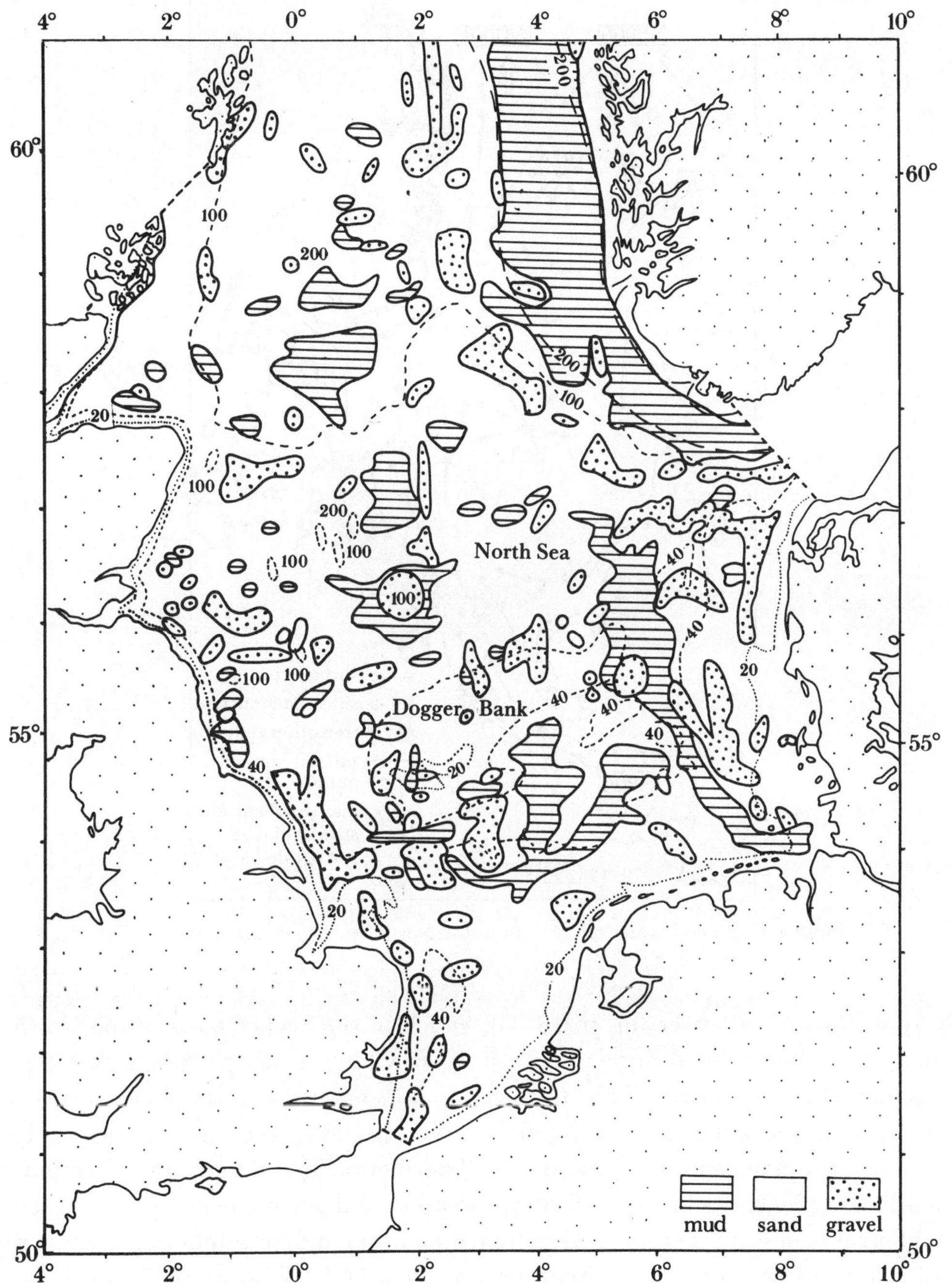

FIGURE 3. Distribution of bottom sediment in the North Sea (after Veenstra 1970).

(1) Wind-forcing can completely reverse the normal flow pattern for a few days, particularly in the shallower areas and where tidal currents are weak (Riepma 1980), but also, for example, in the Straits of Dover. In the Skagerrak strong westerly winds force the water eastwards, reducing the outflow, whereas easterly winds strongly enhance the outflow, an effect that can be felt even at the bottom (Aure & Saetre 1981).

(2) Heating of the surface water during spring and summer results in the formation of

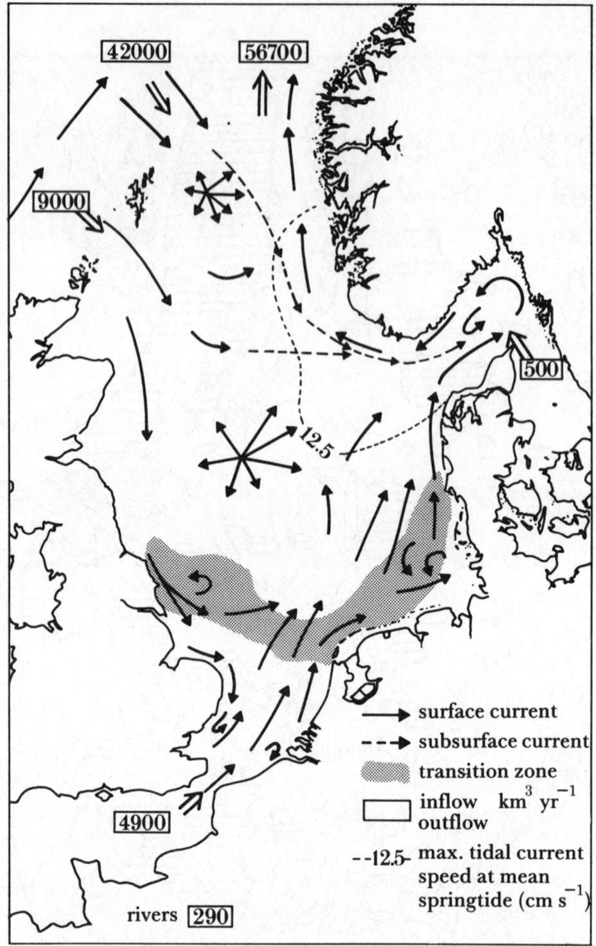

FIGURE 4. General water circulation in the North Sea (from Eisma and Kalf 1987).

relatively warm surface water (up to 18 °C), which in the deeper parts of the North Sea is separated from the cooler deeper water (of *ca.* 6 °C) by a sharp change in temperature (thermocline; Dietrich (1950)). This stratification does not occur in the shallower areas along the coast and in most of the southern North Sea, where tidal currents are strong and the entire water column is regularly mixed (figure 5). The thermocline disappears in autumn when turbulent mixing increases because of stormy weather and cooling of the surface water.

(3) Differences in water density caused by temperature and/or salinity differences result in fronts between water masses of different density (Pingree & Griffiths 1978). Such fronts are common (*a*) in the southern North Sea in a broad zone separating the main mass of North Sea water from the lower salinity water in the south, (*b*) along the Norwegian coast from the Skagerrak to the Norwegian Sea separating the Norwegian Coastal Current water from the more saline North Sea water, and (*c*) on a smaller scale off river mouths and in coastal waters.

(4) Local gyres and other secondary circulations are induced by the bottom topography, the configuration of the coast and, in coastal waters, by differences in density between nearshore and offshore waters. The latter results in a quasi-estuarine coastal circulation whereby the nearshore water moves away from the coast along the surface and offshore water moves shoreward along the bottom (Dietrich 1955; Mittelstaedt *et al.* 1983).

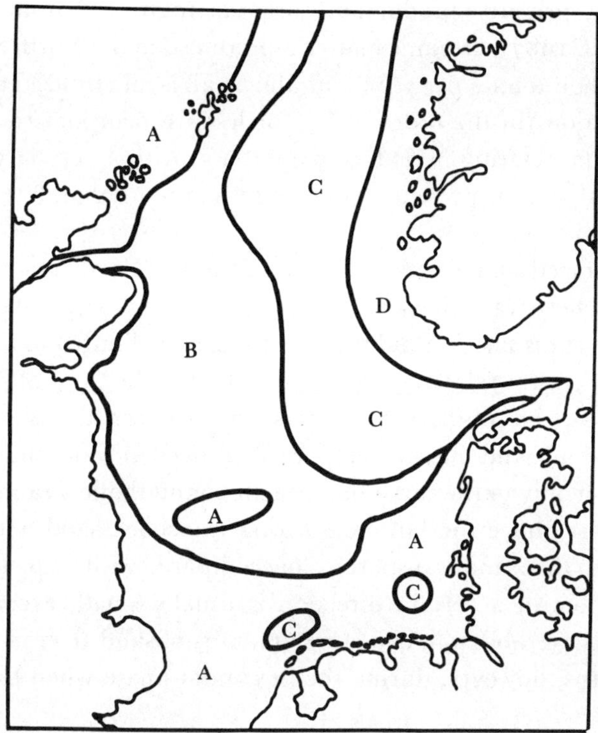

FIGURE 5. Stratification of the water in the North Sea. (A) No stratification all year round; (B) seasonal or permanent temperature stratification (thermocline), no vertical differences in salinity (homohaline); (C) seasonal or permanent salinity stratification (halocline), small seasonal variations; (D) seasonal or permanent salinity stratification (halocline) with strong seasonal variations (from Umweltprobleme der Nordsee 1980).

The effect of wind-induced currents is to enhance mixing, reducing fronts and thermoclines, whereas the formation of fronts and thermoclines indicates reduced mixing. Fronts show a complicated flow structure, usually including downwelling, upwelling, flow parallel to the front, and the formation of large eddies. The water masses separated by fronts often not only show differences in salinity or temperature or both, but also show differences in suspended matter concentration, colour, concentration of dissolved materials, flora and fauna. Fronts are often marked by streaks of foam on the surface. On the whole the North Sea water is renewed every 1–2 years but the renewal time varies in the different parts from more than three years in the German Bight and the western part of the central North Sea to less than six months at the eastern side north of Stavanger (Maier-Reimer 1977).

3. Sediment transport

At present, hardly any sand and no gravel is being supplied by rivers to the North Sea and only a very small amount comes from erosion of cliffs (Eisma *et al.* 1979). Virtually all sand and gravel is being reworked from older deposits by bottom currents and waves. Many sands and gravels contain a more recent admixture of organic matter and carbonate (shell) fragments left over (mainly) from benthic organisms, and probably to a lesser extent, from plankton. Suspended matter (less than *ca.* 125 μm diameter) is supplied to the North Sea to a total of, roughly, 40 million tonnes per year (dry mass), coming from the Channel, from the North Atlantic Ocean, from the Baltic, from rivers, from erosion of cliffs and the sea floor, from the atmosphere and from primary production (organic matter, diatom frustules, carbonate particles; Eisma & Kalf, 1987). Organic matter is produced in the North Sea in large quantities (in the order of 70 million tonnes per year) but almost all is mineralized and consumed so that only a very small fraction (in the order of 1% or less) is incorporated in bottom sediments (chiefly in mud deposits). During periods of plankton growth a very large amount of organic particles, living or dead, can be present comprising up to more than 90% of the total amount of suspended material.

Gravel is not transported under present conditions in the North Sea except in very shallow areas or on the beach where waves have a large impact. In any case, gravel is being transported only over relatively short distances. Sand transport is usually limited to those areas where tidal currents and wave action are relatively strong, i.e. the beaches and offshore in most parts of the southern North Sea, the southern Skagerrak and between the islands in the north, but during heavy storms (fine) sand may be temporarily moved at more than 100 m water depth. Sand goes in suspension only in the surf along beaches or on shallow sand banks, but is usually transported over or just above the bottom. During transport, sand is temporarily stored in ripples of varying size so that sand transport is slow compared with suspended matter transport. Resultant sand transport in a certain direction is usually small, except near to the coast, although tidal currents are able to move large amounts of sand (e.g. in the Southern Bight); almost all of this returns, however, during the next tidal phase when the current goes in the opposite direction.

Suspended matter is moved through the entire North Sea, roughly following the direction of water transport. Most suspended matter is concentrated nearshore, concentrations reaching 100 mg l^{-1} or more, because most of the sources are located nearshore (rivermouths, cliffs, the shallow sea floor) and because the nearshore water circulation concentrates suspended matter near to the coast, at the same time reducing the dispersal of suspended matter from the nearshore sources in an offshore direction. About one third of the suspended matter in the North Sea, which is approximately the same amount as that which comes in from the North Atlantic, flows out into the Norwegian Sea, which is not necessarily the same material that came in from the ocean. The remainder is deposited in a number of isolated areas as indicated opposite (figure 6). In these areas the suspended matter is concentrated by tidal mechanisms or in large tidally or topographically induced eddies. Wave activity near to the bottom is small, either because of the water depth or because the area is sheltered against the wind. Mud deposited in these areas is therefore not easily stirred up again. A balance of the supply and the outflow plus deposition of suspended matter in the North Sea is given in table 1. The figures in this table are approximate; because of the large variability in transport conditions and the limited number of data, budget calculations can only approximately give the amounts of suspended matter that are involved.

Bottom sediment and suspended matter:

(*a*) form a habitat for organisms that live in or on the bottom or attach themselves to suspended particles (bottom fauna, micro-organisms);

(*b*) contain food (in the form of organic matter) for pelagic or benthic organisms that filter suspended matter from the water or consume bottom sediment;

(*c*) provide the right conditions for higher organisms that feed on benthic fauna, use certain types of bottom sediment for spawning grounds, etc.;

(*d*) provide a carrier for substances that are adsorbed on to the particle surfaces, that are transported that way through the North Sea and become concentrated in bottom deposits.

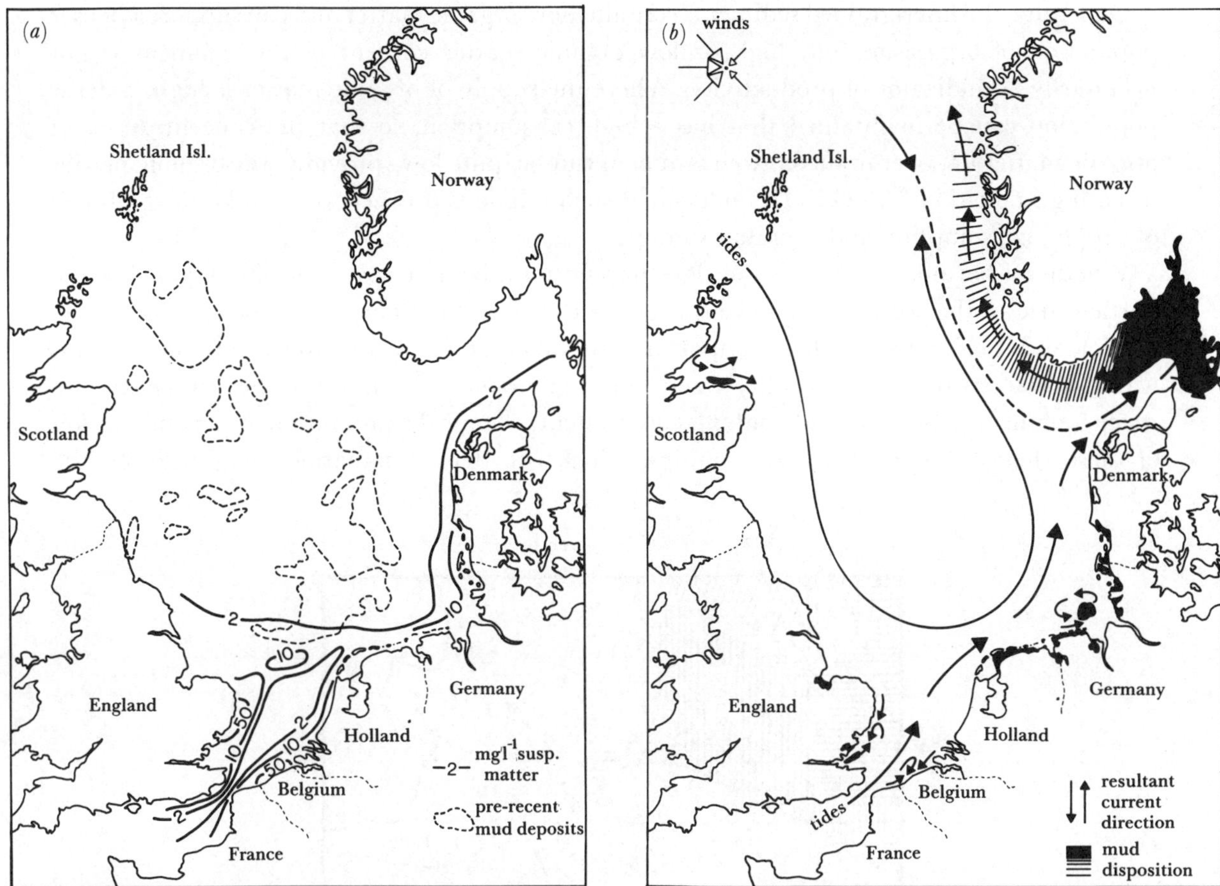

FIGURE 6. (*a*) Distribution of suspended matter (in mg l^{-1}) and (*b*) resultant transport directions, location of recent mud deposits (black) and older mud deposits (from Eisma & Kalf 1987).

TABLE 1. ANNUAL SUPPLY, OUTFLOW AND DEPOSITION (MEGATONNES)

supply	
North Atlantic Ocean	10
Channel	10
Baltic	0.5
rivers	4.8
seafloor erosion	9–13.5 (+?)
coastal erosion	0.7
atmosphere	1.6
primary production	1
total	37.6–42.1
outflow + deposition	
outflow	11.4 + < 3
deposition	
estuaries	2.5
Waddensea + The Wash	3.5
Outer Silver Pit	1–4 (?)
German Bight	3–7
Elbe Rinne	?
Oyster Grounds	?
Kattegat	8
Skagerrak	4–7 (+?)
total	33.4–46.4

Generally, the finer-grained sediments contain more organic matter and can support a denser population of organisms, but high or low organic matter content of the sediment is not necessarily an indicator of productivity. Where the supply of organic matter is high, a dense population can be maintained that has a high consumption, so that the concentrations of organic matter left over in the bottom sediment can be quite low (provided the organic matter has a high nutritional value). To understand such relations it is necessary to know the fluxes of supply, consumption and eventual storage.

Whether a bottom sediment is the right habitat for a benthic species is often related to the particle size of the sediment (Umweltprobleme der Nordsee 1980). It is not necessarily the particle size itself, however, that is important, but other factors related to it: water movement, water characteristics (e.g. turbidity), and also organic-matter content of the sediment, pore space, sediment structure, mud content, clay content, chemical composition, or a combination of these (figure 7). Intricate relationships, which still defy explanation, exist between the

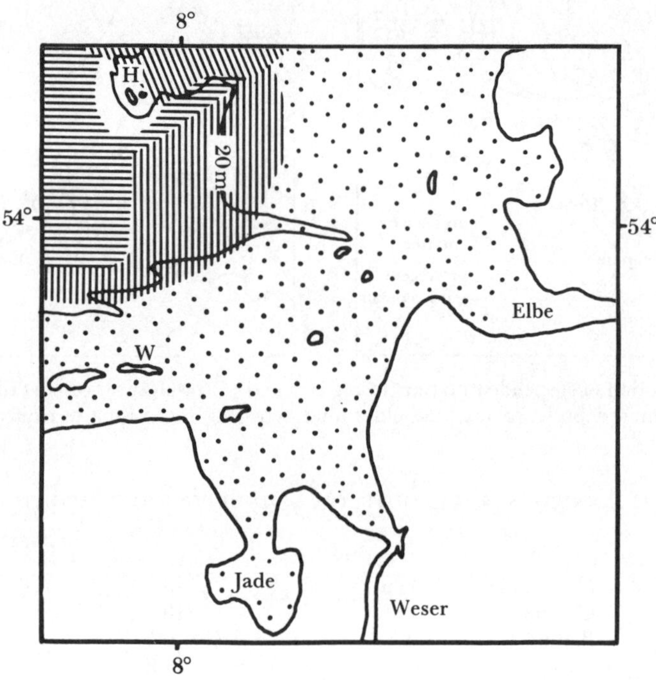

Figure 7. Distribution of bottom fauna communities in the German Bight (from Umweltprobleme der Nordsee 1980). Pointed area: *Macoma balthica* community, coastal sands; vertical lines: *Abra alba* community, mud; horizontal lines; *Echinocardium cordatum–Amphiura filiformis* community, fine muddy sand; oblique lines: *Venus striatula* community, coarse sand and gravel.

presence, absence or the abundance of benthic organisms on the one hand, and sediment characteristics as median diameter and sorting on the other. This can be related to the fact that most benthic organisms have pelagic larvae that must settle within a few weeks. If the right type of bottom sediment is not encountered within that period, they will settle in an area that is not optimum but may be still good enough for a large number of individuals to survive. Also, factors quite unrelated to sediment characteristics may determine whether benthic animals will be present or not; coastal sands in the southern North Sea are very poor in species as well as in specimens not because of the sand or the surf – elsewhere beach sands can be much more inhabited – but because the water becomes very cold during the winter, or because it has a lower salinity, or because of its high turbidity.

4. Nutrients

Phytoplankton, to be able to grow, need water, carbon dioxide, light, and microamounts of nutrients (phosphate, nitrogen compounds, for diatoms also silica) and probably also small amounts of trace metals and some organic compounds such as vitamins. Water and carbon dioxide (as bicarbonate) are present in abundance in the North Sea but light and to some extent nutrients are restricted (little is known about the influence of trace-metal concentrations and organic compounds). The intensity of the sunlight shows seasonal variations and its penetration in the water is limited by reflection, scattering and absorption. Particles in suspension restrict the penetration of light and the growth of the plankton itself, when it becomes abundant, can reduce the light penetration. In the North Sea the low light intensity during the winter months strongly reduces photosynthesis by phytoplankton. Low winter temperatures are not a limiting factor since diatoms can live and grow on and between ice. When light is sufficiently available, the concentration of nutrients usually limits phytoplankton growth.

The nitrogen compounds – nitrate being the most important – come into the North Sea from the North Atlantic, from rivers and from the atmosphere in rainwater (Cushing 1973; Johnston 1973; Umweltprobleme der Nordsee 1980). High concentrations occur where the influx of river water is high (coastal waters along the English and the Belgian–Dutch–German coast; figure 8*a*), but are lower elsewhere in the North Sea. In most of the North Sea there is a delicate balance between the supply of nitrogen from the three sources mentioned, the degree of mixing of surface water and deeper water, the depletion by growing phytoplankton and the amount of nitrate brought back into solution by mineralization of organic matter (including the upward flux of nitrate and other nutrients from the bottom sediments).

The concentration of phosphate, which is not transported through the atmosphere except as aerosols which contribute only a minor quantity, shows a similar distribution as nitrate with high concentrations in the North Atlantic inflow and near river mouths (Firth of Forth, Thames estuary, Belgian–Dutch coast, German Bight; figure 8*b*). The distribution of dissolved silica, which is supplied from the same sources as phosphate, is very similar. Conditions for photosynthesis (phytoplankton growth) are therefore favourable in the coastal areas but growth tends to be limited there (*a*) because of the high suspended-matter concentrations (up to 100 mg l^{-1}) limiting light penetration into the water, and (*b*) dispersal by mixing with offshore water poorer in phytoplankton and nutrients, which is probably more important than the turbidity of the water as a limiting factor. Outside the coastal areas phytoplankton growth in most parts of the North Sea is limited during most of the year by the formation of a thermocline restricting the mixing of surface water and deeper water. In the surface water above the thermocline nutrients are used up but in the deeper water high nutrient concentrations develop because of mineralization and sinking down of organic matter produced in the surface water. In the surface water a kind of steady state develops whereby phytoplankton populations increase to their limits and nutrient concentrations become low, while nutrients that become available by mineralization of the organic matter that is produced are quickly used up again so that their concentration in solution remains low. Only part of the organic matter that is

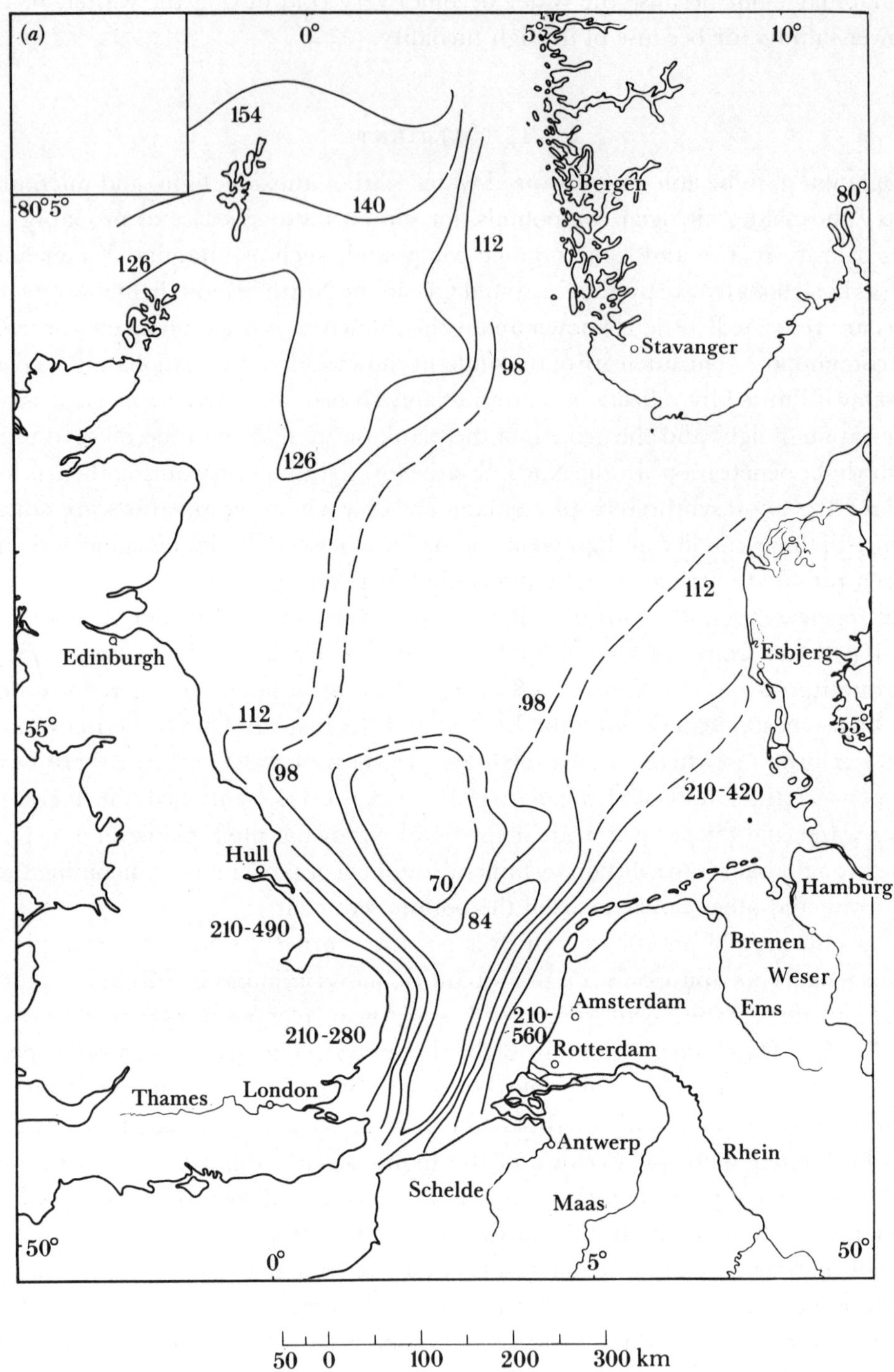

FIGURE 8*a*. For description see opposite.

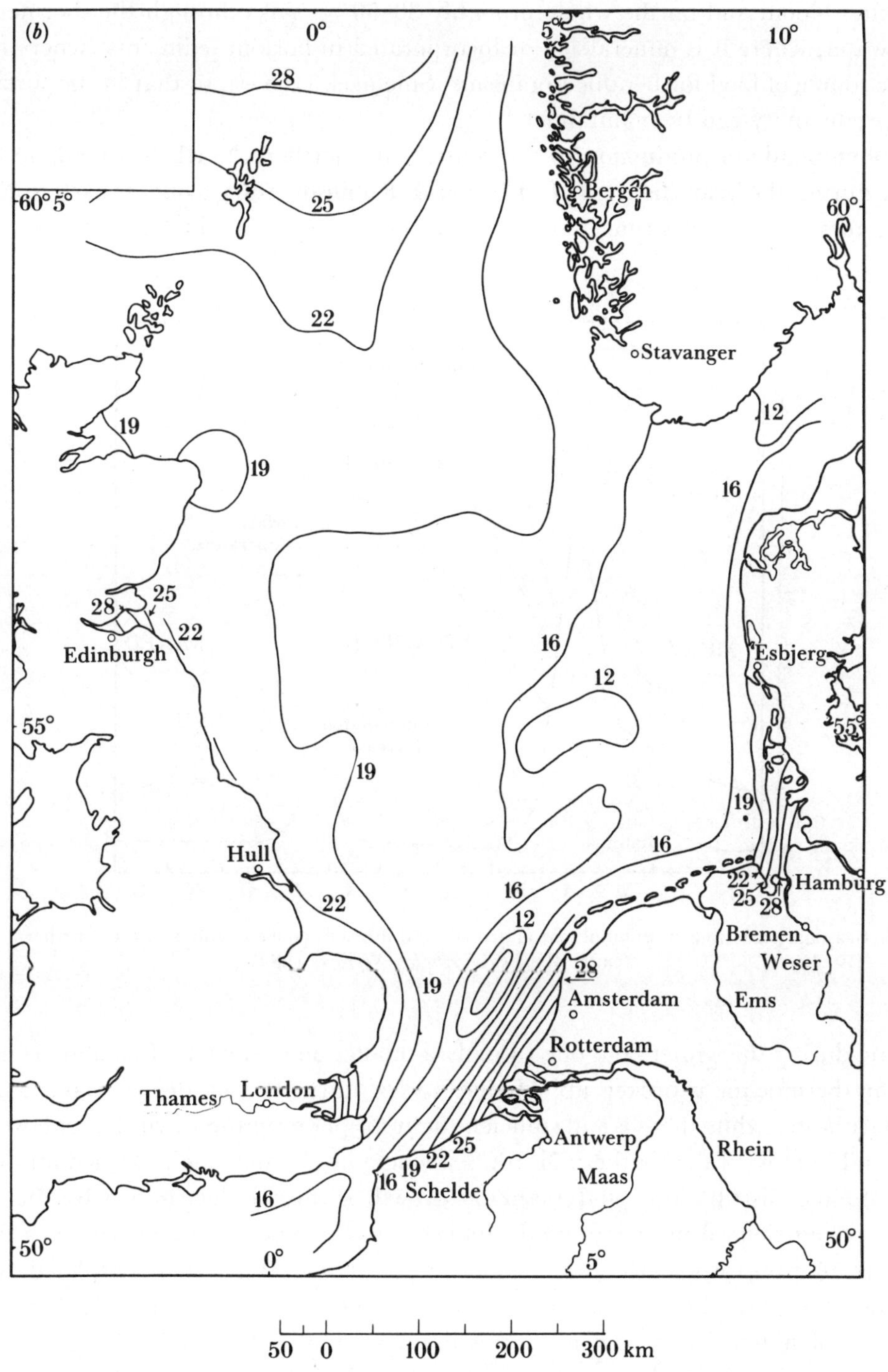

FIGURE 8. Distribution of NO_3-N (*a*), and PO_4-P (*b*), both in micrograms per litre in the North Sea surface water in winter (from Johnston 1973).

produced in the surface water (normally in the order of 1% but much higher after a period of plankton bloom and on the whole probably 30–50%) sinks through the thermocline into deeper water, where it is mineralized or incorporated in bottom sediments (where it forms a valuable source of food for benthic organisms; Smetacek (1984)), so that in the surface water a high productivity can be maintained.

The phytoplankton production in the central and northern North Sea tends to have two maxima during the year (figure 9; Colebrook & Robinson 1965): one in early spring when light becomes increasingly stronger and nutrient concentrations are high because of mixing and

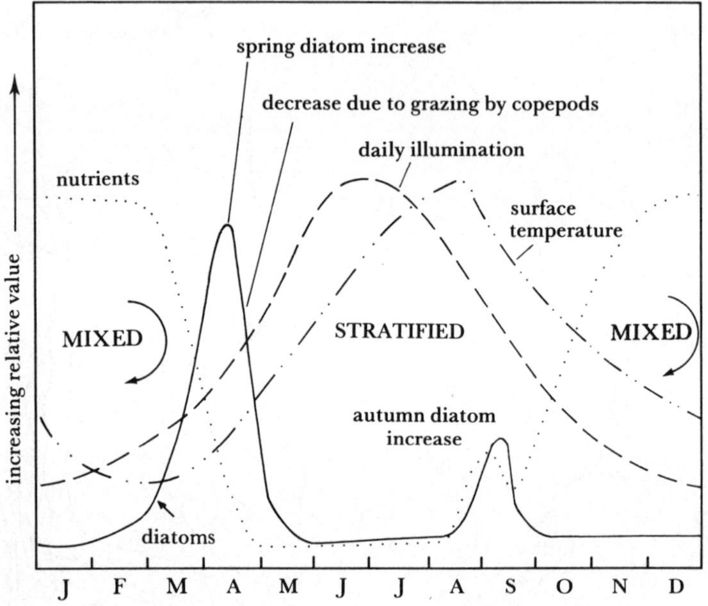

FIGURE 9. Seasonal variations in phytoplankton growth and nutrient concentrations in the (northern) North Sea (after *Waterkwaliteitsplan Noordzee* 1985).

advection during the winter, and one (usually a smaller one, which is often absent) in autumn when the thermocline is broken up. Mixing makes nutrients from deeper water available in the surface water while there is still sufficient light for photosynthesis. In the shallower coastal waters and in most of the southern North Sea where no thermocline is formed and haloclines have a limited distribution, phytoplankton growth is largely determined by the degree of advection of nutrients from rivers and the limiting factors indicated above. Diatoms are usually the dominant group as long as silica is sufficiently available. When that is depleted to very low levels and other nutrients are still there, other phytoplankton species, mainly flagellates, take over. Phytoplankton, living or dead, forms the basis of all other life in the sea. Diatoms are eaten by the larger zooplankton species (copepods), which in their turn are consumed by other pelagic organisms and by fish. The smaller flagellates are mostly eaten by smaller zooplankton species, which are primarily the food of jellyfish. The dead organic matter is consumed (*a*) by organisms like bacteria and copepods that live on organic matter in suspension, (*b*) by organisms that filter suspended matter from the water (consuming living and dead organic matter indiscriminately, the particle size determining whether a particle can be ingested or not), and (*c*) by benthic organisms that eat bottom sediment.

The supply of phosphate and nitrate (but not of silica) to the North Sea has increased considerably during the past decades, because of the human population increase and the increased use of detergents and fertilizers in the countries around the North Sea (van Bennekom *et al.* 1975; Gerlach 1984). The increased supply chiefly reaches the North Sea by way of the rivers and to a lesser extent from direct discharges of organic waste or through the air. The effect of this eutrophication up to now has not been easy to assess. The changes in the populations of phytoplankton and of other marine organisms that have occurred can be related to other factors such as hydrographical changes (e.g. changes in storm frequency, temperature anomalies) and changes in the zooplankton populations that graze on phytoplankton. Some shifts may be related to changes in nutrient supply: along the Dutch coast the silica supply has decreased somewhat during the past 50 years, whereas phosphates and nitrates increased so that diatoms, which were dominant 50 years ago, have been replaced by flagellates which are dominant now. Around 1930 phosphate supply to the coastal waters was a limiting factor for plankton growth, whereas at present mixing with nutrient-poorer offshore water is most likely to be the principal limiting factor. It is also possible that pollutants are (partly) responsible for the observed changes (Postma 1985).

Another possible effect of eutrophication is the lowering of the concentrations of dissolved oxygen in the water (Gerlach 1984). Under aerobic conditions mineralization of organic matter involves uptake of oxygen that is dissolved in the water. In areas with high phytoplankton growth periodically both high and low dissolved oxygen concentrations can be expected; high because of the production of oxygen during photosynthesis so that even supersaturation may develop, and low because of oxygen being used up during the decomposition (mineralization) of organic matter produced by phytoplankton. The actual concentration of dissolved oxygen in the water is the result of the relation between supply (from the atmosphere and/or from photosynthesis) and consumption. In the North Sea, as in almost all other sea areas, the advection of dissolved oxygen through the water (by currents, waves and turbulence) has always been sufficient to maintain, even in the near-bottom waters, sufficiently high dissolved oxygen levels for most organisms to be able to live there (i.e. at least a 30% saturation). Very low oxygen concentrations have now occurred in the German Bight and along the Danish west coast, resulting in a high mortality of bottom fauna, which is probably mainly related to discharges of organic waste or to increased nutrient supply from the Elbe river. In bottom sediments that are little disturbed the decomposition of the organic matter in the sediment usually results in anaerobic conditions (with dark blue or black colouring because of the formation of iron sulphide) below an oxidized layer of variable thickness (ranging from 1 mm to more than 10 cm). The diffusion of dissolved oxygen from the overlying water through the pore-water into the sediment is too slow to compensate for the oxygen consumption at deeper levels. Where deep-burrowing organisms occur in the bottom sediment, their activities bring oxygen-rich water deep into the sediment.

5. Other trace constituents

The natural levels of many trace metals and organic compounds, present in concentrations of a few milligrams per litre or (much) less, are not known because the analytical techniques to determine them reliably in solution in seawater were developed when pollution had already been occurring for many decades. Natural levels of trace metals in sediments and suspended matter can be estimated from fine-grained sediments dating from before *ca.* 1850 (figure 10),

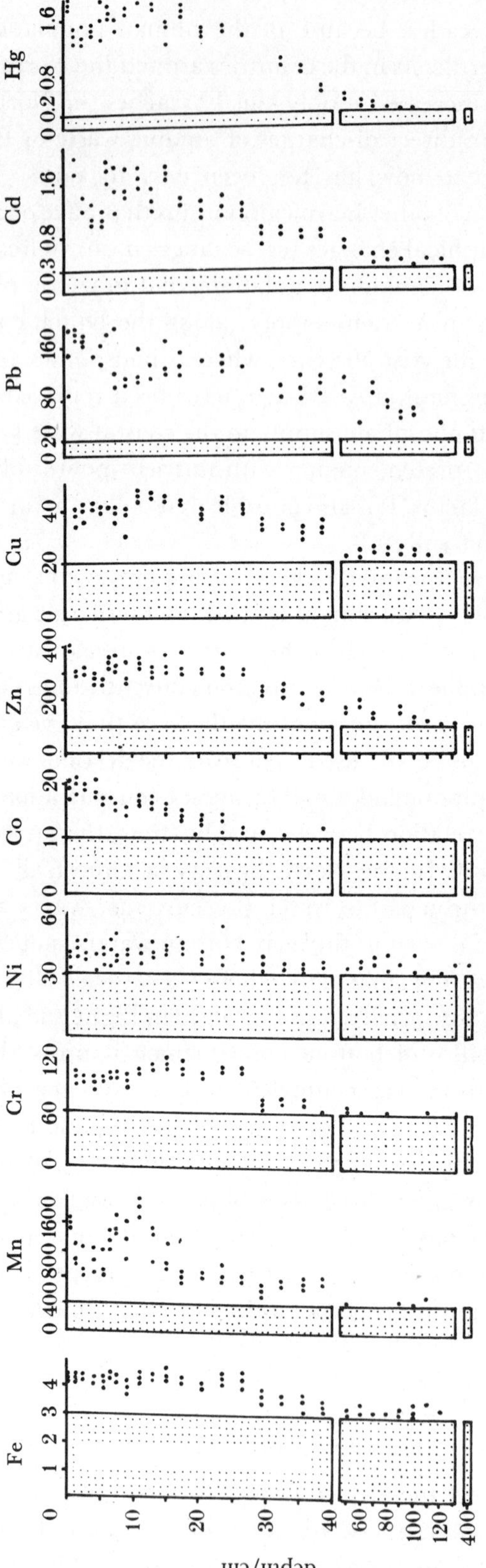

FIGURE 10. Distribution of trace elements in a sediment core from the German Bight (from Förstner & Reineck 1974). During the first part of the 19th century the concentrations were still the natural concentrations (geochemical background). Trace-element concentrations are shown in parts per million by mass of dry sediment.

but the partitioning ratio between the amount in solution and the amount in particulate form is not well enough known to reconstruct natural concentrations in solution. For practical purposes the concentration in North Atlantic water can be taken as the natural concentration in seawater, whereas estimates of natural concentrations in river water are based on the present partitioning ratios. Compared with 1850 the concentrations of many trace metals in sediments have increased by 100–700 %. Other trace constituents, in particular many organic compounds, are purely artificial and only reached the North Sea during the last 100 years (many, much more recently).

For many trace constituents the interaction between the dissolved and the particulate phase and of both phases with organisms is strongly related to the form in which the constituent is present; in inert particles, as irreversibly adsorbed material that can only be desorbed through (partial) conversion of the particulate material, reversely adsorbed material that can be exchanged, and in solution as well as several types of chemical species (oxide, hydroxide, sulphide, ionic, molecular; Salomons & Förstner 1984). Concentration changes in the water, changes in pH, temperature, turbidity and in particle characteristics may result in mobilization of a trace constituent from particles into solution or removal from solution into particulate form. There are two important mechanisms whereby trace constituents are concentrated: one is by adsorption on to particulate matter and deposition of this on the sea floor, and the other is by uptake and concentration in organisms. In both ways the concentration can increase by several orders of magnitude (figure 11). Dead organic matter plays a large role in the adsorption of trace constituents on to particles; all surfaces in seawater are quickly covered by an organic coating and organic matter is very surface-active. The mechanisms involved, in particular those resulting in high concentrations in organisms, are badly known, and the toxic as well as the potentially toxic effects of the concentrated substances are usually difficult to assess. An example is the strong decline in the number of seals in the Dutch Wadden Sea. The study of dead seals indicated that the decline may be related to poisoning with mercury, to parasites, to oil-caused skin infections, to physical disturbance by ships, planes, target practising and tourists chasing the mothers from their young, and to poisoning with organic poisons, notably PCBs. Further research indicated that the effects of mercury were probably neutralized by the uptake of selenium and that the PCBs are the most likely cause (Reijnders & Wolff 1981; *Waterkwaliteitsplan Noordzee* 1985).

Laboratory tests are a logical way to estimate effects of trace constituents (*Waterkwaliteitsplan Noordzee* 1985) but in spite of many such tests, the toxic levels of trace constituents in water or sediments are difficult to estimate because:

(1) there are a great many toxic or potentially toxic substances and a large variety of marine organisms; for many substances and for many marine organisms no tests have been made, while the results of the tests that have been made are difficult to interpret considering the complexity of biological relations in the North Sea and the possibility of synergetic or antagonistic effects (laboratory tests are usually done with only one species and one trace constituent);

(2) for many trace constituents the dispersal patterns and the amounts supplied to the North Sea are not very well known;

(3) the relation between bottom sediment composition and bottom fauna is not well known;

(4) there is the possibility that marine organisms and marine ecosystems adapt to changing conditions, including higher levels of trace constituents.

These uncertainties directly affect decisions on what levels of pollutants can be regarded as

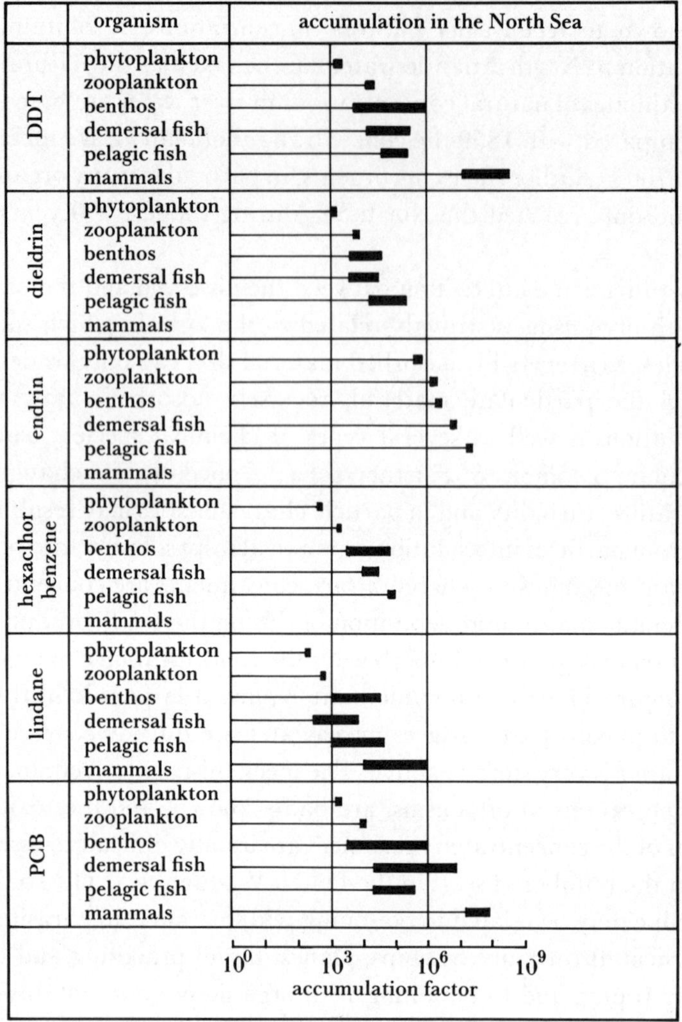

FIGURE 11. For description see opposite.

acceptable. There are only a few instances known from the North Sea where the effects of pollution are clear and where cause can be related directly to effect, such as oil affecting sea birds (but then, whose oil?), copper poisoning fish along the Dutch coast and pesticide poisoning in the Dutch Wadden Sea that could be traced directly to a factory in Rotterdam. In the latter case it took, as in Minamata Bay, a considerable amount of research to establish the link between effect and cause. Prevention of harmful effects caused by trace constituents in the North Sea can therefore only be done on the basis of rather general, and therefore rather vague and unsatisfactory criteria.

6. BIOLOGICAL RELATIONS

Phytoplankton and zooplankton populations in the North Sea are related to water characteristics like temperature, salinity, nutrient concentrations, turbidity, degree of stratification, to the water circulation, to the time of the year and to biotic factors. Within each

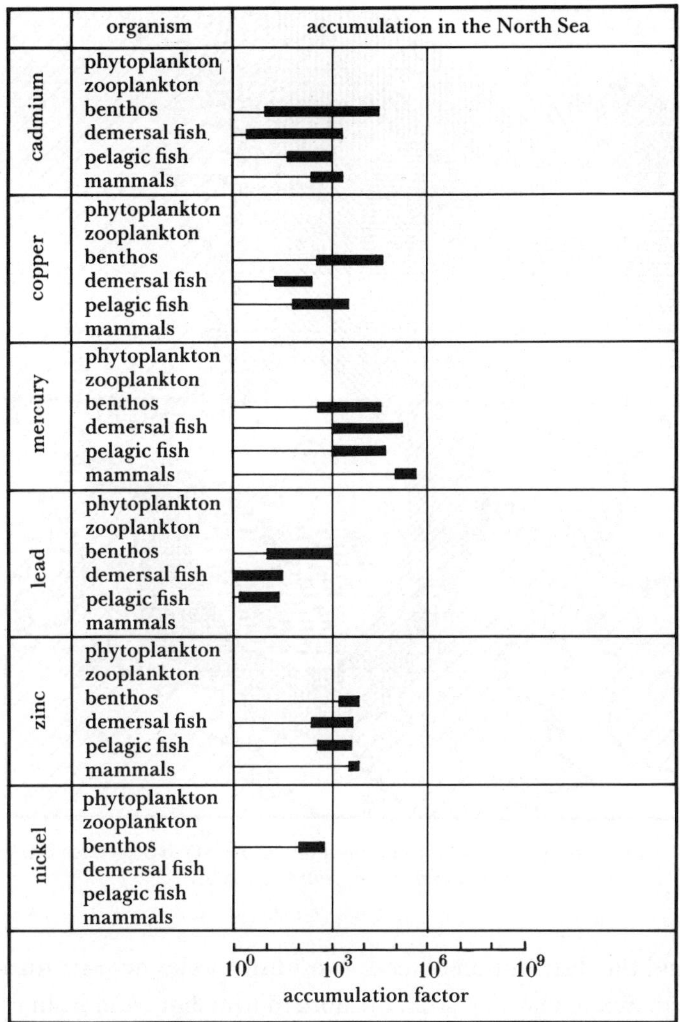

FIGURE 11. Accumulation of some trace metals and organic compounds in different groups of organisms in the North Sea (after *Waterkwaliteitsplan Noordzee* 1985). The accumulation factor is obtained by dividing the concentration in the organisms by the concentration in the seawater.

distributional area, which can only be loosely defined, the phytoplankton shows a patchy distribution (figure 12). The distribution of zooplankton is linked to the phytoplankton distribution (Cushing 1973) but there is usually some time-lag, the phytoplankton population developing first and the zooplankton population following, but the mechanisms leading to an increase or a decrease in zooplankton populations are badly understood. Zooplankton consists of species that are entirely planktonic and species that are only temporarily planktonic, like the larvae of many bottom-dwelling organisms and of fish. Therefore the distribution of bottom fauna and of fish is to some extent related to the same water characteristics as the planktonic populations. Fronts that separate water masses of different characteristics play a large role in these species distributions.

The distributions of the higher organisms at the end of the food chain (fish, mammals and birds) are far more varied and complex than the distribution of phyto- or zooplankton. The distribution of fish is related to the locations of good spawning areas, to areas where the young

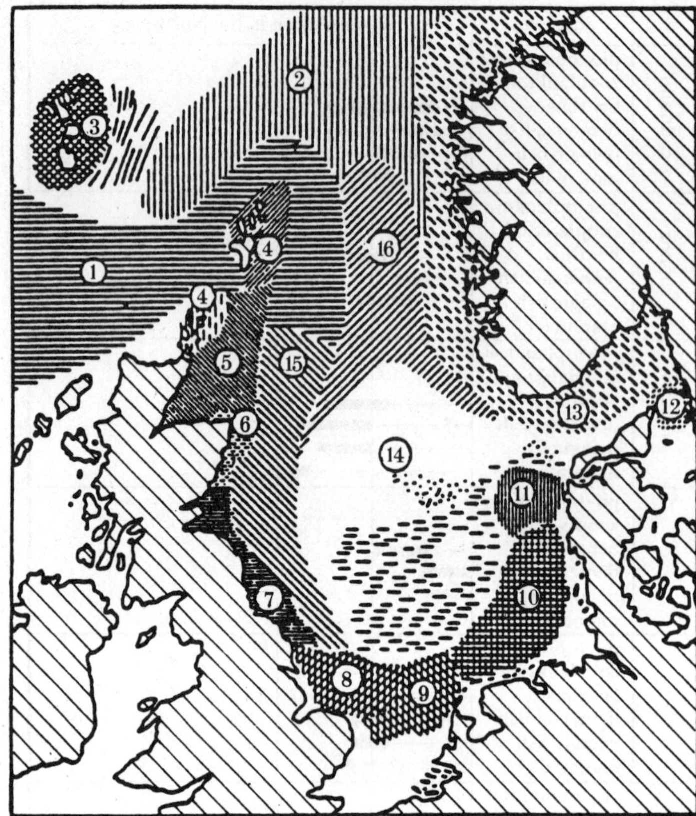

FIGURE 12. Distribution of different phytoplankton communities in the North Sea (after Braarud *et. al.* (1953), from *Waterkwaliteitsplan Noordzee* (1985)).

fish can grow up, and the distribution of food. Most fish species migrate during the year: from one area in the North Sea to another, from offshore to nearshore and inshore waters, and from salt water to fresh water and back. Herring and cod are examples of the first kind, plaice and sole of the second, and salmon and eel of the third. Shallow tidal areas like the Wadden Sea and a number of sandy beaches are very suitable for summer growth of young fish and shrimp (Zijlstra 1972) because of the relatively high temperature, the availability of food based on the local production of algae and other small organisms, and a lower predation. This leads to a seasonal migration: landward in spring, seaward in autumn. Migration to and from fresh water is at present very much reduced because of the destruction of suitable spawning grounds like the gravel beds along many rivers and water pollution.

The distribution of marine mammals is largely determined by their sensitivity to high or low temperatures, by the distribution of the fish that constitutes their food, by the need of some species for certain types of coastal areas to rear their young (e.g. sand flats for common harbour seals) and, at present, also by the degree of water pollution and disturbance. The distribution of bird species is also linked to the availability of food (fish and bottom fauna) as well as to their requirements for nesting (cliffs, dunes and trees). A large number of bird species in the North Sea are migratory; migration occurs to and from areas outside the North Sea and ranges from Greenland and Siberia north and central Africa (Evans 1973).

The general relations between organisms in the North Sea are given in figure 13. To

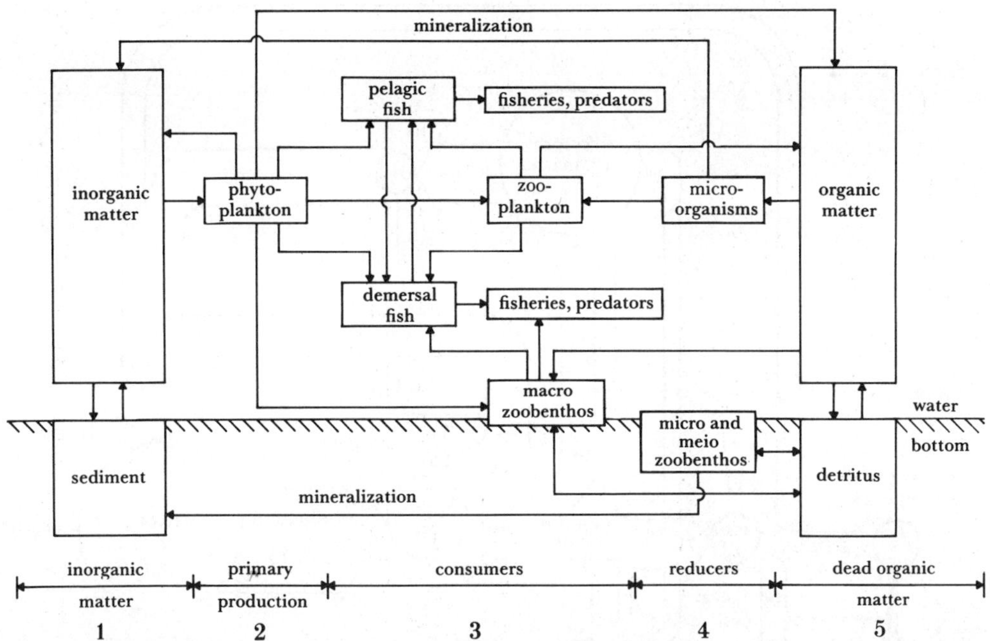

FIGURE 13. General scheme of the relations between the different groups of organisms in the North Sea (from *Waterkwaliteitsplan Noordzee* 1985).

understand these relations better and to get some understanding of the quantities involved, the flux of carbon through this system is considered. An example for the Belgian coast is given in figure 14, which shows the complexity of even a small area (Joiris *et al.* 1982). Data for the North Sea as a whole, however, are not available, mainly because data for the central and northern North Sea and for species that are not of economic interest are lacking. Even the data for biomass production in the different groups, as shown in figure 14, are only rough estimates. Some general relationships, however, can be indicated. There is a short carbon cycle where a large part of the organic matter produced by algae (planktonic or benthic) is consumed by Protozoa and bacteria. For a number of these organisms the substrate is very important as they attach themselves to particles, and they themselves are probably an important food source for bottom fauna. There is also a longer cycle in which organic matter, living or dead, having been consumed by zooplankton, passes to fish and marine mammals (including man). Fisheries interfere in both cycles. In the shallow areas the benthic food chains are important besides the pelagic food chains. In the deeper parts of the North Sea the pelagic cycles dominate and the benthic cycles play a secondary role, although still *ca.* 30% of the available energy in the form of food is used by benthic organisms. At different levels within the food chains there are vast differences in actual biomass and a large variability, while there is not much of a relationship between biomass and trophic level. Also, the reproduction is very different at different levels: the total plankton biomass reproduces itself every 5–10 days (averaged over the year), the biomass of small benthic organisms (meiofauna) every 45 days, the large benthic organisms once or twice a year, and fish only once a year, but there are large variations from year to year. Much of the organic matter produced is 'lost' going from a lower trophic level to a higher one because of respiration and the excretion of organic material, so that only *ca.* 1% of the annually produced organic matter goes into the production of fish. It follows that the effects

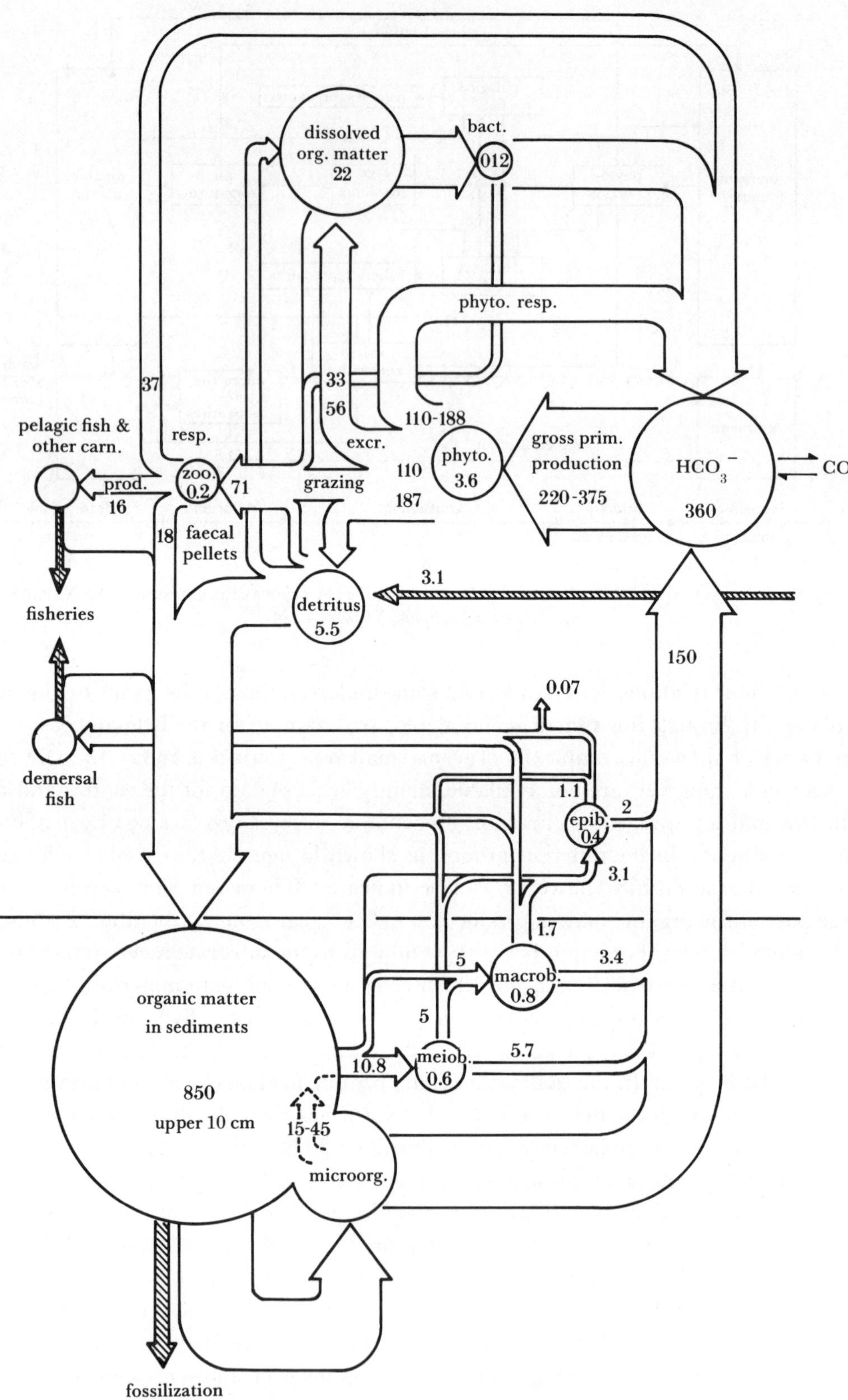

FIGURE 14. Budget of organic carbon off the Belgian coast (from Joiris *et al.* 1982). Numbers within the circles: grams of carbon per square metre present in the water column or in the upper 10 cm of the bottom. Numbers in the arrows; annual flux of carbon expressed in grams of carbon per square metre.

of natural variations or man-induced changes, affecting the phytoplankton production in the first place but also affecting the different trophic levels separately, are difficult to assess. That such changes take place is indicated by the occurrence of a higher biomass of microflagellates and dinoflagellates (more 'red tides' and fluorescing seas) during the past decades compared with the decades before, a lower biomass of diatoms and zooplankton, lower oxygen concentrations near to the bottom in some shallow areas, more demersal fish (living on bottom fauna) relative to pelagic fish, and an increase in the total biomass of fish and in the biomass of shrimp (at least along the Dutch coast). All these changes can be the effect of an increase in the phytoplankton biomass because of increased nutrient supply in combination with a shift in the relative amounts of the supplied nutrients, but such a conclusion can only be tentative because a combination of natural shifts (e.g. in the water temperature), of fisheries (selective removal of certain species), as well as unknown factors at other trophic levels may have caused these changes.

7. Summary

Summarizing, our knowledge on the North Sea is incomplete.

(1) There is a reasonably good idea of the general water circulation through the North Sea and of the composition and the topography of the sea floor, although more detailed knowledge is still inadequate.

(2) There is also a reasonably good idea of the general transport of sediment (bottom transport and in suspension) through the North Sea, but the mechanisms involved are poorly understood, the *in situ* particle characteristics are badly known and on the whole suspended matter transport is much less clear than water transport.

(3) Nutrients have been studied intensively, resulting in a reasonable insight into their origin, distribution and fate but for other microconstituents this knowledge is largely lacking, which is hardly surprising considering that the necessary analytical techniques were developed in the late sixties or even later. Fluxes of nutrients and other trace constituents in the benthic boundary layer need further investigation.

(4) The ecology of the North Sea remains largely obscure because of the complexity of biological relations and the fact that most studies have been limited to commercially important species

(5) Apart from some self-evident or well-studied effects, an assessment of the effects of pollution and other human activities in the North Sea is very difficult to make.

To Dr G. C. Cadee and Dr P. A. W. J. de Wilde I am much indebted for critically reading the manuscript and to H. Hobbelink and R. Nichols for the illustrations.

References

Aure, J. & Saetre, R. 1981 Wind effects on the Skagerrak outflow. In *The Norwegian coastal current*, vol. 1 (ed. R. Saetre & M. Mork), pp. 263–293. University of Bergen.

van Bennekom, A. J., Gieskes, W. W. C. & Tijssen, S. B. 1975 Eutrophication of Dutch coastal waters. *Proc. R. Soc. Lond.* B **189**, 359–374.

Colebrook, J. M. & Robinson, G. A. 1965 Seasonal cycles of phytoplankton and copepods in the north-eastern Atlantic and the North Sea. *Bull. mar. Ecol.* **6**, 123–139.

Cushing, D. H. 1973 Productivity of the North Sea. In *North Sea science* (ed. E. D. Goldberg), pp. 249–266. Boston: MIT Press.

Dietrich, G. 1950 Die natürlichen Regionen der Nord- und Ostsee auf hydrographischen Grundlage. *Kieler Meeresforsch.* **7** (2), 35–69.
Dietrich, G. 1955 Ergebnisse synoptisch ozeanographischen Arbeiten in der Nordsee. *TagBer. Geographentag Hamb.*, pp. 376–383.
Eisma, D. 1973 Sediment distribution in the North Sea in relation to marine pollution. In *North Sea science* (ed. Goldberg, E. D.), pp. 131–150. Boston: MIT Press.
Eisma, D., Jansen, J. H. F. & van Weering, T. C. E. 1979 Sea-floor morphology and recent sediment movement in the North Sea. In The Quaternary history of the North Sea (ed. E. Oele, R. T. E. Schuttenhelm & A. J. Wiggers), *Acta Univ. Ups. Symp. Univ. Ups. ann. quing. Celebr.* **2**, 217–231.
Eisma, D. & Kalf, J. 1987 Dispersal, concentration and deposition of suspended matter in the North Sea. *J. geol. Soc. Lond.* **144**, 161–178.
Evans, P. R. 1973 Avian resources of the North Sea. In *North Sea science* (ed. E. D. Goldberg), pp. 400–412. Boston: MIT Press.
Förstner, U. & Reineck, H.-E. 1974 Die Anreicherung von Spurenelementen in rezenten Sedimenten eines Profilkerns aus der Deutschen Bucht. *Senckenberg. marit.* **6**, 175–184.
Gerlach, S. A. 1984 Oxygen depletion 1980–1983 in coastal waters of the Federal Republic of Germany. *Ber. Inst. Meeresk. Kiel* **130**, 1–87.
Hill, H. W. & Dickson, K. R. 1978 Long-term changes in North Sea hydrography. *Rapp. P.-v. Réun. Cons. int. Explor. Mer* **172**, 310–334.
Joiris, C., Billen, G., Lancelot, C., Daro, M. H., Mommaerts, J. P., Bertels, A., Bossicart, M., Nijs, J. & Hecq, J. H. 1982 A budget of carbon cycling in the Belgian coastal zone: relative roles of zooplankton, bacterioplankton and benthos in the utilization of primary production. *Neth. J. Sea Res.* **16**, 260–275.
Johnston, R. 1973 Nutrients and metals in the North Sea. In *North Sea science* (ed. E. D. Goldberg), pp. 293–307. Boston: MIT Press.
Lee, A. 1970 The currents and watermasses of the North Sea. *Oceanogr. mar. Biol. ann. Rev.* **8**, 33–71.
Maier-Reimer, E. 1977 Residual circulation in the North Sea due to the M2-tide and mean annual wind stress. *Dr. hydrogr. Z.* **30** (3), 69–80.
Mittelstaedt, E., Lange, W., Brockmann, C. & Soetje, K. C. 1983 Die Strömungen in der Deutschen Bucht. *D. Hydr. Inst. publ.* no. 2347. (141 pages.) Hamburg.
Oele, E., Schüttenhelm, R. T. E. & Wiggers A. J. (eds) 1979 The Quarternary history of the North Sea. *Acta Univ. Ups. Symp. Univ. Ups. Ann. Quing. Celebr.* **2**, 1–248.
Pegrum, R. M., Rees, G. & Naylor, D. 1975 *Geology of the North-West European continental shelf*, vol. 2 (*The North Sea*). (225 pages.) London: Trotman Dudley.
Pingree, R. D. & Griffiths, D. H. 1978 Tidal fronts on the shelf seas around the British Isles. *J. geophys. Res.* **83** (C9), 4615–4622.
Postma, H. 1985 Eutrophication of Dutch coastal waters. *Neth. J. Zool.* **35**, 348–359.
Reijnders, P. J. H. & Wolff, W. J. (eds) 1981 *Marine mammals of the Wadden Sea.* rep. no. 7, Wadden Sea Working Group. (64 pages.)
Riepma, H. W. 1980 Residual currents in the North Sea during the INOUT phase of JONSDAP '76. First Results. *'Meteor' Forsch. Ergebn.* A, no. 22, pp. 19–32.
Salomons, W. & Förstner, U. 1984 *Metals in the hydrocycle.* (349 pages.) Berlin: Springer-Verlag.
Smetacek, V. 1984 The supply of food to the benthos. In *Flows of energy and materials in marine ecosystems: theory and practice* (ed. M. J. Fasham), pp. 517–548. New York: Plenum Press.
Umweltprobleme der Nordsee 1980 *Der Rat von Sachverständigen für Umweltfragen. Sondergutachten, juni 1980.* (503 pages.) Verlag Kohlhammer: Stuttgart/Mainz.
Veenstra, H. 1970 Sediments of the southern North Sea. In *ICSU–SCOR Working party 31 Symposium, Cambridge*, vol. 3, pp. 13–23.
Waterkwaliteitsplan Noordzee 1985 Rijkswaterstaat, State Printing Office. (85 pages + 5 additional reports.)
Zijlstra, J. J. 1972 On the importance of the Waddensea as a nursery area in relation to the conservation of the southern North Sea fishery resources. *Symp. zool. Soc. Lond.* **29**, 233–258.

Discussion

R. JOHNSTON (24 *Hilton Drive, Aberdeen, U.K.*). There is no evidence of abnormal shortage of silicate in the open waters of the main part of the North Sea. How does Dr Eisma explain the development of the induced deficiency of silicate in coastal waters off Germany and The Netherlands?

D. EISMA. Although there has probably been a slight reduction in the silicate supply to the southern North Sea in the past decades, the relative deficiency of silicate in the coastal waters off Germany and The Netherlands is caused by the two- to fourfold increase in the supply of phosphate and nitrogen compounds. This has resulted in a shift from diatom predominance to a predominance of flagellates. The effect of this shift is reduced by the limitations on primary productivity imposed by the turbidity of the coastal waters and the mixing of coastal water with offshore water having a lower nutrient content.

J. I. G. Cadogan, F.R.S. (*British Petroleum International Co. plc, London, U.K.*). Dr Eisma said that the assessment of the effects of pollution versus normal changes is difficult to make. Nevertheless does he consider that fixed emission standards are appropriate or should the North Sea be zoned, with different standards being set for different areas?

D. Eisma. Different areas in the North Sea do not have the same sensitivity to pollution. So in a large accident in the North Sea involving oil, tidal flats like the Wadden Sea are more liable to be damaged and more difficult to tidy up than the open waters of the central North Sea, while continuous discharges of small amounts of oil will have similar effects in both areas. Some degree of zoning therefore seems logical, whereby the regulations are adapted to the characteristics of the zoned areas.

Phil. Trans. R. Soc. Lond. B **316**, 487–493 (1987)
Printed in Great Britain

The history and future of North Sea oil and gas: an environmental perspective

By F. G. Larminie

British Petroleum Co. plc, Britannic House, Moor Lane, London EC2Y 9BU, U.K.

The history of the development of the hydrocarbon resources of the North Sea is reviewed in an environmental context. The development of impact assessment techniques and practices and the evolution of monitoring of the physical, chemical and biological environment offshore and onshore, with reference to platforms, subsea pipelines, pipeline landfalls and terminal construction and operation is discussed. A brief account of the development of environmental protection management practices and their application to the design, construction, operation and management of major production projects follows. The paper concludes with a look at the environmental conditions likely to be established as the industry moves into the northern North Sea and areas such as the West Shetland Basin, and their significance for the petroleum industry.

Introduction

The organizers of this discussion meeting asked me for a background on oil and gas in the North Sea Basin, with particular emphasis on the environmental aspects of the development of this new hydrocarbon province. Because the first offshore gas discovery was made more than 20 years ago in 1965, followed by the first commercial oil discovery in 1969, it occurred to me that many younger scientists might well be unaware of the historical background to the oil industry as we know it today. I decided to present a highly selective annotated chronology of the main scientific, legal and technological events of this development. The first part of the paper constitutes this historical summary. The second and third sections discuss the potential impact and future developments of North Sea oil and gas development respectively.

Historical summary

1815 William Smith published the fifteen sheets of his 'Geological Map of England and Wales with part of Scotland'; the map series was based on observations made while working as a surveyor (and later as an engineer) on the construction of the English canals. As a consequence of this pioneering work, Smith is generally regarded as the father of English geology.

At about the same time, two French zoologists, Cuvier and Brongniart, published the first French geological map of the distribution and structural relations of the major rock units of the Paris Basin and north central France.

Up to the late 1920s knowledge of the geology of Europe was based mainly upon surface mapping supplemented by sub-surface information from mining, tunnelling and water-well drilling.

From the 1930s this was supplemented by data from exploratory drilling for oil.

In the early 1930s serious oil exploration in northern Europe started in Austria, Germany, The Netherlands and the United Kingdom, and by 1939 small oilfields had been discovered in all of these countries.

1939 D'Arcy Exploration Company (as the exploration arm of British Petroleum was then known) discovered oil at Eakring in England and this was followed by the discovery and commercial development of a whole series of small fields mainly in the east Midlands in England, but also including one at Kimmeridge in Dorset, England.

1959 The discovery of a huge onshore gasfield at Groningen in The Netherlands by Shell was a major breakthrough. Gas at Groningen was in a Lower Permian sandstone capped by evaporites (including salt). From drilling in the east Midlands, eastern and northeast England the stratigraphic succession was known to be similar and the rocks at these horizons were probably laid down in the same depositional basin, part of which now lay under the southern North Sea between the two countries.

It followed that if there were structures large enough to warrant offshore development there was a good chance that reservoir sands and caprock conditions similar to those of The Netherlands would also exist beneath the seabed in the southern North Sea.

1962 On the basis of this hypothesis, a group of companies performed a reconnaissance seismic reflection survey.

1963 Following the above survey, these companies implemented an aeromagnetic survey of most of the area south of 58°N, together with some additional seismic reconnaissance. The results were encouraging and the technology was available to explore for, produce and develop oil or gas fields in the prevailing conditions (water depths, wind, waves etc.) in the southern North Sea.

Hiatus: drilling was not possible because bordering states had not yet agreed the ownership of the continental shelf outside their territorial waters and states could not grant title to offshore tracts. The Geneva Convention on the Law of the Sea resulted from the first UN Law of the Sea Conference in 1958, and it established the sovereign rights of states bordering the continental shelf as: 'the... seabed and the subsoil of the marine area adjacent to the coast to a depth of 200 metres or beyond that limit to where the depth of the superjacent waters admits the exploitation of the natural resources.'

1964 Her Majesty's Government passed the Continental Shelf Act in April 1964 and ratified the Geneva Convention in the following month. The sub-division of the North Sea sector of the European Continental shelf was then settled between the riparian states either on the basis of median lines or by mutual agreement.

1965 Drilling started and that year British Petroleum discovered the first commercial hydrocarbons in its 48/6 well, the discovery well of the West Sole gasfield. In the next three or four years the southern North Sea was extensively explored and some dozen gas fields discovered. Success came early in the exploration of this basin, but it is important to note that the industry drilled over one hundred dry holes (absence of reservoir sands was the main reason for failure). Results suggested that the southern basin was predominantly a gas province and in the search for oil the focus shifted northwards to a thicker accumulation of sediments in a deep basin lying in much deeper water between Scotland and Norway. Drilling in this basin started in Norwegian waters.

1969 In November of this year Phillips made the first commercial oil discovery, the Ekofisk field, in the Norwegian sector.

1970 British Petroleum made the first commercial oil discovery in the British sector with its 21/10-1 well, the Forties Field discovery well. (This was not the first oil discovery; that honour belongs to Amoco's Montrose discovery.)

1971 In July, Shell discovered the Brent oil field in the East Shetland Basin.

1977 In July, British Petroleum made a significant discovery at Clair in very deep water in the West Shetland Basin.

The southern North Sea gasfields were in relatively shallow water and as previously noted the technology for their development was already available. When the industry moved northwards in the North Sea, conditions were vastly different and the main challenge lay in the combination of environmental factors which included such things as: deep water; extreme wind and wave conditions; air and water temperature; bottom sediment character; long distance from shore. Oil development had not previously been attempted in this combination of extreme marine environmental conditions. After the fact, some idea of the magnitude of the engineering design problems faced can be shown from a comparison of just two major features of the West Sole and Forties platforms:

	West Sole	Forties Charlie
deck load	2 kt	12 kt
support jacket	*ca.* 0.6 kt	*ca.* 16 kt

Another critical factor was the design of foundations because soil mechanics data were sparse. But solutions were found and there are now steel jackets, concrete gravity platforms, tethered semi-submersibles (Buchan), tension leg platforms (Hutton), Shell's underwater manifold systems (North Cormorant), sub-sea pipelines; in short the whole panoply of oil and gas production systems in place and producing.

Therefore in 20 years the United Kingdom offshore oil and gas industry has developed from nothing in 1964 to an annual average North Sea oil production rate of *ca.* 2.5 million barrels† of oil per day and over 10^8 m^3 of gas per day. (There were major developments in the Norwegian, Dutch and Danish sectors during this period and these effectively parallel events in the United Kingdom sector, but time and space preclude a detailed account here.)

How much of this oil produced has gone astray?

The following figures are taken from the United Kingdom Department of Energy 'Brown Book' (1985). From 1979 to 1984 inclusive, the total stabilized crude oil production was 574.3 Mt. There were 335 spills, resulting in 1.782 kt of spilled oil. These figures include one spill of 980t and five spills averaging 55 t each, and if these isolated (and relatively minor) events are excluded the average size of oil spill over the most recent six year period for which figures are available is in the range of 1–3 t. This input is not significant. However, in addition there were long-term, low-level (chronic) hydrocarbon inputs from, for example, the discharge of drill cuttings and oil-based drilling muds. The impact and significance of these are discussed by other speakers in this meeting.

† 1 barrel ≈ 0.159 m^3.

Potential impacts

The potential impacts are many and varied and the discovery of oil and gas in the North Sea basin exposed the paucity of our knowledge of the physical and biological characteristics of the seabed and the superincumbent waters. Added to this, oil is where one finds it which severely limits the options for locating production, transport, storage and export facilities. The design of the best possible system must seek to ensure the safety and integrity of all components, minimize the environmental impact, and resolve potential conflicts with other users and to this end, the requirements of good engineering practice and protection of the environment are to a large extent synonymous.

Sea and sea-bed

Existing uses include fisheries, shipping, gravel extraction and recreation. Possible conflicts include:

(a) location of platforms and other production installations;

(b) location of sub-sea pipelines;

(c) interruption to seabird migration by flares on platforms and major pollution incidents such as a well blow-out during the annual auk (Alcidae) migration across the North Sea;

(d) pollution as a result of an accident to a platform, pipeline or other production installation.

Coastline

Existing applications include residential use, recreation (e.g. beaches, marinas and golf-courses) commerce, nature reserves and sites of special scientific interest, defence and agriculture. Possible conflicts include:

(a) sensitive shorelines such as important coastal habitats, pipeline landfalls, impact of helicopter overflights on seabird colonies;

(b) amenities, for example Forties pipeline landfall at south end of Cruden Bay golf course;

(c) disruption of traditional activities in harbours and pre-empting berthing space;

(d) pollution from pipeline failure, offshore blow-out, or terminals which can be accidental or chronic (including both liquid effluent and emissions to the atmosphere).

Environmental management

It was apparent from the outset that the extreme conditions existing in the northern North Sea would necessitate particular attention to the environmental aspects of the development at all stages of the operation. Nobody had developed oilfields or laid large diameter pipelines in such water depths and at such distances from shore, and reliable baseline data on the physical and biological environment was in short supply. In consequence a lot of work had to be done to get information on such factors as wind speed, wave heights (for example the '100 year' wave), amplitudes, velocities and frequency, bottom-sediment conditions (for platform foundations and pipe-laying), bottom currents, rates of biological colonization of structures (for platform design and maintenance to obviate potential fouling problems), the impact of platform emplacement and operation on the benthic fauna, and the wind and current systems which would determine the movement of oil (essential for oil-spill trajectory modelling), to name but a few.

In my Company, the techniques of environmental impact analysis (EIA) were formulated and applied to the development of the Forties field, starting at the design stage of the project, and the importance of environmental baselines, not only for the offshore part of the development (platforms and submarine pipelines), but also for the onshore elements (e.g. pipeline landfall, overland pipelines and storage and export terminals), was recognized from the outset.

The EIA was supported by a comprehensive physical, chemical and biological monitoring programme during construction, commissioning and operation, aimed at determining the nature and extent of change consequent upon the development. From an initial 'in house' requirement this philosophy was extended to major multi-participant projects in which British Petroleum was involved (e.g. Sullom Voe terminal in Shetland).

Over time, monitoring techniques have evolved, and this is particularly so in the case of biological monitoring, where there have been major scientific advances related to the development of 'effects monitoring'. This involves the detection of changes in an organisms's response to petroleum hydrocarbons at the molecular or cellular level, and the results are of great scientific interest as they are providing an insight into the defensive physiology of marine invertebrates in response to environmental perturbations (both natural and man-made). This is a reversal of the usual relation between basic and applied science, and in this case research originally commissioned for applied industrial purposes has opened up a wider field of basic interest to science.

Having done the impact assessment and the monitoring programme, there is an essential third element in the process, namely to assess performance against prediction as objectively as possible. This can be achieved by an audit–review of the project when it has been in operation for some time, and the procedures of impact assessment, monitoring and audit–review are the essential components of environmental protection management. These procedures, well conducted, will ensure that the environmental impact of industrial activities is comprehensively assessed and appropriate technical and operational procedures adopted to minimize the adverse impact. This approach places the emphasis on good management techniques, together with sound science, and is an integral part of the management of many of British Petroleum's industrial activities world-wide. I speak only for my own company regarding such an approach to integrating environmental conservation and management, and other companies or consortia will have their own approaches. The orderly development of the North Sea oil and gas province to date, with minimal adverse consequences for the marine environment, could not have been achieved unless these or similar practices were the norm throughout the offshore oil and gas industry.

What next?

Forecasting, a risky process at the best of times, is even more fraught with difficulty given the present state of the crude-oil market.

If exploration picks up again the following considerations will be relevant:

as of 1985 the total proven and probable reserves in the United Kingdom sector of the North Sea were 15×10^9 barrels (2×10^9 t);

by the end of 1984 a total of 5.2×10^9 barrels had been produced;

in the development of the North Sea offshore oil province the big discoveries came early in the search, followed by a progressive reduction in average size with time;

it has recently been estimated by the United Kingdom Offshore Operators Association

(UKOOA) that about 80 new fields could be developed by the year 2000 and in British Petroleum's opinion, apart from finds in the as yet largely unexplored, recently licensed areas, the next generation of fields is likely to average 100 million barrels or less;

the 'frontier areas' west of Shetland and Scotland extending into the Atlantic towards Rockall, recently licensed, are aptly named (further offshore, in very deep and stormy waters) and largely unexplored;

reservoirs within the North Sea basin proper will be smaller and more complex, including gas condensates where a high proportion of hydrocarbon liquids must be separated from the gas before each is piped ashore;

exploration will extend further north, and this trend is already evident in the Norwegian sector;

solutions to the problems may include putting more of the production facilities on the seabed, more remote control and less manual intervention (e.g. remote controlled, unmanned platforms being developed for British Petroleum's new southern North Sea gasfields).

Should the UKOOA estimates prove correct, development of the 80 new fields referred to above would require a capital investment programme well in excess of the total sum expended on all the development to date. At the time of writing (February 1986), given the economic circumstances attendant upon the fall in oil prices, it is a moot point when this development will take place.

References

Cuvier, G. & Brongniart, A. 1811 *Essai sur la geógraphie minéralogique des environs de Paris, avec une carte géognostique et des coupes de terrain.*

Smith, W. 1815 *A map of the strata of England and Wales with a part of Scotland exhibiting the collieries, mines and canals, the marshes and fenlands originally overflowed by the sea and the varieties of soil according to variations in the substrata.*

United Kingdom Department of Energy 1985 *Development of the oil and gas resources of the United Kingdom 1985* (the 'Brown Book'). London: HMSO.

Discussion

R. B. Clark (*Department of Zoology, University of Newcastle upon Tyne, U.K.*). What will be done with redundant platforms after exhaustion of an oilfield?

F. G. Larminie. The short answer is nobody knows. SERC funded a study by the Institute of Offshore Engineering at Heriot Watt University, U.K. on alternative uses of offshore installations, and the question is currently the subject of active debate between the oil industry, the fishing industry and the U.K. Department of Energy. In the southern North Sea the smaller, lighter, gas production platforms in relatively shallow water can be removed and British Petroleum has already removed a redundant platform from its West Sole gasfield.

J. K. Rudd (*Amoco Europe and West Africa, Inc., Tottenham Court Road, London, U.K.*). With his extensive experience of monitoring around platforms, and with the benefit of hindsight, what would the speaker do now?

F. G. Larminie. Much as before, but in the established production areas there is now a body of knowledge which would reduce the need for baselines, and it is my opinion that a combination of bottom sediment studies, biological effects monitoring of selected benthic organisms and chemical analysis of the superincumbent waters would provide an effective framework for a monitoring programme. Outside the established areas, for example west of Scotland and north of 62° N, more baseline work will be required and here too the monitoring programme will be tailored to address the specific site-related impact of the installation.

M. L. Tasker (*Nature Conservancy Council*, 17 *Rubislaw Terrace*, *Aberdeen*, *U.K.*). With the move of the oil and gas industry into deeper water, new techniques of oil production will be used. Some of these involve remote controlled systems based on the seabed. Is it possible to foresee any new dangers to the environment arising from such systems?

F. G. Larminie. The most obvious possibility is 'out of sight, out of mind', but I tend to think that the performance and integrity of remote-controlled systems is likely to be even better than that of manned installations. They will be scrutinized by remote imaging systems and will probably be far more regularly and closely monitored than is the case with the sub-sea part of existing installations. The effect of an oil spillage would be much the same as that from a seabed blow-out at a conventional drilling–production system, and existing proven practices would be used to kill a wild well.

Phil. Trans. R. Soc. Lond. B **316**, 495–509 (1987)
Printed in Great Britain

A survey of inputs to the North Sea resulting from oil and gas developments

By D. R. Bedborough[1], R. A. A. Blackman[2] and R. J. Law[2]

[1] *Petroleum Engineering Division, Branch* 4, *Department of Energy, Thames House South, Millbank, London SW*1*P* 4*QJ, U.K.*

[2] *Aquatic Environment Protection Division* 2, *Directorate of Fisheries Research, Ministry of Agriculture, Fisheries and Food, Fisheries Laboratory, Remembrance Avenue, Burnham-on-Crouch, Essex CM*0 8*HA, U.K.*

The annual input of petroleum hydrocarbons to the North Sea has recently been estimated to be between 100 and 170 kt and is derived from a variety of sources.

Although there is uncertainty about the size of inputs from some sources, there is general agreement that the atmosphere, rivers and land run-off (including coastal sewage), and coastal oil industry activities combined with shipping, remain sources of major inputs. However, the size of annual inputs from the offshore oil and gas exploration and exploitation activities has recently increased to about 20 kt and these activities now form one of the major sources of petroleum hydrocarbons to the North Sea. This increase is almost entirely due to the use of oil-based drill-muds and the consequent discharge of drill cuttings contaminated with residual mud.

At present, experience in the United Kingdom has shown that this input of fresh, unweathered oil rapidly enters otherwise uncontaminated offshore sediments, producing strictly local effects around the point-source discharges. The nature and composition of this input differs from the majority of the inputs to coastal waters and sediments, and from the diffuse atmospheric input to offshore waters.

Of the 140 kt of materials other than oil discharged annually to the North Sea from oil and gas developments in the United Kingdom, 98–99% arise from drilling operations, but the vast majority of inputs from this source are biologically inert or derivatives of natural products. Surveys indicate that, of the remaining materials, less than 50 t of the more toxic products (i.e. those with a 96 h LC_{50} to *Crangon crangon* of less than 1 part/10^6) are discharged into United Kingdom waters annually.

The largely uncontaminated offshore North Sea waters and sediments remain little affected by offshore oil and gas developments, but if these activities enter already contaminated estuarine and coastal waters, the contamination and effects from this source will be harder to distinguish.

Introduction

Estimates of the global input of petroleum hydrocarbons to the seas have been provided by the United States National Academy of Sciences (N.A.S. 1975, 1985) but valid estimates of inputs to a particular body of water such as the North Sea require the compilation of detailed local information. This is no easy task, and it is fortunate that recent interest in the sources, fates and effects of oil in the North Sea has led to the gathering of such information, which is made available both by national governments and international regulatory bodies such as the Paris Commission.

Sources of petroleum hydrocarbon inputs to the North Sea

The Eighth Report of the Royal Commission on Environmental Pollution (1981) provided figures for the estimated inputs of petroleum hydrocarbons to United Kingdom waters from various sources (table 1). This table is taken from table 2.2 of that report. At about the same time, Whittle *et al.* (1982) contributed a paper to the Royal Society Discussion Meeting on 'The long-term effects of oil pollution on marine populations, communities and ecosystems',

Table 1. Estimated annual inputs of petroleum hydrocarbons to United Kingdom waters†‡ (kilotonnes) (from Royal Commission on Environmental Pollution (1981))

natural seeps	0.1–0.3
atmosphere	8–9
run-off from land (including urban run-off and inland municipal waste)	5–10§
coastal discharges of municipal wastes (sewage)	7§
coastal refineries	8
other coastal effluent	11
oil terminal operations	0.1
accidental losses from tankers at sea	10
operational discharges from tankers at sea	?
losses from general shipping	?
offshore production	0.5
discharges from ballast water treatment facilities	0.5
rounded totals	50–56‖

† Defined as: the North Sea (west of the median line), the English Channel (full extent), the Celtic Sea (south-west to the 100 fathom¶ line), the Irish Sea (full extent) and Scottish coastal waters (west and north to the 100 fathom line). Total area so defined = 715000 km^2.

‡ Compiled from estimates by the United Kingdom Petroleum Industries Association and Government Departments.

§ Including an unknown proportion of hydrocarbons of biogenic origin.

‖ No data for two items.

¶ 1 fathom = 1.83 m.

which also provided best estimates for such inputs. Table 2 is taken from table 7 of their paper and presents the position in about 1980. Both apply to approximately the same area of United Kingdom waters. There is generally good agreement between the two estimates, the differences being almost entirely due to their different estimates of the atmospheric input and the fact that the Royal Commission on Environmental Pollution report (1981) makes no estimates for operational losses from ships.

More recently the Institute of Offshore Engineering at Heriot-Watt University has conducted a survey of contaminant inputs (I.O.E. 1985) which includes inputs of petroleum hydrocarbons. Table 3 is taken from table 7.6 of their summary and updates the position to about 1982–83. Some idea of the probable relative inputs owing to operational tanker discharges and general shipping losses can be gained from table 4, taken from table 7.5 of the Institute of Offshore Engineering report, but there is much uncertainty about the actual quantities entering the North Sea from these sources. The derivation of the 23 kt of oil from offshore oil and gas production (bottom line of table 3) is shown in table 5 which is taken from table 7.4 of the Institute of Offshore Engineering report. It is difficult to compare in detail these estimated inputs to the whole North Sea, which include those originating from North Sea states other

TABLE 2. ESTIMATED ANNUAL INPUTS OF PETROLEUM HYDROCARBONS TO UNITED KINGDOM WATERS† (KILOTONNES)

(After Whittle *et al.* 1982.)

	best estimates	
natural seeps	*< 0.3*	
offshore oil activity	*3.41–3.56*	
formation and production water		0.75–0.9
displacement water		0.05‡
accidental spills		0.1‡
platform runoff		0.01
drill cuttings		2.5
transportation	*7.55–9.0*	
tankers: large spills		1.9–2.2§
small spills		0.5–0.1
non-tanker spills		< 1.0
operational losses from ships		4.0–5.0
shore terminals		
refinery terminals, coastal storage depots, crude loading terminals		0.1–0.2
oily water discharges		0.5
refineries	*6.0–6.5*	
municipal and industrial effluents	*15.51–16.91*	
sewage effluent to tidal waters		0.5–1.4
sewage sludge dumped at sea		6.5–7.0
industrial waste dumped at sea		< 0.01
industrial effluent		8.5
urban runoff to tidal waters		
rivers	< 7.5	
atmosphere	30.0‖	
(*dredge spoil*)¶	(2.8)¶	
total	70.3–73.8	

† Area considered for inputs: west and north to 200 m depth, south to latitude 49° N and east to median line, excluding west of Eire. Total area considered approximately 750000 km².

‡ Plus large spills from major incidents.

§ Plus massive spills from major incidents.

‖ Maximum estimate 250.

¶ Not included in inputs total, redistribution rather than input.

than the United Kingdom, with those previously presented for United Kingdom waters, because of the different types and sizes of area receiving them. However, given that all the United Kingdom offshore production inputs cited in tables 1 and 2 were to the North Sea, that little of the United Kingdom coastal and atmospheric inputs will enter Celtic Sea and northwest Scottish waters, and that the Institute of Offshore Engineering (1985) assumes that 50% of the total United Kingdom inputs enter the North Sea, two general statements can be made about relative North Sea inputs with time. Firstly, despite differences in estimated size of inputs, it is agreed that the atmosphere, rivers and land run-off (including coastal sewage), and finally coastal oil industry activity combined with losses and discharges from shipping, form three major sources of petroleum hydrocarbons. Secondly, the input from North Sea oil and gas developments has changed from a minor to a major source and this is almost entirely because of the increase in the amounts of oil discharged on drill-cuttings from the use of oil-based drill-muds. This is illustrated by the most recent figures from the United Kingdom Department

TABLE 3. ESTIMATED ANNUAL INPUTS OF PETROLEUM HYDROCARBONS TO THE NORTH SEA† (KILOTONNES)

(After I.O.E. 1985.)

natural seeps	0.3–0.8‡
atmospheric	19§
rivers, land runoff (including inland municipal waste)	40–80
coastal sewage discharges	3–14
coastal refineries	6.0
oil terminal operations (including reception facilities)	0.8
other coastal industrial effluent	9
accidental losses from tankers at sea	5–12
operational discharges from tankers at sea	?
losses from general shipping	?
offshore production	23
total	107–165

† North Sea defined as waters north of the Strait of Dover, the Skagerrak and the area to 61° N to a line at 4° W.

‡ Estimated for an area of 5.3×10^5 km^2 of continental shelf at 0.5–1.5 t 1000 km^{-2}.

§ Includes United Kingdom inputs to non-North Sea waters.

TABLE 4. INPUTS OF OIL TO THE NORTH SEA IN 1982 FROM SHIPPING AND TANKER TRANSPORTATION OPERATIONS AS REPORTED TO THE PARIS COMMISSION IN 1983 (KILOTONNES)

(After I.O.E. 1985.)

	U.K.	Norway	Denmark	Netherlands	Germany
losses from tankers	3.0–10.0	0.56	0.0005	1.65	0.32
losses from shipping (including operational discharges from tankers)	?	2.0	1.5	?	0.2
reception facilities	0.5	n.a.	0.002	0.046	0.012
oil terminals	0.5	1.5	0	n.a.	0.29
inland navigation	?	n.a.	0	4.0–9.0	5.7

n.a., Not available.

of Energy showing that the total quantity of oil discharged on cuttings by the United Kingdom has increased from 5.78 kt in 1981 to 19.6 kt in 1985. These also show that the earlier figures for this input in tables 1 and 2 were underestimates.

There are several points to be made about inputs from these four major sources. It is to be expected that coastal waters receive a greater annual deposition of petroleum hydrocarbons from the atmosphere than more distant North Sea waters. This input is diffuse and its composition and environmental effects are largely unknown because of changes brought about by processes such as photolysis and photo-oxidation while in transit. Inputs from rivers to the North Sea can be considered more as point sources and the effects are thought to be restricted to coastal waters. This is because most of their oil load is highly modified, degraded and adsorbed onto suspended solids by the time it reaches the sea and the bulk of this material will sediment out and be trapped in the estuary or be confined to adjacent coastal waters. However, the effects there of river-borne oil are augmented by the many, direct, industrial point sources of oil into estuaries and coastal waters including those from coastal oil industry activity. General land run-off, most shipping losses and operational discharges from shipping may be considered as a diffuse coastal input or as a series of minor, local point sources. The input from offshore

TABLE 5. TOTAL COMBINED SECTOR NORTH SEA OIL INPUTS FROM OFFSHORE OPERATIONS BY SOURCE, 1982–1983 (KILOTONNES)

(After I.O.E. 1985.)

source	kilotonnes	comment
production and displacement water	3.0	increasing annually
oil on cuttings	20.0†	reflects annual usage of oil-based muds and drilling activity
drainage	0.1	relatively constant
atmospheric inputs from flaring	0.2–1.0	approximate figure, decreasing
spills	0.4–2.0	highly variable from one year to the next, range makes no allowance for major incidents such as prolonged blowout
total North Sea	23.7–26.1	

† No input from Dutch Sector drilling activity is assumed.

oil and gas developments is very different because it enters the sea almost entirely as unweathered oil attached to particles which pass directly to otherwise uncontaminated offshore sediments around a series of discrete point sources. The low-toxicity alternative base-oils now in common use also differ significantly in composition from the residues of crude, refined and lubricating oils that characterize most of the other inputs because of the absence of higher molecular mass aromatic fractions. They also differ from the fresh crudes and fuel oils typical of large spillages because of a reduced content of lower molecular mass aromatics. Therefore the environmental effects of discharged base-oils in sediments can be expected to differ from those of other oil inputs. If large-scale oil and gas development moves into coastal and estuarine waters this point-source contamination could be added to areas already likely to be under general stress from other oil inputs and other contaminants, such as heavy metals and persistent organohalogens. On the other hand the biological populations in these areas are likely to be more adapted to the presence of oil than those offshore.

A more detailed examination of the sources and nature of the inputs from offshore oil and gas developments to the whole North Sea can be made by considering the United Kingdom sector activity, because these inputs are largely due to drilling activities and the United Kingdom has the greatest number of offshore installations and the largest drilling programme.

UNITED KINGDOM INPUTS TO THE NORTH SEA FROM OFFSHORE OIL AND GAS DEVELOPMENTS

Sources of information

The pollution control authority for the United Kingdom offshore oil and gas industry is the Petroleum Engineering Division (PED) of the Department of Energy (DEn). A close liaison is maintained between DEn and the two Government fisheries departments, the Department of Agriculture and Fisheries for Scotland (DAFS) and the Ministry of Agriculture, Fisheries and Food (MAFF). The Petroleum Engineering Division of DEn receives information of all discharges of oil from the United Kingdom offshore industry into the marine environment because of its powers under the Prevention of Oil Pollution Act 1971. This Act covers the discharges of oil in produced water and storage displacement water, oil spills and oil on drill cuttings.

There are no statutory regulations covering the discharge of chemicals which are part of normal operations, but surveys of chemicals used offshore during 1982 and 1984 have been carried out and when combined with information from the voluntary 'Notification Scheme for the Selection of Chemicals for Use Offshore' give an indication of the quantities, types and toxicities of chemicals used and discharged in both drilling and production activity.

Growth of offshore oil and gas activity

Before discussing the various discharges we shall briefly consider the growth in the United Kingdom offshore oil and gas activity over the past five years, bearing in mind that gas was first produced offshore in 1967 and oil in 1975. Gas production has risen from 37.4×10^9 m^3 in 1981 to 43.0×10^9 m^3 in 1985 and oil production has risen from 87.7 mt to 127.5 mt over the same period. The great majority of discharges arise from drilling operations and the number of wells drilled annually has increased from 211 to 273. Only a small fraction of the discharges from production activity arises from gas production; the majority is associated with oil production and during the past six years the number of oil production platforms has increased from 26 to 36.

(*a*) *Oil*

Oil from offshore activity enters the marine environment from three sources.

(i) *Oil spills.* Table 6 gives details of offshore oil spills from platforms reported to the Department of Energy from 1981 to 1985. Each year there have been two or three spills in the range of 30–80 t but the majority of spills are in the range of 1–3 t. However, there is no discernible pattern of oil spills and it is not possible to predict their future numbers or sizes.

TABLE 6. OFFSHORE OIL SPILLS FROM INSTALLATIONS REPORTED TO THE DEPARTMENT OF ENERGY 1981–1985

year	number	total amount/t	number of producing platforms
1981	71	104	26
1982	42	162	28
1983	62	186	32
1984	47	130	34
1985	87	310	35

(ii) *Oil associated with produced water and storage displacement water.* When oil is first produced from a geological reservoir the water content is usually low but over the life of any production platform the water content gradually increases. When the water content exceeds a certain level (usually that figure acceptable for transport via a pipeline) water is separated from the oil and discharged offshore. Some offshore installations possess facilities for storing oil in sub-sea cells with an oil–water interface. When oil is transferred to these the displaced storage water is discharged.

Section 3 of the 1971 Prevention of Oil Pollution Act makes it an offence for any offshore installation to discharge any oil or any mixture containing oil. However, it is not practicable to remove oil totally from production water and so installations are exempted from the provisions of the 1971 Act under section 23. Each exemption has a schedule attached which sets out the requirements for sampling and analysis of the relevant discharge. Platforms are expected to comply with a monthly average oil in water standard of less than 40 mg l^{-1} and records of analyses for oil water and volumes of water discharged are returned to DEn each month. These returns, combined with a recent survey of projected production water discharges from each platform, enable a picture of past and future discharges to be constructed.

Figure 1 shows this picture for all fields currently in production from 1979 to 1996. Data for 1979 to 1984 are actual recorded figures, for 1985 to 1989 are reliable projections and for 1990 onwards are speculative because of problems in forecasting. New fields not yet in production have not been included.

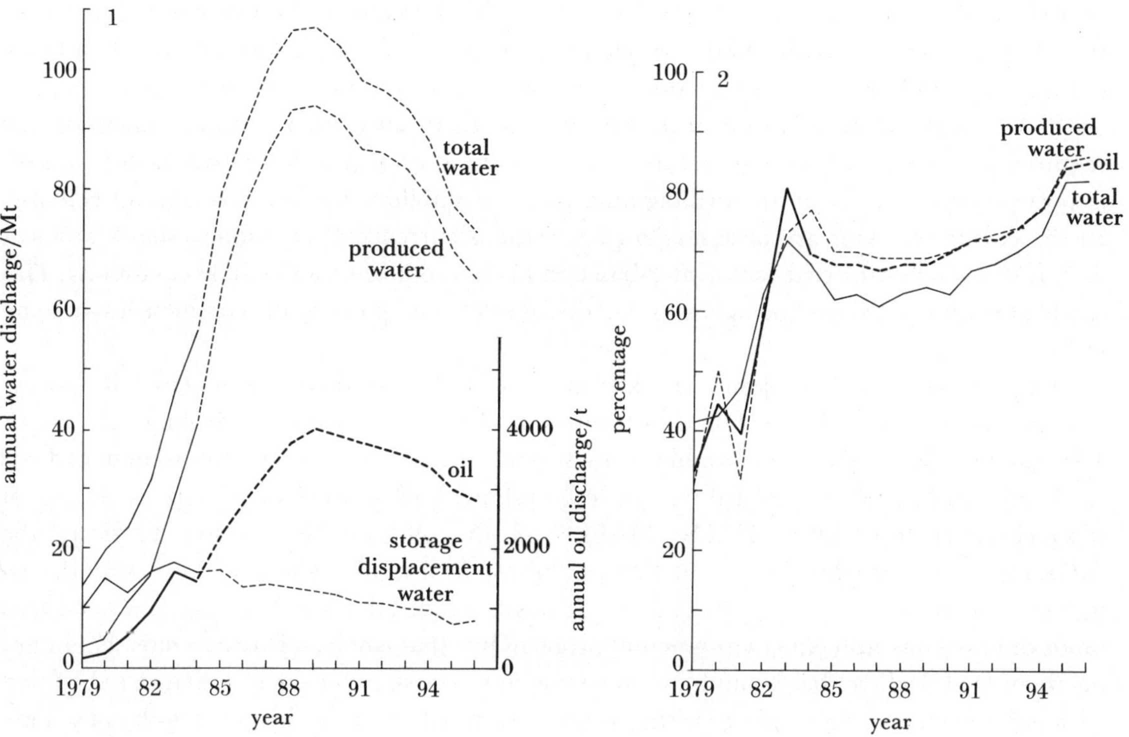

FIGURE 1. Projection of oil and water discharged from United Kingdom North Sea oil operations.

FIGURE 2. Projection of United Kingdom East Shetland Basin discharges as a percentage of all United Kingdom North Sea oil production discharges. The left ordinate is the East Shetland Basin discharge as a percentage of total North Sea discharge.

The indications are that water production will peak around 1989–1990 although the tail-off may not be as steep as shown when new platforms come into production. A peak figure of around 110 Mt per year of water can be expected, i.e. around twice the 1984 figure. It should be noted that current installed capacity for water treatment is around 220 Mt per year.

The projected increase in the discharge of oil associated with the water will be larger, rising to a probable figure of around 4 kt per year (about three times the 1984 figure); this is because of the concentration of oil in discharged water increasing as the water flow rate approaches the control equipment design capacity. At present, two thirds of platforms are operating with their control equipment at between 10 % and 50 % of design capacity, the median figure being

30 %. It has been assumed that the average concentration will be 40 mg l^{-1} for produced water and 10 mg l^{-1} for storage displacement water. The latter input will decrease slowly from a current annual figure of around 18 mt to around 9 mt.

Within the North Sea, much of the discharged water comes from the East Shetland Basin. This is a well-defined and relatively concentrated area of activity which tends to produce about three times more water per barrel of oil than other areas because of its formation characteristics. Figure 2 summarizes the importance of the East Shetlands area which is now producing around 70 % of water and associated oil and could produce an even higher percentage of discharged oil beyond 1989.

(iii) *Oil on drill cuttings*. In 1984, Section 1 of the 1971 Act was amended such that the definition of oil was changed from 'crude oil, fuel oil, lubricating oil and heavy diesel oil' to 'any oil produced directly or indirectly from crude oil'. This alteration was required to enable the 1971 Act to apply to diesel-oil and refined mineral-oil used in the formulation of oil-based muds (OBM). Drilling muds are traditionally regarded as mixtures of clays, water and chemicals circulated to the drill bit to lubricate the systems, carry away rock cuttings, maintain the required pressure at the bit end and to provide an aid to geological formation evaluation. An important requirement for any drilling mud is that it stabilizes the borehole against collapse. In the North Sea OBMS are necessary owing to the occurrence of claystones, shales and salt formations which can react with water-based muds to produce unstable hole conditions. This can lead to protracted drilling operations, reduced safety and possibly the complete loss of some wells.

The past five years have seen a marked increase in the use of OBM, although this appears now to have levelled off and should in future reflect the general level of drilling activity. In 1981 the main base-oil used to formulate OBM was diesel-oil; concern about the amount of diesel oil being discharged to sea led to the establishment of a working group made up of representatives from DEn, DAFS, MAFF and the offshore oil industry to study the environmental effects around production platforms. As a result, new guidelines were drawn up to control the discharge of oil on cuttings. It was concluded that the discharge of water-based muds did not constitute an environmental problem but that whole, oil-based muds should not be discharged. If diesel-based muds are to be used it is necessary for the installation to be fitted with equipment to remove oil from the cuttings. In practice, since the new regulations came into effect no diesel-based muds have been used. Where an acceptable low toxicity base-oil OBM is to be used the installation and operation of efficient equipment to separate cuttings from mud is considered to be adequate.

Table 7 shows the number of wells drilled using OBM from 1981 to 1985. From 1981 to 1984

TABLE 7. WELLS DRILLED WITH OIL-BASED MUD (OBM) ON THE UNITED KINGDOM SECTOR 1981–1985

year	1981	1982	1983	1984	1985
total wells drilled	211	229	223	290	273
percentage drilled using OBM	36	50	65	76	77
percentage using low toxicity mud	2	15	39	71	77
weight of diesel discharged/t	5500	6000	6200	1400	0
weight of low toxicity oil discharged/t	280	2600	8300	18400	19600
total oil discharged/t	5780	8600	14500	19800	19600
oil per well/t	76	75	100	90	93

an average figure of 177 g oil per kilogram of dry cuttings was assumed and this has been substantiated by the figures for 1985 when an average of 180 g kg^{-1} was recorded. The total quantity of oil discharged on cuttings is determined primarily by the amount of drilling activity and it is not possible to determine what this will be in the future.

(iv) *Total oil inputs from installations.* Total inputs of oil to the North Sea from the United Kingdom sector offshore oil and gas activity are summarized in table 8. It can be seen that the main input of oil (90 %) is from drill cuttings with reported oil spills representing only about 1 % of the oil released into the sea.

TABLE 8. TOTAL INPUTS OF OIL TO THE NORTH SEA FROM UNITED KINGDOM SECTOR OFFSHORE OIL AND GAS ACTIVITY

year	1981	1982	1983	1984	1985
total input/t	*6409*	*9689*	*16386*	*21350*	*22072*
percentage from:					
oil spills	1.6	1.6	1.1	1.0	1.4
produced water	8.2	9.6	10.4	6.7	9.8
drilling	90.2	88.8	88.5	92.3	88.8
total stabilized crude oil produced from offshore fields/mt	*87.7*	*100.1*	*110.5*	*120.8*	*127.5*

(*b*) *Chemicals*

In 1979 the United Kingdom Government established its 'Notification Scheme for the Selection of Chemicals for Use Offshore' (CNS) to monitor and advise on the use and discharge of chemicals from oil and gas activities on the United Kingdom continental shelf. The aim of the voluntary scheme is to prevent damage being caused to the marine environment by discharges other than oil from offshore installations.

Surveys of chemicals used in 1982 and 1984 were carried out to determine the quantities and types of chemicals discharged. An attempt has been made to separate drilling operations from production operations; here production refers to chemicals added after the production manifold (prior to and downstream of phase separation) and also to chemicals added for the treatment of injection water or cooling water.

(i) *Production chemicals.* In 1984, 22 kt of production chemicals were used compared to 9 kt in 1982, the main increase being about 7.5 kt of gas treatment chemicals (mainly monoethylene glycol, triethylene glycol and methanol). Gas platforms were not included in the 1982 survey. Production chemicals constituted only 5–10 % of all chemicals used and typically only one fifth by mass were discharged to the sea. Production chemicals therefore constituted only 1–2 % of all discharges. Table 9 lists the type of products discharged because of offshore production

TABLE 9. PRODUCTS DISCHARGED AS A RESULT OF OFFSHORE PRODUCTION OPERATIONS (TONNES)

year	1982	1984
scale inhibitors	950	1200
corrosion inhibitors	550	220
biocides	180	500
demulsifiers	30	600
oxygen scavengers	30	120
gas treatment	—	520

operations. The main differences between 1982 and 1984 would appear to be the increased discharges of biocides, demulsifiers and oxygen scavengers. However, in the 1982 survey the use of onboard chlorine generators was not covered and this accounts for the majority of the increased discharge of biocides in 1984. The increased discharge of oxygen scavengers and demulsifiers is due to increased rates of water injection and water production.

There has been a marked increase in the proportion of production chemicals used and discharged for which full toxicity data have been made available; this has increased from 33 % of the discharged materials by mass in 1982 to 70 % in 1984. No toxicity or product composition data have been made available for about 20 % of the mass discharged but given that the scheme is voluntary it will be difficult to improve this situation. Nevertheless, from the information received, it seems unlikely that any production chemicals used and discharged had potentially harmful toxic properties (i.e. with a 96 h LC_{50} to brown shrimp, *Crangon crangon*, of less than 1 part/10^6).

It is not possible to predict future discharges of production chemicals other than to say that they can be expected to increase as more treated water is injected into reservoirs and more residually contaminated water is produced with the oil, even though oil production itself is expected to decline.

(ii) *Drilling fluids*. The number of wells drilled increased from 223 in 1982 to 290 in 1984 and the quantity of chemicals used increased by roughly the same proportion. However, the quantity of chemicals discharged remained relatively constant at 133 kt in 1984 compared with 136 kt in 1982, i.e. the percentage of chemicals used which were discharged dropped from 81 % to 57 %. This reflects the increased use of oil-based muds: from 50 % of all wells drilled in 1982 to 76 % of all wells drilled in 1984. United Kingdom regulations allow the discharge of whole, water-based muds but not of whole, oil-based muds which have to be returned to the shore for reprocessing. The discharges of the base-oil component were excluded from both surveys because they are covered elsewhere.

Of the other drilling fluid components discharged, 127 kt in 1984 and 131 kt in 1982 are inert or derivatives of natural products and are considered to have little or no environmental impact apart from local effects of physical smothering, sediment grain size alteration and organic enrichment. Note that even the discharge of clean rock cuttings would lead to the first two of these effects and that barites and bentonite, which together account for around 75 % by mass of all drilling discharges, are virtually insoluble and inert.

There are no important differences between the types and quantities of drilling fluid chemicals discharged during 1982 and 1984, and table 10 lists the majority of chemicals discharged as a result of offshore drilling operations. Polymeric viscosifiers such as carboxymethyl celluloses and starches are generally biodegradable and have little or no environmental effect beyond organic enrichment. The main lignosulphonate used is ferrochrome lignosulphonate with 3–4 % Cr^{3+}. Chromium is strongly adsorbed onto the clay minerals in the drilling mud and probably has little or no effect outside the immediate settlement zone. Of the total 5–6 kt of 'non-inert' materials, full composition and toxicity data are available for about 50 % by mass but no information on 25 % by mass. However, because of statutory requirements, although toxicity data are not available on all the individual mud components, they are available on the complete mud systems which must have a 96 h LC_{50} to *Crangon* of greater than 6000 parts/10^6. Therefore, although it would be desirable to bring more individual drilling products into the CNS, such an omission is not as serious as it might at first appear

TABLE 10. PRODUCTS DISCHARGED AS A RESULT OF OFFSHORE DRILLING OPERATIONS (TONNES)

year	1982	1984
weighting agents and inorganic gelling products	*112400*	*111600*
inorganic chemicals	*12560*	*11000*
polymeric viscosifiers	*4280*	*3680*
lignosulphonates, lignites, etc.	*700*	*630*
minor additives:		
surfactants, detergents	250	200
de-foamers	65	400
biocides	150	20
corrosion inhibitors	150	20
drilling lubricants	190	200
oxygen scavengers	160	65
dispersants	30	95
pipe release agents	25	15
scale inhibitors	10	10

in estimating potential environmental damage. From the toxicity data available, it can be predicted that the great majority of the 'non-inert' materials have a 96 h LC_{50} to *Crangon* of greater than 1 part/10^6.

As with the discharge of oil associated with drill cuttings it is not possible to predict future discharges of drilling chemicals, which will be related to overall drilling activity.

One tentative conclusion from the two surveys of production and drilling chemicals is that the quantity of more toxic chemicals (i.e. those with a 96 h LC_{50} to *Crangon* of less than 1 part/10^6) discharged from offshore oil and gas activity on the United Kingdom continental shelf during 1982 and 1984 was less than 50 t per year. Therefore although the quantity of non-oily materials discharged offshore to the United Kingdom continental shelf is large, the effects are expected to be restricted in extent and severity. Similarly, the oil inputs from the United Kingdom offshore oil and gas industry enter North Sea waters and sediments from a series of point-sources and have strictly localized effects compared with coastal inputs. This can be shown by considering the results of surveys carried out by MAFF, monitoring hydrocarbon concentrations in United Kingdom waters and sediments.

HYDROCARBON CONCENTRATIONS IN UNITED KINGDOM WATER AND SEDIMENTS

Water

Hydrocarbon concentrations in uncontaminated offshore water are generally less than 2 $\mu g\ l^{-1}$. Law (1981) found total hydrocarbon concentrations in subsurface waters of 0.7–1.3 $\mu g\ l^{-1}$ (mean 0.9 $\mu g\ l^{-1}$) Ekofisk crude oil equivalents near the median line of the North Sea between 55 °N and 57 °N, south of the Auk and Ekofisk oil fields. In a more recent survey in 1985 carried out between Auk, the Forties field and the Norwegian coast, concentrations between 0.7 and 1.3 $\mu g\ l^{-1}$ (mean 1.0 $\mu g\ l^{-1}$) were found. All these values were determined by fluorescence spectroscopy and blank-corrected. In the southern North Sea concentrations were somewhat higher, 1.3–2.7 $\mu g\ l^{-1}$, with a mean of 1.7 $\mu g\ l^{-1}$ (Law 1981). In coastal areas and around industrialized estuaries higher concentrations are found. A series of transects were sampled during a cruise in 1984 directly out to sea from four east-coast estuaries, those of the rivers Tyne, Tees, Humber and Thames. Concentrations along these transects fell from 9.4 to 0.8 μg l^{-1}, 83 to 0.4 μg l^{-1}, 15 to 0.5 μg l^{-1}, and 11 to 1.5 μg l^{-1} respectively. Concentrations typical of offshore water were not reached within 40 miles of the coast off the Humber, but were established 20 miles off the other three estuaries. By comparison, preliminary results from a study of the dispersion of oily water discharges from a North Sea production platform in 1985 show the maximum hydrocarbon concentration 250 m down-current of the platform to be 60 μg l^{-1} at 10 m depth.

Sediments

In clean, sandy, offshore sediments on the Dogger Bank in the central North Sea total hydrocarbon concentrations are 0.4–2.0 μg g^{-1} dry mass Ekofisk crude-oil equivalents (Law & Fileman 1985). This is similar to the range found found in clean, sandy sediments in water greater than 100 m deep in the Western Approaches of 1.1–2.0 μg g^{-1} dry mass (Law 1981); a general background figure for clean sand would therefore appear to be less than 2 μg g^{-1}. Finer sediments tend to have higher concentrations, and so in muddier deposits background concentrations may reach 5 μg g^{-1}.

Sediment hydrocarbon concentrations vary widely in areas affected by oil pollution. During investigations of a chronic oil pollution problem associated with a refinery discharge Dicks & Iball (1981) found aliphatic hydrocarbon concentrations ranging from 100 to 5620 μg g^{-1} dry mass in an estuarine saltmarsh. Sediment samples collected intertidally from the outer Humber estuary and the north Lincolnshire coast following the *Sivand* oil spill in 1983 ranged from 16 to 1300 μg g^{-1} dry mass Ekofisk crude oil equivalents at sites believed to be unaffected by the spill but subject to chronic low-level input. Discharges of drill cuttings from offshore operations in which oil-based muds are used may lead to very high sediment concentrations close to discharging platforms. Davies *et al.* (1984) reviewed the effects of the use of oil-based drilling muds in the North Sea and compiled results from surveys done by a number of different groups around North Sea installations. They found that hydrocarbon concentrations in sediments within 250 m of a platform could be 1000 times the background concentration, but concentration gradients away from the platform were very steep and normal background concentrations were usually reached 2–3 km from the platform. Very high concentrations could be found beneath and directly alongside platforms: greater than 50000 μg g^{-1} dry mass alongside Beatrice (Davies *et al.* 1984) and 160000 μg g^{-1} dry mass diesel equivalents underneath Murchison (Law *et al.* 1982). Although the effects of such accumulations on benthic animals are obviously very marked, major biological effects occur in only a very limited area, generally within 500 m of the platform (Davies *et al.* 1984).

Most oil and gas developments to date have been in offshore areas largely unaffected by other hydrocarbon inputs. In the future it seems likely that more fields will be developed in coastal and even estuarine environments and it will then be more difficult to link effects and causes than at present.

Coastal areas

Data obtained during two recent pre-development surveys carried out by MAFF will serve to illustrate the greater complexity of sediment hydrocarbon distributions in coastal areas, whether in the North Sea or elsewhere. The first survey was carried out in the English Channel in 1982, south and west of the Isle of Wight. Sediment hydrocarbon concentrations ranged from 0.3 to 370 μg g^{-1} dry mass Ekofisk crude oil equivalents (figure 3).

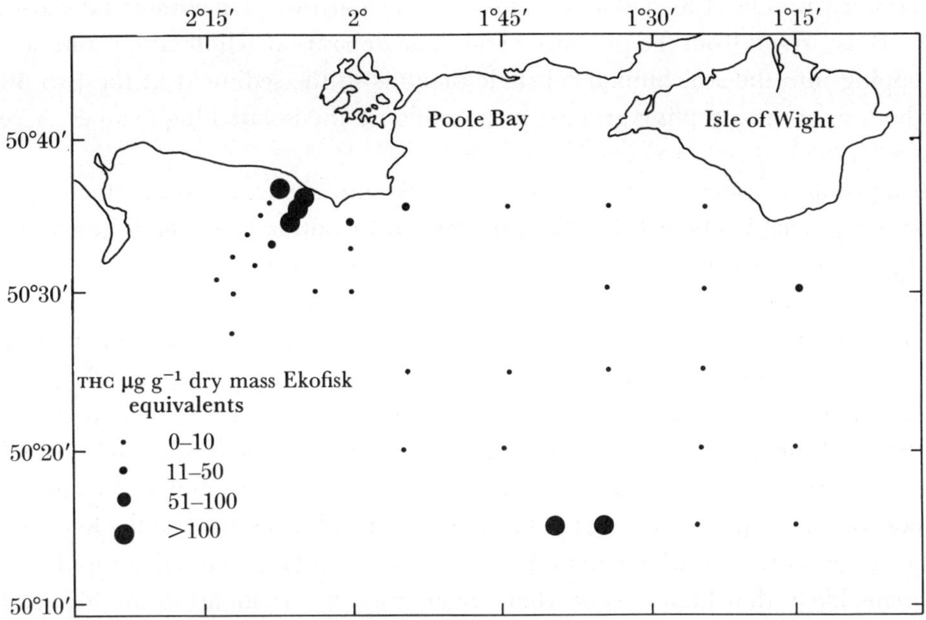

FIGURE 3. Total hydrocarbon concentrations in surface sediments from the English Channel, south and west of the Isle of Wight, in 1982.

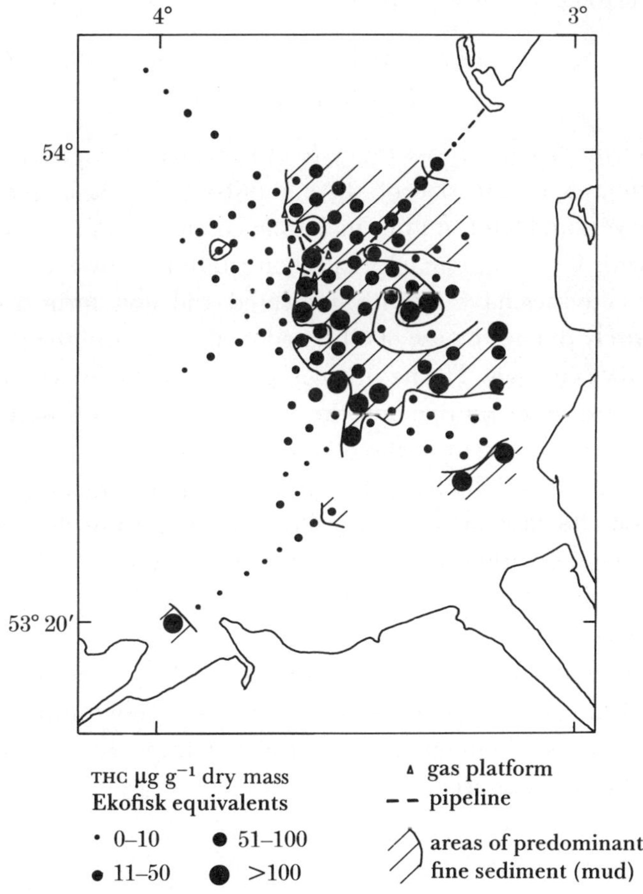

FIGURE 4. Total hydrocarbon concentrations in surface sediments from the eastern Irish Sea in 1982–1983.

The high concentrations (120–370 μg g^{-1}) found in the northwestern corner of the grid, close to the Dorset coast, result from the presence of oil shale deposits at Kimmeridge and of a shale shelf outcropping into the sea. Similar oil shale deposits in the sediment at the two offshore sites at the bottom of the sampling grid are responsible for the isolated high values there (160 and 280 μg g^{-1}).

The second pre-development survey was done in 1982 and 1983 around the (then) proposed Morecambe Bay gas field. The sediment hydrocarbon distribution is related to the occurrence of fine mud particles as the dominant size fraction of the sediment (figure 4).

The gas field is situated at the western edge of a large mud patch extending down the eastern margin of the eastern Irish Sea, with sandy sediments occurring to the west of the gas field. Concentrations of sediment hydrocarbons ranged from 3.3 to 410 μg g^{-1} dry mass Ekofisk crude oil equivalents. The association between fine sediment and high concentrations of hydrocarbons is evident but the source of the hydrocarbons is more difficult to establish. The eastern Irish Sea is subjected to hydrocarbon inputs from a large number of different sources, including most, if not all, those discussed in the first part of this paper. A large proportion of the hydrocarbons reaching the eastern Irish Sea either enter adsorbed onto particulate material or rapidly become so, and it seems likely that high hydrocarbon concentrations are found in muddy sediments because this is where these are being deposited. The distribution is complex, and interpretation of monitoring data for sites at any distance from the discharge point would be much more difficult in an already contaminated coastal area like the eastern Irish Sea, than in comparatively clean offshore waters.

Conclusion

The inputs of petroleum hydrocarbons to the North Sea arise from a wide variety of sources. Although recent estimates of the size of these inputs varies, it is generally agreed that atmospheric return, rivers and land run-off (including coastal sewage), and coastal oil industry activities combined with shipping, remain major contributors. However, inputs from offshore oil and gas industry activities have recently increased and now form a major source. This increase is almost entirely due to the use of oil-based drill-muds and the consequent discharge of mud residues with rock cuttings. The nature, composition and effects of this input differ from those of inputs from most other sources and the United Kingdom situation is that its effects are restricted to small areas adjacent to the point-source discharges.

Of the materials other than oil discharged to the sea from offshore oil and gas developments in the United Kingdom, 98–99% arise from drilling operations but the vast majority of these are biologically inert or derivatives of natural products. Surveys of the materials used and discharged indicate that less than 50 t per year of the more toxic products are discharged into United Kingdom waters.

The largely uncontaminated offshore waters and sediments of the North Sea remain little affected by offshore oil and gas developments. If such activities move into already contaminated estuarine and coastal waters, contamination and effects from this source will be harder to distinguish from those of other sources.

References

Davies, J. M., Addy, J. M., Blackman, R. A. A., Blanchard, J. R., Ferbrache, J. E., Moore, D. C., Somerville, H. J., Whitehead, A., & Wilkinson, T. 1984 Environmental effects of oil-based drilling muds in the North Sea. *Mar. Pollut. Bull.* **15**, 363–370.

Dicks, B., & Iball, K. 1981 Ten years of saltmarsh monitoring – the case history of a Southampton Water saltmarsh and a changing refinery effluent discharge. In *Proceedings, 1981 Oil Spill Conference.* (Publication 4334) pp. 361–374. Washington, D.C.: American Petroleum Institute.

I.O.E. 1985 *Input of contaminants to the North Sea from the United Kingdom* (compiled by W. C. Grogan). Final Report prepared by the Institute of Offshore Engineering, Heriot-Watt University, Edinburgh, for the Department of the Environment, London.

Law, R. J. 1981 Hydrocarbon concentrations in water and sediments from UK marine waters, determined by fluorescence spectroscopy. *Mar. Pollut. Bull.* **12**, 153–157.

Law, R. J., Blackman, R. A. A. & Fileman, T. W. 1982 Surveys of hydrocarbon levels around five North Sea oil production platforms in 1981. ICES CM 1982/E:14, (6 pages). (Unpublished manuscript.)

Law, R. J. & Fileman, T. W. 1985 The distribution of hydrocarbons in surficial sediments from the central North Sea. *Mar. Pollut. Bull.* **16**, 335–337.

N.A.S. 1975 *Petroleum in the marine environment.* Washington, D.C.: National Academy of Sciences.

N.A.S. 1985 *Oil in the sea: inputs, fates and effects.* Washington, D.C.: National Academy of Sciences.

Prevention of Oil Pollution Act 1971 (26 pages.) London: HMSO.

Prevention of Oil Pollution Act 1971 (Application of section 1.) Regulations 1984. (Statutory instrument 1984, no. 1684.) (2 pages.) London: HMSO.

Royal Commission on Environmental Pollution 1981 *Eighth report: oil pollution of the sea.* (Cmnd 8358) London: HMSO.

Whittle, K. J., Hardy, R., Mackie, P. R. & McGill, A. S. 1982 A quantitative assessment of the sources and fate of petroleum compounds in the marine environment. *Phil. Trans. R. Soc. Lond.* B **297**, 193–218.

Discussion

R. E. Jones (*School of Ocean Sciences, University College of North Wales, Menai Bridge, U.K.*). Do the monitoring techniques record the quantity of short-chained hydrocarbons discharged and is this fraction accurately incorporated in the total hydrocarbon budget?

R. A. A. Blackman. The monitoring techniques used by MAFF to measure total hydrocarbons in water and sediment do not record short chain hydrocarbons. The figures for the budget of total petrogenic hydrocarbons entering the North Sea from the various sources are derived from a wide variety of monitoring methods, very few of which accurately record the short chain hydrocarbons. Such compounds are generally considered to be relatively ephemeral, except under continuous discharge conditions, and of low acute toxicity, although less is known of their chronic and sublethal effects.

J. G. Parker (*Shell Exploration and Production, Aberdeen, U.K.*). There is considerable generation of hydrocarbons in the sea due to planktonic activity. Has there been an assessment of the contribution from this source, and if so, how does it compare with the input from the offshore industry?

R. A. A. Blackman. Whittle *et al.* included estimates for biogenic hydrocarbons in United Kingdom waters in their paper (1982, p. 204). Phytoplanktonic production was then estimated to contribute about 100 kt annually, while total marine biogenic input to United Kingdom waters (from phytoplankton, zooplankton, benthos and fish) was estimated to be about 165 kt annually. These figures are similar to a 'standing stock' of petrogenic hydrocarbons in the North Sea of about 110 kt, based upon a background subsurface water concentration of 2 $\mu g\ l^{-1}$ (Whittle *et al.* 1982, p. 215) but biogenic hydrocarbons consist of a smaller range of, for the most part, simple straight and branched-chain compounds that are easily degradable and exert no toxic effects at these low levels.

Phil. Trans. R. Soc. Lond. B **316**, 511 (1987)
Printed in Great Britain

Effects of North Sea oil and gas developments on fisheries

[Abstract only]

By R. Johnston
24 Hilton Drive, Aberdeen AB2 2NP, U.K.

North Sea oil and gas developments have introduced pressures on the fishing industry, on fishing operations, and on the marine ecosystem including commercial fish species. Operational incompatibilities and competition for personnel, facilities on land and space at sea may arise.

Among the pollutants are various accidental and operational inputs of hydrocarbons, dispersants used in clean-up, and process chemicals. Effects can be assessed on various bases: toxicity for specific organisms, impairment of food chains, and perturbation of natural populations.

Much is being done to resolve dissonances between the industries and to pre-empt operational incompatibilities.

Numerous measures have been introduced to minimize any pollution threats. Research continues to explore and quantify the effects on the marine ecosystem.

Phil. Trans. R. Soc. Lond. B **316**, 513–524 (1987)
Printed in Great Britain

Seabirds and North Sea oil

BY G. M. DUNNET

*Department of Zoology, University of Aberdeen, Tillydrone Avenue,
Aberdeen AB9 2TN, U.K.*

The history of large and well-publicized incidents of seabird mortality resulting from oil pollution from tanker accidents (such as the *Torrey Canyon* in 1967) and from unattributed oil slicks at sea, gave rise to real fears that the development of North Sea oilfields would result in serious mortality and declines in the large populations of seabirds breeding and wintering in and around the North Sea. The oil industry recognized the problem and attempted to minimize pollution risks in all exploration, production and transport operations.

Preliminary maps were prepared showing the distribution of vulnerable concentrations of breeding and wintering birds to facilitate contingency planning, and, particularly in Shetland, Orkney and the Moray Firth, extensive and long-term programmes are established to monitor:

(1) the numbers of breeding birds;
(2) wintering concentrations (particularly of sea ducks);
(3) the distribution and abundance of seabirds at sea;
(4) the numbers, and percentage oiled, of birds found dead on beaches.

The feared increase in oil pollution incidents has not materialized. The few accidents associated with offshore production have had little effect on seabirds. Tanker accidents have been few, but have had large, temporary and local effects (e.g. the *Esso Bernicia* at Sullom Voe in 1978–79). Breeding seabird populations in the area have increased, although in recent years, in some places, some species may have declined, but these declines cannot be attributed to oil activities in the North Sea. The Beached Bird Survey suggests that chronic oil-induced mortality is at a relatively low level.

INTRODUCTION

Seabirds and oil do not mix. Since the early years of this century oil pollution at sea has continuously caused the contamination and death of seabirds, many of which are washed up on beaches as conspicuous indicators of environmental contamination offshore. Given the widespread public interest in birds, these events receive a great deal of attention, and since the early 1920s the Royal Society for the Protection of Birds (R.S.P.B.) has been actively concerned about the effect of oil pollution on seabirds, and seabird populations (R.S.P.B. 1979). In these early years the source of the oil polluting the sea was largely the discharge of oily wastes from oil burning ships, and from accidents involving ships, especially oil tankers. The dramatic increase in the size of oil tankers in the past thirty years aggravated those concerns, and the grounding of the fully laden *Torrey Canyon* off the Scilly Islands in April 1967, probably killing 30000 seabirds (Bourne *et al.* 1967; Bourne 1970), firmly and clearly focussed attention onto the oil industry.

It is not surprising that when exploration for oil and gas began in the North Sea in the 1960s leading to discoveries of oil in the northern North Sea in the 1970s, ornithologists were very apprehensive about the consequences of these developments for seabirds. In the North Sea, particularly the northern North Sea, oil exploration and production would occur in very difficult environments: far offshore, in deep water and subject to severe weather conditions, especially in winter. The oil industry had little previous experience of operating in such conditions, and a largely new technology had to be developed (Sanders 1972). The risk of oil pollution at sea resulting from these developments was considered to be large. Although few quantitative data were available it was known that around the northern part of the North Sea large concentrations of breeding seabirds assembled to huge mixed-species colonies, although many species were more widely and more evenly distributed along the coastline. While some of these species stayed throughout the year others left the North Sea for the winter. Ringing recoveries indicated that there was a substantial influx of seabirds into the North Sea in winter from populations breeding further to the north. Several species of sea ducks also gather in various parts of the North Sea in winter. Although most seabirds had been studied to some extent at their breeding colonies, very little was known about the overall distribution of breeding colonies, less about the numbers of breeding seabirds, and almost nothing about their distribution and movements at sea, even in the breeding season.

Perceived threats to seabirds from oil industry operations

Concern about the well-being of seabirds and their populations related to a number of perceived threats from offshore operations. Below I list the major threats and give a brief commentary of the degree to which they have been realized during North Sea operations up to now.

(i) *Tanker accidents*

It was felt that tankers moving to and from the major crude oil terminals at Sullom Voe in Shetland and Flotta in Orkney, especially during bad weather and the short days of winter, would have a high risk of collision, breakdown or running ashore. A major incident such as the grounding of the *Torrey Canyon* or of the *Amoco Cadiz* in Brittany in 1978 (Conan 1982) could cause enormous mortality of seabirds in the northern North Sea at any time of year. No incident of such magnitude has occurred during North Sea operations so far. The largest single tanker incident relates to the *Esso Bernicia* at the Sullom Voe terminal in December 1978. On this occasion over 1100 t of Bunker C fuel oil were spilt into the sea and over 3700 oiled bird corpses were picked up (Heubeck 1979).

The provision of good navigational aids and the maintenance of tight operational procedures are largely responsible for the generally good record of tanker operations. The latest world-wide data on accidental oil spills from tankers show a reduction in the number of spills (to 42%) and in the amount of oil spilled (to 41%) between 1974–79 and 1980–85 (Advisory Committee on Pollution of the Sea 1986). However, the impact of an accident involving a loaded supertanker is still a legitimate fear.

(ii) *Spills at terminals and offshore loading facilities*

At all offshore production installations and loading sites, as well as at oil terminals, elaborate contingency plans have been designed and implemented to cope with spills that may occur during 'normal' operations. These plans include the provision of a wide range of equipment and sometimes major onshore structures (such as the spur booms at Sullom Voe), a body of trained personnel and detailed procedures to be followed in the event of a spillage. In this context particular concern was expressed in relation to the Beatrice oilfield located well inside the Moray Firth and only a few miles offshore from some of the largest seabird colonies in Britain. Records are kept for all spillages and remedial action taken at the time, and so far none of these has caused large mortality of seabirds.

(iii) *Leaks and fractures in pipelines*

New technology was developed to lay and maintain large diameter pipelines in deep water. These now total hundreds of miles of underwater pipes in the North Sea and so far they have been remarkably free from trouble. Although a number of incidents are recorded of pipelines being damaged by external action, they have proved to be safe and efficient. No significant seabird mortality can be attributed to this cause.

(iv) *Blowouts at wells and production installations*

Once again fears relating to this source of oil pollution at sea have been largely unrealized. The only exception was the major blowout on the Ekofisk field in April 1977 when the spilt oil was dispersed rapidly by wind and rough seas but caused little, if any, observed seabird mortality.

(v) *Flares*

The flaring of unwanted gases at production platforms offshore was considered likely to attract birds particularly in misty weather with the obvious probability that the birds would be killed. Sage (1979) considered this to be a major hazard especially for migrating passerine birds in autumn and in spring. However, detailed observations and records from the North Sea Bird Club members have indicated that while some mortality of passerines does occur, this does not reach the large numbers predicted even in poor visibility at the height of the migration season, and the impact on both landbird and seabird populations may be regarded as minimal.

(vi) *Illegal spillages of oil from shipping*

Oil continues to be discharged into the sea illegally from shipping and the first indication of such events is commonly the occurrence of dead or contaminated living seabirds on the beaches. The source of this oil can rarely be attributed to any particular ship or category of ships, although attempts have been made to relate such spillages to oil operations in the North Sea. In particular, between December 1978 and April 1979 there were large numbers of oiled birds on beaches from the Moray Firth to Shetland (Richardson *et al.* 1982). This coincided with the opening of the Sullom Voe terminal before the completion of the ballast water treatment facilities there. It was inferred that tankers approaching the terminal may have discharged dirty ballast water at sea, and it was also inferred that other shipping may have taken the opportunity to discharge oil contaminated water under these circumstances. Since then there has been a great increase in the proportion of tankers using segregated ballast and strict rules for the quantity of ballast on board when ships arrive at terminals. In the beached-bird surveys, which I mention below, dead and dying oiled birds continue to be found on beaches around the North Sea although the numbers of these have shown no detectable trend over the past ten years (Stowe 1982*a*; Heubeck 1987) despite a number of major incidents such as that which occurred in the Skagerrak and the southeastern coastlines of the North Sea in the winter of 1980/81 (Mead & Baillie 1981). The source of this oil was never confirmed.

These risks remain, and I cannot overemphasize the need for the most careful vigilance and tight controls over all aspects of the operations together with highly developed contingency plans, with the full range of equipment and trained personnel necessary to implement them. It cannot be denied that good fortune has played a part in the generally good environmental record of the North Sea operation to date. There is no room for complacency; the highest standards must be maintained.

Studies and actions undertaken by ornithologists

Acutely aware of the general lack of detailed knowledge about seabirds onshore at their breeding colonies, or offshore at sea, ornithologists in Great Britain quickly developed a number of areas of work designed to improve our understanding of the distribution and abundance of breeding seabirds, their distribution and mobility at sea and detailed studies of a variety of parameters relating to population dynamics. It was clearly important to identify species and areas specially vulnerable to spilt oil at sea. Several of the more important of these are listed below.

(i) *Operation Seafarer and subsequent surveys*

The Seabird Group, established in 1965, undertook as one of its first major commitments a survey of the distribution and numbers of seabirds breeding around the coast of Great Britain and Ireland. The main fieldwork was completed in the breeding seasons of 1969 and 1970 and the results were published in book form in 1974 (Cramp *et al.* 1974). Named by the late James Fisher after a poem called 'The Sea Farer', composed probably in the seventh century, this study pioneered methods for locating and counting (or estimating) the numbers of twenty-four species of breeding seabirds with widely differing patterns of breeding biology, breeding habitats and attendance at colonies. Though criticized in detail as better methods were evolved, this still serves as a baseline against which subsequent changes in distribution and abundance can be measured. Arising from the experience gained in Operation Seafarer, methods of counting the numbers of breeding birds of most species have now been standardized and the R.S.P.B. developed a programme of monitoring seabirds by organizing counts at a series of sample colonies around the British coastline (Stowe 1982*b*). Further, intensive sampling of seabird colonies was designed and done in the northern North Sea particularly in relation to the Sullom Voe terminal in Shetland (Richardson *et al.* 1981), the Orkney area in relation to the Flotta terminal and in the Moray Firth area in relation to the Beatrice field (Mudge & Aspinall 1985). The results of these intensive surveys have now been analysed and interpreted and I comment on these below.

In 1979 the Norwegian Seabird Project was established to determine the size of the breeding population of seabirds in Norway, to map the distribution of seabird concentrations along the Norwegian coast, and to perform annual counts on wintering seabirds including sea ducks. Although this programme did not cover the entire Norwegian coastline the results for the period 1979–84 have now been published by Røv *et al.* (1984). Breeding seabirds are rather scarce around the other coastlines of the North Sea but data on their distribution and abundance are available for Heligoland and Denmark.

(ii) *Preliminary maps of seabird concentrations*

In 1974 the Nature Conservancy Council brought together all the then existing information to produce maps which showed, for the British coastline of the North Sea, breeding concentrations of seabirds and the areas of known concentrations of seabirds, particularly sea ducks in winter. These maps were entirely preliminary in nature and drew arbitrary lines based on presumed foraging distances out to sea from the main breeding concentrations. Areas of known concentrations of sea ducks were also identified. These maps were produced to identify areas of seabird concentrations with special vulnerability in the event of an oil spill. A second edition was produced in 1980.

(iii) *Distribution and abundance of birds at sea*

In the early 1970s the Natural Environment Research Council supported pilot studies of seabirds at sea. This work was carried out by Dr W. R. P. Bourne from the University of Aberdeen and pioneered the development of methods for quantitatively measuring the distribution and abundance of seabirds at sea under the very difficult conditions of observing from a ship. Some early results of this work are presented in Bourne (1976, 1980). These early studies have now been developed in the North Sea by the Nature Conservancy Council in a continuing programme partly funded by the oil industry. Their results from the North Sea have been published in a final report entitled *Seabird distribution in the North Sea* (Blake *et al.* 1984). This report provides a large amount of information on the distribution and indices of abundance of different age-categories of about 30 species of seabird in the North Sea during the period 1979 to 1982. Although these maps provide useful data on seasonal variations in the distribution patterns of different age categories of different species in the North Sea, what they cannot do is provide important information on the mobility of individual birds between the various concentrations of birds and from season to season.

A similar study of the distribution of seabirds offshore in the Moray Firth has been carried out by Mudge *et al.* (1984) for Britoil. This work is on a much smaller scale than the more extensive Nature Conservancy Council programme, and has produced important data showing variations in seasonal patterns of fifteen different species of seabirds. In addition, Barrett (1983) has described the patterns of distribution and mobility of the large flocks of sea ducks occupying this area.

(iv) *Beached-bird surveys*

Records of oiled birds on beaches have been collected for many years, but since 1966 the Royal Society for the Protection of Birds, in conjunction with the Seabird Group, has organized and standardized a national beached-bird survey. From 1969 the February surveys have been incorporated into an international beached-bird survey involving the countries bordering the southern North Sea, and in 1971 the survey was made more systematic by concentrating on five predetermined weekends from September to March (Stowe 1982*a*). Extensive data are now available from these surveys, and more intensive, specialized versions of the survey have been instigated in the northern North Sea, especially in Shetland and Orkney. Recent analyses (Stowe 1982*a*; Heubeck 1987) conclude that in general the data show no statistical trend, but that frequencies of oiled birds are lower than the high levels of 1978–80. However, this level of oiled birds is still a matter for concern in the southern North Sea (Camphuysen 1984; Vauk 1984).

(v) *Ringing recoveries*

The ringing of seabirds, mainly at their breeding colonies, has been actively pursued, and sometimes specially encouraged, in the United Kingdom and in other parts of Europe. The recovery of ringed birds during beached-bird surveys have given us a considerable amount of new information on the provenance of birds occurring dead or oiled or both on different North Sea beaches during winter. The British Trust for Ornithology is carrying out an analysis of the pattern of recoveries of ringed birds throughout the North Sea, especially noting the proportion of ringed birds that have been found oiled from various places. An initial analysis of this type is presented by Mead & Bailey (1981) but we look forward to seeing the full analysis in the near future. However, like the mapping projects of birds at sea, these data give us brief information of a one-off nature, and tell us little of the overall mobility of the birds.

The number of seabirds using the North Sea

A number of estimates have been produced over the years of the number of seabirds breeding round the North Sea. Evans (1973) estimated that the number of seabirds (19 species) breeding 'around the North Sea' amounted to 1.28 million pairs. This tally excluded species such as the storm petrels, and had a number of question marks against some of the species which are rather difficult to count. At a seabird conference in Aberdeen in 1977 Bourne (1978) provided 'provisional' numbers of breeding seabirds of the northern Atlantic and northwest Pacific Oceans from which it is possible to extract the numbers associated with the North Sea. He gives as a total for 'North Sea and Baltic' 1.987 million breeding birds, though he makes the point that in some areas counts of individual auks have been equated arbitrarily with pairs. Evans (1984) gave an extensive table of the estimated number of breeding pairs of seabirds in northwest Europe from which with the Norwegian data in Barrett & Vader (1984) it is possible to extract approximate numbers relating to the North Sea. I give this total in table 1, but it

Table 1. Estimates of the number of breeding pairs of seabirds around the North Sea

(Derived from Evans (1984) in Croxall *et al.* (eds) (1984). Note totals for England and Scotland include birds breeding on their west coast and are not associated with the North Sea.)

England	436911
Scotland	2318996 + *ca.* 350000
Belgium	10240
Holland	294786
West Germany	126384
Denmark	308953
Sweden	530674
southwest Norway	*ca.* 150000
totals	4176944 + 350000 = 4526944

is important to note that the totals for England and Scotland will include seabirds nesting on the west coasts of these countries and which therefore are not properly related to the North Sea. In the original table there are large margins of error given for some species (such as the British storm petrel, *Hydrobates pelagious*) and on the orders of magnitude of some populations (for example Leach's storm petrel, *Oceanodroma leucorhoa*) and I have taken where appropriate the minimum rather than the maximum estimate. From this table the total number of breeding pairs is 4.53 million, and is substantially higher than the earlier estimates.

It is obviously difficult to get precise numbers for the breeding populations of seabirds, and we need not ever expect complete accuracy. However, as I have discussed elsewhere (Dunnet 1982*a*), the delayed maturity that is common in seabirds (many species do not breed until they are three to five years old, and some not until they are about ten) indicates substantial numbers of pre-breeding birds related to these breeding populations. Indeed the pre-breeding population may amount to approximately 50 % of the breeding population in terms of numbers (Dunnet 1982*a*, table 2). However, many of these pre-breeding birds will not be in the North Sea. It is well known for kittiwakes, *Rissa tridactyla* (Coulson 1966), and for fulmars, *Fulmarus glacialis* (Dunnet 1982*b*), that young birds may spend the first two or three years of their lives as far from their native colonies as the Grand Banks of Newfoundland and the Barents Sea in the Arctic. Precisely when they all return to 'prospect' for breeding sites is not clear but undoubtedly during the breeding season there will be a number of pre-breeding individuals of most seabird species in attendance at breeding colonies; to estimate this number is at present virtually impossible.

The situation in winter is even more complex. Many of our breeding seabirds are migrants, leaving the breeding area completely for the winter. Others are partial migrants with some components of the population departing while others remain in the vicinity of their colonies. Others may be resident, or undergo only local movements, throughout the year. In addition, large numbers of birds that breed farther north move down into the North Sea for winter. While ringing recoveries and mapping of the distribution of birds at sea tell us something about these winter concentrations and species distributions, these methods are not capable of giving us accurate estimates of the number of birds present at this time.

The results of seabird studies in the North Sea in relation to oil developments

A large amount of data has been accumulated over the past 20 years and subjected to many reviews. Bourne (1968) discussed the general problem of oil pollution and bird populations. At an International Conference of the Sea held in Rome, Tanis & Mörzer-Bruyns (1968) and Clark (1968) reviewed the impact of oil pollution on seabirds in Europe. In 1977 the British Ornithologists' Union organized a conference on 'The changing seabird populations of the North Atlantic' at Aberdeen and several papers were given relating specifically to oil pollution whereas many more were on general population studies of seabirds (Anon 1978). The Royal Society for the Protection of Birds prepared a major review of the topic which it submitted to the Royal Commission on Environmental Pollution in 1979. I prepared reviews of the situation in 1980 (Dunnet 1980) and again at a Royal Society Discussion Meeting (Dunnet 1982*a*). At the annual conference of the Seabird Group in 1982 there was further discussion on the general problem and Clark (1984) provided a further review. In May 1984 the United Kingdom Offshore Operators' Association organized a symposium of this subject at Aberdeen, the results of which are published by Clark *et al.* (1984). Most recently the International Council for Bird Preservation has published its *Status and conservation of the world's seabirds* (Croxall *et al.* (eds) 1984) in which threats to seabirds are discussed generally and in relation to specific areas including northwest Europe.

With so many recent reviews it may not be surprising that there is not much new left that can be said here today. I will however try to provide a general survey of such conclusions as are possible about the effects of the North Sea oil operations on the seabird populations there.

Monitoring breeding populations

Although extensive data sets exist from programmes of monitoring the numbers of breeding seabirds, few have been formally analysed and published. However, several have been presented at recent conferences, and many exist in the form of annual reports sometimes with analyses of a run of several years.

In general there has been a continuous increase of most species of seabird around the North Sea, but there are signs of patchy declines in some areas of kittiwakes and puffins, *Fratercula arctica* (Stowe 1982*a*). None of these changes can be associated with the activities of the oil industry.

An important problem in interpreting these data is the lack of knowledge of 'natural' variability in the numbers of breeding birds at any colony from year to year. My own data on the total number of fulmars nesting on the Orkney Island of Eynhallow each year from 1958 to 1985 inclusive show large deviations in most years from the overall trend line of population change (Dunnet *et al.* 1979). This study of a breeding population on an uninhabited island, little disturbed by people, may be considered to relate to a 'natural' population not subjected to any abnormal or unusual mortality causes. These data show how difficult it would be to interpret changes in numbers over any small run of years in terms of unusual rates of mortality, or of departures of the population from the long term trend line.

In recent years monitoring at breeding colonies has been extended to include collecting information on reproductive output. Some very interesting results have been obtained. For example at Vedoy, Rost, guillemots, *Uria aalge*, have had only four good breeding years since 1971 and these years have been characterized by a large number of chicks of good mass being produced. In all the other years a few chicks with low masses were produced. However, even in good years the numbers produced were far fewer than in the years from 1958 to 1963 (Røv *et al.* 1984). Similarly for the Arctic tern, *Sterna paradisaea*, in Shetland, there has been very variable reproductive success over the past eight years. Breeding production was very bad in 1984 and 1985, poor in 1978–79 and 1980 and good in only 1982 and 1983 (Heubeck & Ellis 1986). These Shetland results were not due to the weather during the breeding season, because in all years the auks managed to breed reasonably well. In 1985 kittiwakes in Shetland had disastrous breeding success compared with previous years (Heubeck & Ellis 1986). Highly variable breeding success is a feature of many species of marine fish, and has also been demonstrated to occur in the eider, *Somateria mollissima*, breeding on the northeast Scottish coast (Milne 1974). However, in general the many intensive studies of breeding biology of seabirds have shown a tendency for overall breeding success to vary only a little between years. The present situation therefore may indicate some new situation developing, and this is commonly thought to be related to food supply and probably to the increasing industrial fisheries for small fish (Furness 1982). It certainly is not a phenomenon that can be attributed to the activities of the oil industry.

Birds at sea

The most recent work on the distribution and abundance of birds at sea has been published by Blake *et al.* (1984), Mudge *et al.* (1984) and Barrett (1983). In these studies maps have been produced which show the crude densities of different age categories of the various seabird species in sample areas of the North Sea. Different maps are produced for different seasons and some information is available on the foraging range of different species from their breeding colonies. This is largely new and useful information and is undoubtedly helpful in the context of oil spill contingency planning, although it tells us nothing about the mobility of individual seabirds, also an important feature in relation to possible oil spills.

Barrett (1983) has produced valuable new data about sea ducks in the Moray Firth over two winters. Eight species and a total of over 30000 ducks concentrate there each winter. The seasonal and daily pattern of movements and the locations of feeding and roosting concentrations show considerable stability within and between seasons. Kinnear (1976) and Huebeck (1983) have described the distribution of moulting flocks of eiders in Shetland; at this stage in their annual cycle they are flightless and are especially vulnerable to oil pollution.

Beached bird survey

The Beached Bird Survey continues with monthly counts carried out in each of the winter months. The general pattern of a winter peak in the numbers of dead birds picked up, and in the proportion of these which are oiled, and a summer low is well known. A recent detailed study of the Shetland data (Heubeck 1987) covers the period 1979 to 1985. After the high levels of bird mortality and the high proportions contaminated with crude oil in Shetland and Orkney during 1979 and early 1980 (Richardson *et al.* 1982), the proportion of corpses with oil has never exceeded 26% since February 1982 which compares favourably with other areas of Great Britain. The proportion of auks found oiled in Shetland in recent years was also low and like that found in similar areas of Britain (Stowe 1982*a*; Underwood & Stowe 1984). Further, the incidence of oiled seabird corpses near the Sullom Voe terminal has been low since the winter of 1979/80, and this is important because a high proportion of the Shetland population of eider and black guillemot, *Cepphus grylle*, form moulting and breeding concentrations within a 20 mile radius of the oil terminal (Ewins & Kirk 1985; Heubeck 1983).

However, the Shetland data show an increase in the numbers of guillemots found dead on beaches. Breeding populations of guillemots have increased in northern Scotland and on the North Sea coast in recent years (Stowe 1982*b*) but the increase in the number of corpses has been too great and too sudden to be accounted for by this. There may be a change in the distribution of guillemots remaining to moult in the Shetland waters after the breeding season (Robertson 1985). An important point is that a few of these guillemots are contaminated by oil, and mortality may possibly be induced by a shortage of food. To conclude, over the six years from March 1979 to February 1985 the incidence of oiled corpses declined from a high in 1979–80, and has remained low ever since.

We await the full analysis of ring recoveries from the beached bird surveys which is currently being undertaken by the British Trust for Ornithology. This information will relate winter mortality in different parts of the North Sea to the colonies of origin of the affected birds, and will enable predictions to be made of the possible changes in breeding numbers which could result from the presumed lowered recruitment. For example in the winter of 1980–81 approximately 60000 birds may have been killed by oil in the southern North Sea. Rings on some of these birds showed that many were immature and came from breeding colonies in Orkney, Shetland, northeast Scotland and Saltee Island. Given that patterns and processes of recruitment to breeding populations are largely unknown (Ollason & Dunnet 1983; Dunnet 1982*a*), and the lack of precision of monitoring breeding numbers, it is virtually certain that any consequential reductions in the local breeding populations would not be detectable; it is almost certain that present monitoring is insufficiently precise to detect such changes.

References

Anon. 1978 The Changing Seabird populations of the north Atlantic. Proceedings of conference, Aberdeen University 1977. *Ibis* **120**, 103–136.

Advisory Committee on Pollution of the Sea 1986 *ACOPS Yearbook 1985–1986*. London.

Barrett, J. 1983 *Moray Firth seaducks: winters 1981–82 and 1982–83*. Report to Britoil plc by The Royal Society for the Protection of Birds.

Barrett, R. T. & Vader, W. 1984 The status and conservation of breeding seabirds in Norway. In *Status and conservation of the world's seabirds* (ed. J. P. Croxall, P. G. H. Evans & R. W. Schreiber). Cambridge: International Council for Bird Preservation. Tech. publ. no. 2.

Blake, B. F., Tasker, M. L., Hope Jones, P., Dixon, T. J., Mitchell, R. & Langslow, D. R. 1984 *Seabird distribution in the North Sea*. Huntingdon: Nature Conservancy Council.

Bourne, W. R. P. 1968 Oil pollution and bird populations. *Field Studies* **2** (suppl.), 99–121.

Bourne, W. R. P. 1970 Special review – after the *Torrey Canyon* disaster. *Ibis* **112**, 120–125.

Bourne, W. R. P. 1976 Seabirds and pollution. In *Marine Pollution* (ed. R. Johnston), pp. 403–502. London: Academic Press.

Bourne, W. R. P. 1978 The seabirds of the eastern North Atlantic. *Ibis* **20**, 117–119.

Bourne, W. R. P. 1980 The habitats, distribution and numbers of northern seabirds. *Trans. Linn. Soc. N.Y.* **9**, 1–14.

Bourne, W. R. P., Parrack, J. D. & Potts, G. R. 1967 Birds killed in the *Torrey Canyon* disaster. *Nature, Lond.* **215**, 1123–25.

Camphuysen, C. J. 1984 Landelijke olieslachtoffer-telligen in Februari in Nederland, 1965–84. *Nieuwsbrief, Nederlands Stookolieslacktoffer-Onderzoek* **5**, 41–55.

Clark, R. B. 1968 Oil pollution and the conservation of seabirds. In *Proc. int. conf. oil pollution of the Sea, Rome, 1967*, pp. 76–112.

Clark, R. B. 1984 Impact of oil pollution on seabirds. *Environ. Pollut.* A **33**, 1–22.

Clark, R. B., Dunnet, G. M. & Addy, J. M. 1984 Seabirds and North Sea oil. *Mar. Pollut. Bull.* **15**, 272–274.

Conan, G. 1982 The long-term effects of the *Amoco Cadiz* oil spill. *Phil. Trans. R. Soc. Lond.* B **297**, 323–333.

Coulson, J. C. 1966 The movements of the Kittiwake. *Bird Study* **13**, 107–115.

Cramp, S., Bourne, W. R. P. & Saunders, D. 1974 *The seabirds of Britain and Ireland*. London: Collins.

Croxall, J. P., Evans, P. G. H. & Schreiber, R. W. (eds) 1984 *Status and conservation of the world's seabirds*. Cambridge: International Council for Bird Preservation. Tech. publ. no. 2.

Dunnet, G. M. 1980 Seabirds and oil pollution. In *Energy in the balance*, pp. 51–64. Guildford: Westbury House.

Dunnet, G. M. 1982*a* Oil pollution and seabird populations. *Phil. Trans. R. Soc. Lond.* B **297**, 413–427.

Dunnet, G. M. 1982*b* Ecology and Everyman. *J. Anim. Ecol.* **51**, 1–14.

Dunnet, G. M., Ollason, J. C. & Anderson, A. 1979 A 28-year study of breeding Fulmars *Fulmarus glacialis* in Orkney. *Ibis* **121**, 293–300.

Evans, P. 1973 Avian resources of the North Sea. In *North Sea science* (ed. E. D. Goldberg), pp. 400–412.

Evans, P. G. H. 1984 Status and conservation of seabirds in northwest Europe (excluding Norway and the U.S.S.R.). In *Status and conservation of the world's seabirds* (ed. J. P. Croxall, P. G. H. Evans & R. W. Schreiber). Cambridge: International Council for Bird Preservation. Tech. publ. no. 2.

Ewins, P. J. & Kirk, D. A. 1985 Autumn and winter distribution and ringing recoveries of Shetland Tysties *Cepphus grylle*. Unpublished report to the Shetland Oil Terminal Environmental Advisory Group.

Furness, R. W. 1982 Seabird–fisheries relationships in the northeast Atlantic and North Sea. In *Marine birds: their feeding ecology and commercial fisheries relationships* (ed. D. N. Nettleship, G. A. Sanger & P. F. Springer). Proc. Pacific Seabird Group Symp. *Can. Wildl. Serv. spec. Publ.*

Heubeck, M. 1979 Seabirds and recent oil pollution. *Shetland Bird Club Rep.* **1978**, 47–51.

Heubeck, M. 1983 *Surveys of Eider moult flocks, 1982 and 1983*. Unpublished report to the Shetland Oil Terminal Environmental Advisory Group.

Heubeck, M. 1987 The Shetland Beached Bird Survey, 1979–1986. *Bird Study* (In the press.)
Heubeck, M. & Ellis, P. 1986 1985 – Disastrous breeding season for Kittiwakes and Terns in Shetland. Unpublished report to the Shetland Oil Terminal Environmental Advisory Group.
Kinnear, P. K. 1976 Eider moult concentrations in Shetland, Autumn 1976. Unpublished report to the Nature Conservancy Council.
Mead, C. & Bailey, R. S. 1981 Seabirds and Oil: the worst winter. *Nature, Lond.* **292**, 10–11.
Milne, H. 1974 Breeding numbers and reproductive rate of Eiders at Sand of Forvie National Nature Reserve Scotland. *Ibis* **116**, 135–154.
Mudge, G. P. & Aspinall, S. J. 1985 *Seabird studies in east Caithness.* Report to Britoil plc by The Royal Society for the Protection of Birds.
Mudge, G. P., Crooke, C. H. & Barrett, C. F. 1984 *The offshore distribution and abundance of seabirds in the Moray Firth.* Report to Britoil plc by The Royal Society for the Protection of Birds.
Ollason, J. C. & Dunnet, G. M. 1983 Modelling Fulmar populations. *J. Anim. Ecol.* **52**, 185–198.
Richardson, M. G., Dunnet, G. M. & Kinnear, P. K. 1981 Monitoring seabirds in Shetland. *Proc. R. Soc. Edinb.* B **80**, 157–179.
Richardson, M. G., Heubeck, M., Lea, D. & Reynolds, P. 1982 Oil pollution, seabirds and operational consequences around the Northern Isles of Scotland. *Environ. Conserv.* **9**, 315–321.
Robertson, I. S. (ed.) 1985 *Shetland bird report 1984.* Lerwick.
Røv, N., Thomassen, J., Anker-Nilssen, T., Barrett, R., Folkestad, A. O. & Runde, O. 1984 Sjøfuglprosjektet 1979–1984. *Viltrapport* **35**, Trondheim.
R.S.P.B. 1979 *Marine oil pollution and birds.* Sandy, United Kingdom: Royal Society for the Protection of Birds.
Sage, B. 1979 Flare up over North Sea birds. *New Scientist,* **81**, 464–466.
Sanders, N. 1972 North Sea oil: can the technology cope? *New Scientist,* **56**, 380–382.
Stowe, T. J. 1982*a* *Beached bird surveys and surveillance of cliff-breeding seabirds.* Report for the Nature Conservancy Council by the Royal Society for the Protection of Birds.
Stowe, T. J. 1982*b* Recent population trends in cliff-breeding seabirds in Britain and Ireland. *Ibis* **124**, 502–510.
Tanis, J. J. C. & Mörzer-Bruyns, M. F. 1968 The impact of oil pollution on seabirds in Europe. In *Proc. int. conf. oil pollution of the Sea, Rome, 1967*, pp. 67–74.
Underwood, L. A. & Stowe, T. J. 1984 Massive wreck of seabirds in eastern Britain, 1983. *Bird Study* **31**, 79–88.
Vauk, G. 1984 Oil pollution dangers on the German coast. *Mar. Pollut. Bull.* **15**, 89–93.

Discussion

Y. Samiullah (*Monitoring and Assessment Research Centre, King's College London, U.K.*). Professor Dunnet's discussions on seabirds and other contributions for fisheries only consider biological effects in terms of overall mortalities. Does he not feel that this could lead to a general complacency within the oil industry, particularly with regard to the possible sublethal and cellular effects of chronic low-level inputs of petroleum hydrocarbons to the marine environment?

G. M. Dunnet. Sublethal effects of pollution are important, especially as early warning of environmental contamination. This approach is used on marine invertebrates in an integrated monitoring programme at Sullom Voe. However, if these effects do not manifest themselves at the population level, they may not be serious in terms of environmental management, nature conservation or the fisheries resource. I think the oil industry is now well informed about the subtleties of biological systems, and certainly there is no room for complacency.

D. P. Stone (*Northern Environmental Branch, Department of Indian and Northern Affairs, Ottawa, Canada*). Is Professor Dunnet aware of any seabird mortality or oiling having occurred which could be attributed to the use of oil-based muds at shallow water drilling sites?

G. M. Dunnet. I am not aware of any oiling or mortality of seabirds that can be attributed to the use of oil-based drilling muds.

M. L. Tasker (*Nature Conservancy Council, Aberdeen, U.K.*). The oil industry can be congratulated on the fact that there has been no major impact by North Sea oil and gas developments on seabirds so far; however, the potential for damage still exists, and it is hoped that the high standards of safety will be maintained.

Research on seabirds at sea in the North Sea has been conducted by the Nature Conservancy Council with oil industry and Government support since 1979. The mobility of seabirds in the breeding season, and some important features of seabird distribution in the autumn and winter, have been examined in detail during the past two years. In general, areas of the eastern North Sea, particularly off Norway, can now be identified as being in need of survey rather than areas in the southern North Sea.

Phil. Trans. R. Soc. Lond. B **316**, 525–544 (1987)
Printed in Great Britain

The impact of oily discharges on the meiobenthos of the North Sea

By C. G. Moore[1], D. J. Murison[2], S. Mohd Long[1] and D. J. L. Mills[1]

[1] *Department of Brewing & Biological Sciences, Heriot-Watt University, Chambers Street, Edinburgh EH1 1HX, U.K.*

[2] *Department of Agriculture and Fisheries for Scotland, Marine Laboratory, Victoria Road, Torry, Aberdeen AB9 8DB, U.K.*

The impact of hydrocarbon discharges on the intertidal and subtidal meiobenthos of the North Sea is examined primarily by a consideration of two field investigations. The first study examines the effects of an oil refinery discharge on intertidal meiofauna in the Firth of Forth, while the second describes the impact of oil platform discharges on the surrounding meiobenthos.

The impact of the refinery effluent is only clearly distinguishable upstream of the discharge, as downstream the effects are confused with those of a second petrochemical discharge. The meiofaunal community is only strongly affected on the upper shore and this appears to be chiefly the result of an organic enrichment effect causing a raising of the redox potential discontinuity (RPD) layer. All meiofaunal taxa examined are sharply reduced in density and species richness within 320 m of the discharge but at 600–900 m from the discharge meiofaunal densities are enhanced or depressed, relative to clean sediments, dependent upon the seasonal pattern of the RPD layer. Farther down the shore the impact is only felt at most by a slight reduction in species richness and subtle change in species abundance patterns on the middle shore for a distance of about 600 m. The meiofaunal responses to the petrochemical discharges seem similar to those described for the macrofauna in the same area, although a small meiofaunal population persists in the most polluted sediments in the absence of macrofauna.

The discharge of drilling cuttings, contaminated with oil-based drilling mud, was found to strongly modify meiofaunal densities within 800 m of the Beryl A Platform. Nematode densities are strongly reduced in the vicinity of the platform and it is thought that the impact on this infaunal taxon may be due to slow degradation within the sediment of toxic fractions of the diesel base of the drilling mud. By contrast copepod densities were greatly enhanced in one survey and the difference in impact is considered to be due to the epibenthic habit of the species involved, enabling them to flourish in conditions of high food or low predation and competition or all three. The species involved seem typical members of meiofaunal communities of organically enriched sediments. Some improvement in meiofaunal densities throughout the period 1984–85 is thought to be possibly the result of a switch from diesel-based to low-toxicity drilling muds.

It is concluded from these and other studies that hydrocarbon discharges into the North Sea are unlikely to be causing extensive damage to meiofaunal communities.

1. Introduction

At a meeting on the environmental effects of North Sea oil and gas developments, interest in the impact on benthic organisms is focused on the implications of drilling platform activities for the surrounding benthos. The effects of the discharge of drill cuttings, contaminated with oil or water-based drilling mud, on the community structure of the macrobenthos is routinely monitored at many platforms and so the identification of broad patterns of response is now possible (Davies *et al.* 1984; Kingston, this symposium). However, metazoans passing through a 500 μm mesh sieve (meiofauna) have hitherto been overlooked in platform monitoring studies and yet there is a *prima facie* cause for concern about possible damage to meiofaunal communities. The great mass of the meiofauna is permanently associated with the sediment, few species having a planktonic larval phase, and thus adverse sedimentary conditions must be tolerated by all life stages. Also, meiofaunal species are dependent upon the sediment for suitable food supplies, very few suspension feeders being known.

The paucity of knowledge regarding meiofauna necessitates a broad approach to the appraisal of potential damage by oily discharges. Knowledge about the impact of oil pollution on meiofaunal communities is very limited and concerns largely the effect of spillages on the total density of the major meiofaunal groups, such as nematodes and copeods, or at best examines one of these groups at the species level. While the investigation of meiofaunal communities at the level of species does present some taxonomic problems, this should not provide justification for overlooking this component of the ecosystem. Indeed, meiofauna offers a number of advantages for employment in effects monitoring (Heip 1980; Platt & Warwick 1980; Raffaelli & Mason 1981).

This contribution aims to assess the impact of platform and other chronic discharges of oily wastes on North Sea meiobenthos, partly by a consideration of the limited amount of published literature but largely by examination of some of the results from two continuing field investigations. One of these concerns a recently initiated programme of meiofaunal monitoring around an oil platform, while the other study examines in more detail the impact of an oil refinery discharge on meiofaunal community structure.

2. Effects of hydrocarbon discharges on intertidal meiobenthos

(*a*) *Refinery impact*: *the Grangemouth field study*

(i) *Study area and methods*

British Petroleum Oil Grangemouth Refinery Ltd is located along the banks of the Forth Estuary, Scotland, beside British Petroleum Chemicals Ltd (Grangemouth). The adjacent shore consists of an extensive mudflat, which receives an oily effluent from the refinery of about 15–25 Ml d^{-1} comprising ballast water, boiler and cooling tower blowdown and process water (figure 1). About 500 m from the refinery discharge point British Petroleum Chemicals discharges an effluent of 6–9 Ml d^{-1}, which includes organic solvents, ammonium salts and small amounts of oil. There are also two small sewage discharges to the mudflats (McLusky 1982). The mudflats in the area have been extensively reclaimed and for the most part only extend upwards to a little above mid-tide level. The highest region of the mudflats is found along the shoreline receiving the petrochemical effluents, which discharge at the top of the mudflats at about the level of mean high water neaps. The salinity in this region of the estuary is generally within the range 19–32‰ (Bagheri & McLusky 1982, C. G. Moore, personal observations). The estuary mouth is to the east of Grangemouth.

The impact of the petrochemical effluents was investigated by surveying the meiofauna and certain physical and chemical environmental parameters in late July 1984. As interest was

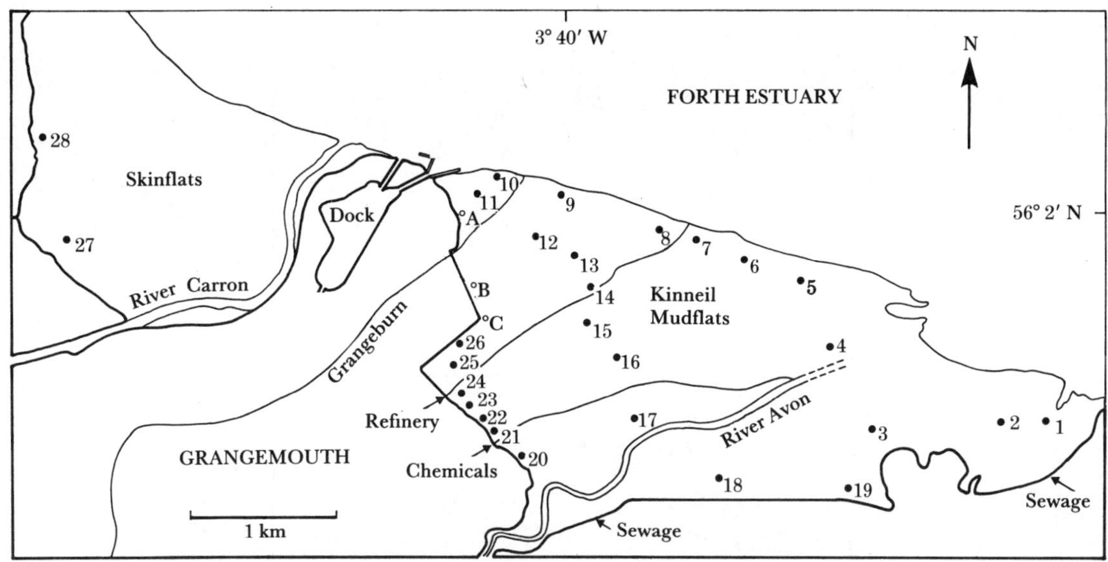

FIGURE 1. Grangemouth mudflats in the Forth Estuary, showing the location of sampling sites and effluent discharges. The sites are located along three longitudinal transects at approximately MLWN (sites 1–10), MTL (11–19) and MHWN (20–28). A, B and C are meiofaunal production study sites.

centred on the impact on community structure at different intertidal heights, a total of 28 sampling sites were located along longitudinal transects at approximately MLWN, MTL and MHWN. Because of the restriction of the upper shore mudflat to the close vicinity of the discharges, two control sites for the MHWN transect were located on a similar mudflat at Skinflats to the west of Grangemouth.

At each site triplicate core samples of area 5.5 cm^2 were taken to a depth of 3 cm for meiofauna. The meiofauna passing through a 500 μm aperture sieve, but retained on a 63 μm sieve, was extracted by Ludox centrifugation (McIntyre & Warwick 1984). Cores were also taken for, *inter alia*, grain size analysis by the dry sieving and pipette methods (Buchanan 1984), determination of the redox potential depth profile (Pearson & Stanley 1979) and hydrocarbon analysis of the surface 3 cm of mud. Total aromatics were determined by UV spectrophotometry using a chrysene standard, fluorescence intensity being measured at 360 nm with excitation at 310 nm (Anon. 1976). Total aliphatics were determined by IR spectrophotometry with a crude-oil standard, absorbance being measured at 2930 cm^{-1}.

(ii) *Results*

Although both petrochemical effluents have cut distinct channels in the mud, much oil is stranded at the top of the shore. There is a clear concentration gradient with distance from the discharge point (figure 2). Total aliphatic hydrocarbon concentrations in dried sediment at MHWN range from 2451 to 5981 μg g^{-1}, total aromatics from 6 to 158 μg g^{-1}. Oil content of the mud decreases with distance down the shore, although a clear concentration gradient in both aliphatics and aromatics with distance from the effluent channel is discernible at MTL but not at MLWN.

The high organic loadings of both discharges have strongly influenced the redox conditions of the sediment at MHWN (figure 3), with the redox potential discontinuity layer (RPD) at a depth

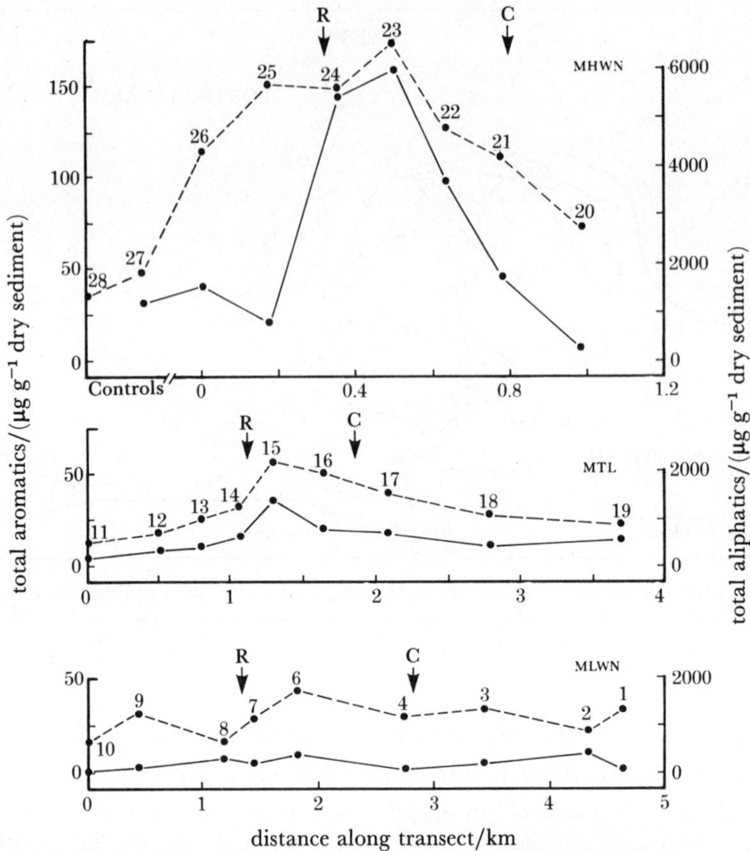

FIGURE 2. Concentration of total aliphatic hydrocarbons (dashed line) and total aromatic hydrocarbons (solid line) at sites (numbered) along transects at three heights on Grangemouth mudflats, 25–27 July 1984. R and C denote positions of British Petroleum Refinery and British Petroleum Chemicals effluent channels.

of about 2 mm, compared with 20–21 mm at a similar tidal level at the control sites and 20–25 mm along the MTL and MLWN transects (apart from a single site).

The petrochemical discharges cause a gross reduction in the density of all meiobenthic taxa recorded at MHWN for a distance of at least 200–300 m from both effluent channels, although there is some recovery of nematode numbers between the discharges, and of nematode and annelid numbers at the ends of the transect (figure 3). Maximum copepod density in this region is only five individuals per 10 cm^2. Along the mid and low tide transects there is no clear evidence of meiofaunal density changes in response to the refinery discharge. Mean densities of nematodes, copepods and annelids are respectively 1219, 274 and 205 individuals per 10 cm^2 at MTL and 1692, 51 and 66 individuals per 10 cm^2 at MLWN.

Meiofaunal species richness exhibits a clear response to the effluents at MHWN, with the minimum of just 7 species being recorded at sites within 180 m of the refinery discharge, rising to 17 species 320 m upstream of the discharge and 20–29 species at the control sites (figure 4). The evenness pattern (reflected also in the Shannon–Wiener diversity index) is quite the opposite to what might be expected. Dominance is reduced in the most polluted sediments, as no species can flourish in such harsh conditions. The pattern of species richness along the MTL transect suggests that this is a more sensitive indicator of perturbation of the community than density. Richness is depressed over a considerable distance in the area around the effluent

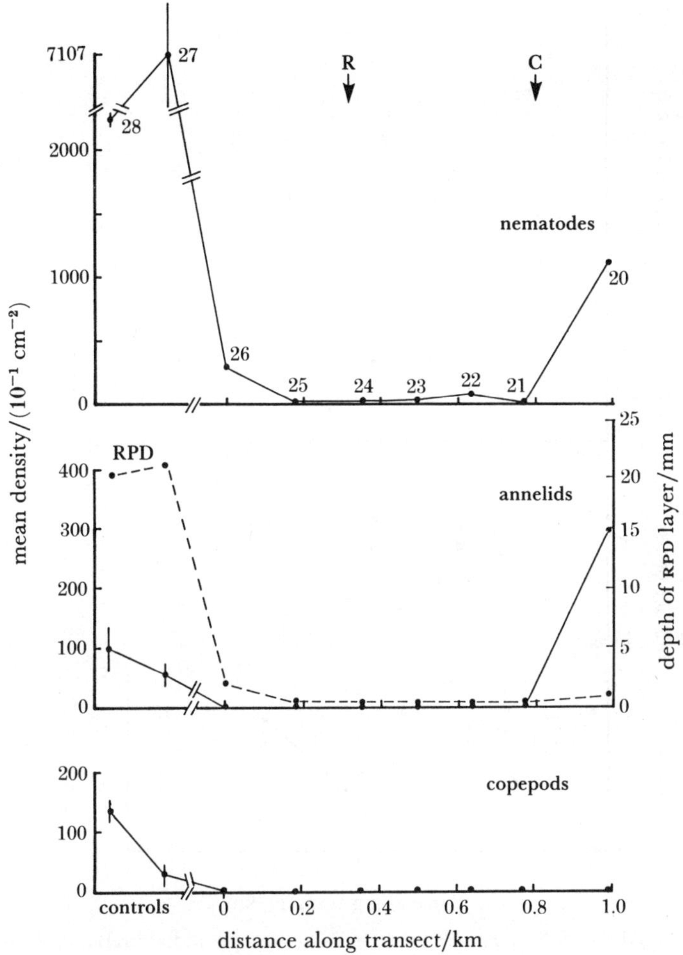

FIGURE 3. Mean density (with standard error bars) of the three dominant meiofaunal taxa (solid lines) and depth of the RPD layer (dashed line) along a transect at approximately MHWN on Grangemouth mudflats, 27 July 1984. R and C denote positions of British Petroleum Refinery and British Petroleum Chemicals effluent channels.

channels. The diversity and evenness measures seem unaffected by the oily discharge (although there is a strong increase in dominance just downstream of the chemical effluent channel).

Seventy-eight meiofaunal species were recorded throughout the area but the major spatial trends in species abundance patterns of the meiofaunal community can be summarized in the ordination in figure 5. The proximity of the sampling sites on the plot reflect their similarity in terms of composition of the community, except that sites at the two ends of the horseshoe configuration are in fact separated by subsequent axes (trends). Shore height is clearly important in explaining compositional differences but along the upper and mid shore transects there are community gradients related to pollution levels. Group A sites are all grossly polluted MHWN sites with a meiofauna consisting chiefly of low density populations of the nematodes *Sabatieria pulchra* and *Rhabditis marina*. The group B site at one end of the MHWN transect is similar in composition but the meiofauna is more species rich and abundant, and is strongly dominated by the nematode *Diplolaimella ocellata*. The group C site is the most distant of the MHWN transect sites from the refinery discharge and most closely resembles the MTL sites. It is dominated by

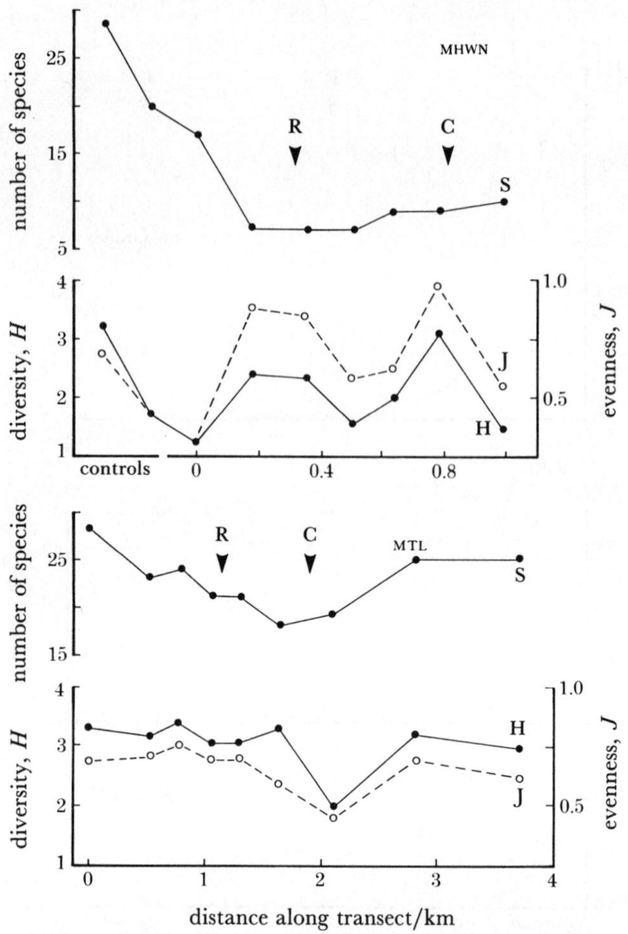

FIGURE 4. Meiofaunal species richness, Shannon–Wiener diversity (bits per individual) and Pielou evenness (J) along transects at approximately MHWN and MTL on Grangemouth mudflats, 25–27 July 1984. Values are based on all specimens of copepods, polychaetes and oligochaetes found in triplicate 5.5 cm^2 cores at each site plus nematodes present in a sample of 200 specimens from each site. R and C denote positions of British Petroleum Refinery and British Petroleum Chemicals effluent channels.

the nematodes *Hypodontolaimus balticus*, *Daptonema setosum* and *Tripyloides gracilis*, and the polychaete *Manayunkia aestuarina*. The MHWN control sites at Skinflats are more similar to the MTL sites than the MHWN transect sites, although of these they most closely approach the group C site but support many species that were virtually absent on the upper shore transect, notably the nematodes *Ptycholaimellus ponticus*, *Metachromadora remanei* and *Microlaimus globiceps*, and all copepod species. The group C site most closely resembles the most atypical, and presumably the most pollution-perturbed, MTL site (group D). This site, just downstream of the chemical effluent channel, is distinct from the other MTL sites by the high density of *Daptonema setosum* and a flourishing population of the copepod *Mesochra lilljeborgi*. The remaining MTL sites show a variation in species composition and abundance apparently related to their distance from the refinery channel, with group E sites lying within about 600 m of the channel and group F sites beyond this distance. The difference in the meiofaunal communities between these groups is rather subtle but is most clearly exemplified by modest enhancement of the polychaete *Capitella capitata* and strong enhancement of the copepod *Platychelipus littoralis* within group E. Group G consists of the lower shore sites.

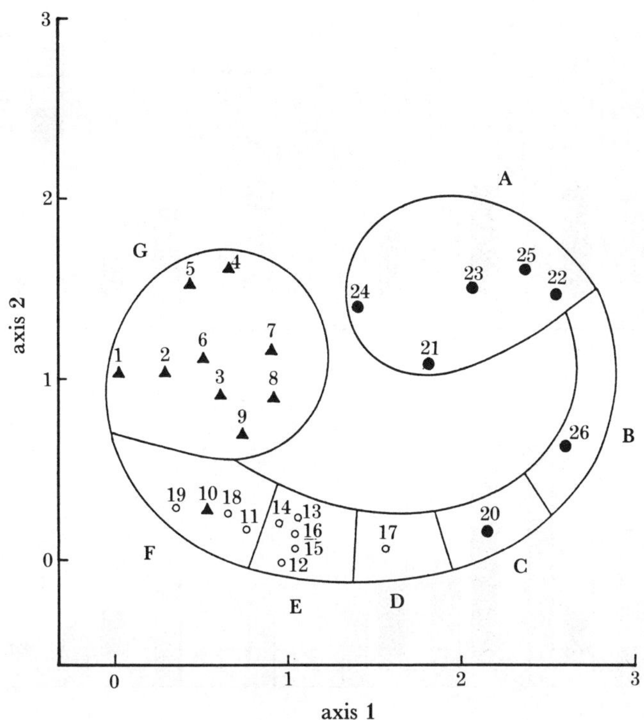

FIGURE 5. Ordination (detrended correspondence analysis) of sites at approximately MLWN (triangles), MTL (open circles) and MHWN (filled circles) on Grangemouth mudflats, based on $\log(x+1)$ transformed densities of nematode, copepod and annelid species. Letters denote site groupings. Axes are scaled in standard deviation units.

The degree of impact of the discharges on the meiobenthos varies seasonally. This is shown by the results of a study, currently in progress, of meiobenthic production levels in the area. Figure 6 shows the temporal pattern of densities of the three dominant taxa at three sites upstream of the refinery channel (figure 1). These sites are at the same tidal level (between MTL and MHWN) and experience strong, moderate and weak hydrocarbon contamination. Oil levels are yet to be measured but the oil pollution gradient is visibly obvious. Densities are higher at the more polluted sites in April. In July the moderately polluted site shows marked enhancement of densities, but with increasing temperature and elevation of the black layer to within a few millimetres of the surface, the densities of nematodes and copepods fall to below those of the cleanest site. This decline is even more severe at the most polluted site, where the meiofauna is almost eradicated in August and September. Thus the meiofauna at the more polluted sites exhibit both enhancement and depression of density, dependent upon the time of year and degree of pollution.

(iii) *Discussion*

It is impossible to firmly link oil levels at Grangemouth with environmental impact as the chemical effluent is also influencing the biota. The effects of the two discharges appear to be fairly similar, varying from toxicity to enrichment. The visual impact of both discharges on the biota is identical, both channels being accompanied by parallel ribbons of dense microalgae (chiefly filamentous Cyanophyceae) a little distance from their banks. Similar blue-green algal cover has been found associated with other refinery effluents by Konig (1968) and Baker (1971). The work of McLusky (1982) on the macrofauna of the area also suggests that the effluents

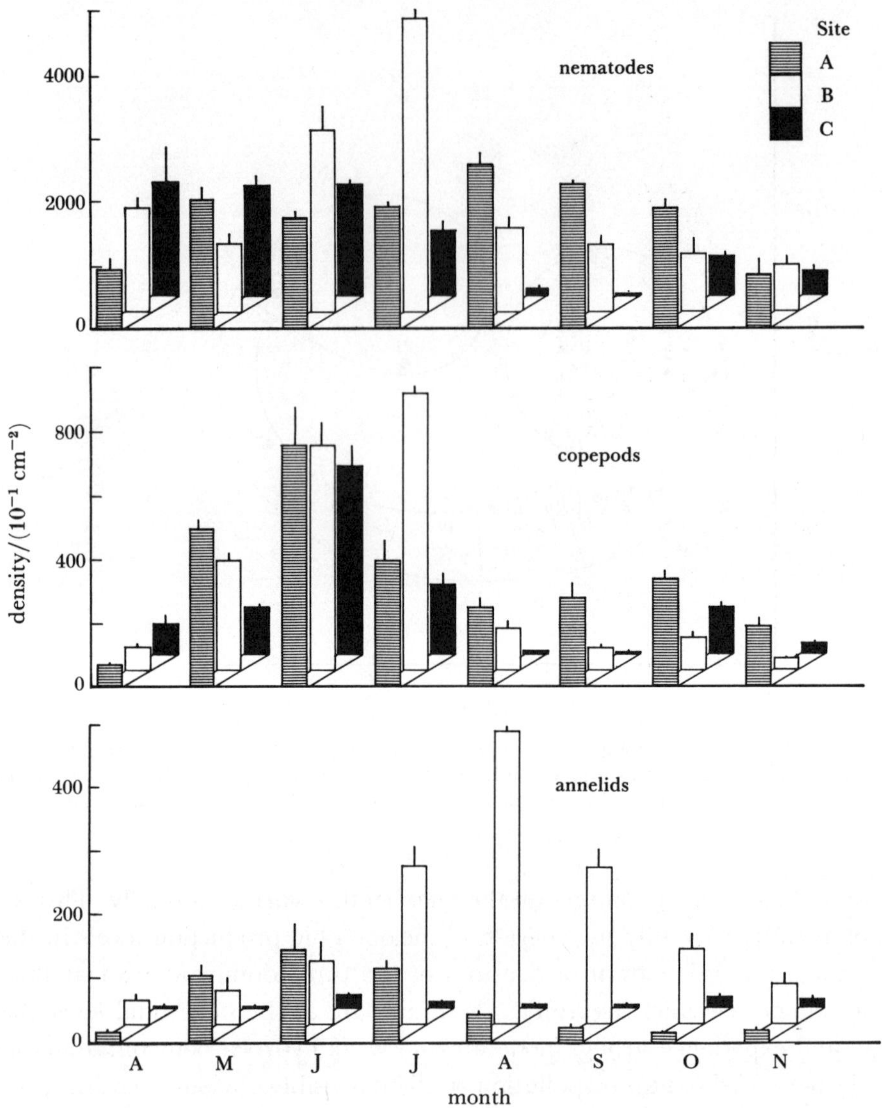

FIGURE 6. Temporal changes in mean density (with standard error bars) of the dominant meiofaunal groups in 1984 at sites between MTL and MHWN on Grangemouth mudflats. The sites experience weak (A), moderate (B) and strong (C) hydrocarbon contamination.

are having similar effects on the biota. Moreover, although the refinery effluent clearly gives rise to extremely high oil concentrations in the sediment, it cannot be assumed that the environmental impact is necessarily associated with hydrocarbon components of the effluent.

Meiofaunal responses to different levels of petrochemical pollution observed at Grangemouth may be summarized as follows:

(*a*) gross pollution: meiofaunal density is very low, most taxa virtually disappear but a few species of nematodes can persist;

(*b*) severe pollution: nematodes and annelids increase in density and species richness but copepods remain very low;

(*c*) moderate pollution: meiofaunal densities may vary from impoverishment to enrichment, dependent upon season;

(*d*) slight pollution: background densities are achieved but species richness is reduced and there is an abnormal species abundance pattern;

(*e*) minimal pollution: background densities, species richness and species composition.

It is interesting to compare this meiofaunal response to that of the macrofauna at Grangemouth, which has been monitored for several years by McLusky (1982). The patterns are very similar, McLusky identifying a similar series of response levels in terms of density and diversity. Moreover, the response of the meiofaunal community appears to broadly correspond to that of the macrofaunal community in the same region of the mudflat.

Because of the difficulty in distinguishing the impacts of the chemicals and refinery effluents downstream of the refinery, the effects of the refinery effluent are more clearly discernible in the upstream direction. The effect on the meiobenthos appears to be largely due to the consequences of organic enrichment of the sediment, causing strongly reducing conditions, although there appears to be also a localized toxic or physical effect by constituents of the effluent. McLusky (1982) also noted the resemblance of the macrofaunal response to one typical of organic enrichment.

There were subtle differences in community structure at MTL but the discharge was only strongly felt on the upper shore, where the impoverishment of the fauna can be largely explained by the reduced nature of the sediments. However, at sites within 320 m of the discharge redox conditions differed little and yet the fauna was clearly richer at 320 m. Thus there also appears to be, as expected, a strong toxic or physical response gradient in the immediate vicinity of the discharge (cf. Bagheri & McLusky 1982).

At sites between 600 and 900 m from the discharge the seasonal fluctuation in redox conditions seems to exert a major role in controlling density. The process of organic enrichment producing impoverishment of macrofauna through raising of the RPD depth is a well-known mechanism (Pearson & Rosenberg 1978) and this is exacerbated on the upper shore at Grangemouth by desiccation of the mud. However, it is known that certain meiofaunal species can tolerate such extreme consequences of organic enrichment. For example Bouwman *et al.* (1984) recorded extremely high densities of nematodes on an organically polluted upper shore mudflat in the Ems-Dollart Estuary, The Netherlands, which also suffered desiccation, with a thin (*ca.* 2 mm) aerobic layer, and an absence of macrofauna. It is difficult to compare the two studies because of the lack of detailed knowledge regarding the physical and chemical sedimentary conditions and the duration of tidal cover but clearly the habitats are somewhat different, with the Dollart area being more equivalent to Forth mudflats below MHWN in the upper reaches of the estuary. This is shown by a comparison of the salinity régimes and fauna in the Dollart area (Bouwman 1983; Bouwman *et al.* 1984) with those throughout the Forth Estuary (C. G. Moore, unpublished results).

McLusky (1982), employing a 250 μm mesh sieve, reported an azoic area within 250 m of the refinery discharge. Although such a macrofaunal azoic area may vary in extent from year to year, no macrofauna or even juvenile annelids were present in the meiofaunal samples taken in this area in late July 1984. The meiofauna samples within 165 m of the refinery discharge, however, contained a total of 11 meiofaunal species with densities from 29 to 42 individuals per 10 cm^2. Such an impoverished meiofauna is ecologically insignificant; however, the extent of seasonal density variation is unknown.

(*b*) *Oil spill and experimental studies*

The only other studies of the effects of oil on the meiofauna of muddy shores concern accidental or experimental coating by crude oil. Boucher (1985) monitored the meiofauna of a mudflat heavily contaminated by the *Amoco Cadiz* spill and failed to record any marked changes in meiofaunal density until two years after the spill. No evidence is presented to relate these changes to an oil effect. Neither Naidu *et al.* (1978) nor Fleeger & Chandler (1983) found experimental spraying of crude oil to adversely affect the meiofauna, both studies reporting an enhancement of certain taxa. Even though Fleeger & Chandler (1983) sprayed oil to a depth of 2 cm only a slight and non-significant early decrease in species richness was recorded. Decker & Fleeger (1984) examined recolonization of azoic mud that had been experimentally mixed with oil. In comparison with unoiled controls there was little effect of the presence of oil on the copepod community, even at a concentration of 3810 $\mu g\ g^{-1}$ total aromatics. Only one species was apparently affected, showing initially depressed densities, but after 60 days, enhanced densities. Similarly, copepod species richness was initially lower in heavily-oiled mud but higher than in unoiled controls after 60 days. Polychaetes and nematodes showed depressed densities in very heavily oiled sediment (3810 $\mu g\ g^{-1}$) but there were no temporally consistent differences between the densities of these taxa in unoiled sediment and sediment with an oil content of 1330 $\mu g\ g^{-1}$ total aromatics.

The effects of Grangemouth refinery effluent on the meiobenthos are clearly stronger than those recorded in the above investigations. This difference is probably mainly due to the extreme superficiality of the oxidized sediment layer at Grangemouth, which was not noted in the other studies.

From the available evidence it appears that the meiofauna of intertidal estuarine muddy sediments is surprisingly resilient to oil pollution, although the effects of high concentrations of refined oily products are unknown. So far as density changes are concerned the Grangemouth study and that of DeLaune *et al.* (1984) indicate that oil impact on the mudflat meiofauna is no more adverse than that on the macrofauna. It is too early to gauge the relative sensitivities of the two size groups concerning more subtle aspects of community structure such as diversity and species composition, which this and other studies show to be more sensitive to low pollution levels than simple density.

In contrast to the resilience of mudflat meiofauna, sandy shore communities have generally been reported as suffering impoverishment as a result of oil spills, particularly when the spill is of high aromatic content. Thus Rützler & Sterrer (1970), Wormald (1976) and Giere (1979) recorded drastic reductions in meiofaunal densities after spills of fuel oil or fuel–crude oil mixtures and reductions of one or more major taxa have been reported by Boucher (1980), Fricke *et al.* (1981) and McLachlan & Harty (1982) after spills. The general picture to emerge from these studies is that nematodes are somewhat more tolerant than interstitial harpacticoids.

3. Effects of hydrocarbon discharges on subtidal meiobenthos

(*a*) *Oil platform impact: the Beryl field study*

(i) *Study area and methods*

The Beryl field is situated 180 km southwest of the Shetland Isles. The Beryl A platform is located in 115 m of water on a bed of silty sand. Drilling commenced at Beryl A in 1976, with 37 wells completed up to May 1982 using mostly diesel-based drilling muds. After a 16 month interval drilling recommenced in September 1983 using low-toxicity drilling mud to cut a further four wells, although there was another break in drilling from January to June 1984.

As part of a more comprehensive series of surveys around North Sea oil platforms the Department of Agriculture and Fisheries for Scotland (DAFS) Marine Laboratory has monitored sediment hydrocarbons and heterotrophic hydrocarbon mineralization rates at Beryl since 1979. McIntosh *et al.* (1983) and Massie *et al.* (1985) should be consulted for methods and detailed analysis of the aromatics. In 1982 macrobenthos monitoring was included in the programme (Moore 1983) and in the surveys of 1984 and 1985 biological monitoring was extended to encompass a limited investigation of the meiobenthos.

In May 1984 single Craib cores (area 24.6 cm^2) of at least 6 cm in length were taken at nine sites for redox depth profile determination (Pearson & Stanley 1979) and meiofaunal analysis. The meiofauna retained on a 45 μm aperture sieve was extracted by the decantation method (McIntyre & Warwick 1984) and subsampled using an Elmgren splitter (Elmgren 1973). In May 1985 duplicate cores (area 25.0 cm^2) of at least 6 cm were taken by Scottish Marine Biological Association (SMBA) multiple corer at 11 sites for meiofaunal and redox analysis. Meiofauna extraction was as for 1984 except that a Ludox centrifugation stage was added (McIntyre & Warwick 1984). Grain size data are available for three sites in 1984, when core samples of the surface 4–6 cm were taken from a Smith–McIntyre grab and graded by dry sieving and pipette analysis (Buchanan 1984).

(ii) *Results*

The dumping of cuttings contaminated with drilling mud has led to very high levels of sediment hydrocarbons within 800 m of the platform (figure 7), although concentrations of these and naphthalene mineralization rates fluctuate with the level of drilling activity (McIntosh *et al.* 1983).

The sediment samples taken in 1984 indicate that sediment structure is affected by the dumping of cuttings for a distance of at least 200 m. At sites 800 and 4800 m south of the

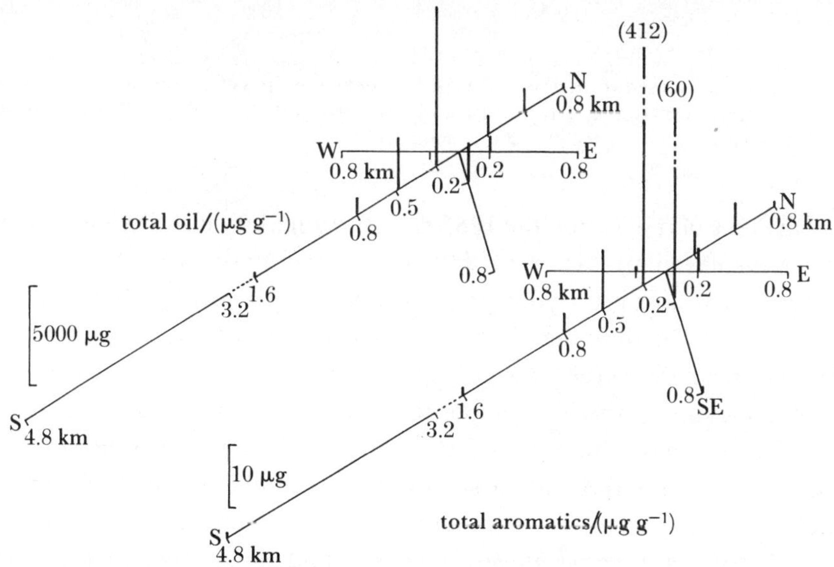

FIGURE 7. Concentration of total oil and total aromatics in sediments around the Beryl A oil platform in May 1985.

platform sediment particle size distribution was unimodal with a median diameter of 91–99 μm and 13–22 % silt–clay content. There was an additional mode in the clay range, 200 m north of the platform, reducing the median particle diameter to 60 μm and enhancing the silt–clay fraction to 51 %. Owing possibly to the reduction in permeability of the sediment around the platform, or through a mechanism of organic enrichment, the redox potential of the superficial sediments is reduced within 800 m (figure 8), with the RPD rising to 1 cm 200 m southeast of the platform in May 1985.

The pattern of meiofaunal density with distance from the platform is shown in figure 8 for the 1985 survey. The meiofauna was dominated strongly by nematodes and copepods, together representing 94–97 % of the total numbers. Both taxa show a clear response to platform pollution. Nematode density is sharply depressed within 800 m of Beryl, falling from a density of 1821–2246 individuals per 10 cm² at 3200–4800 m south of Beryl to 497 individuals per 10 cm² at a point 200 m southeast. Copepod density shows the opposite trend rising from 22–45 individuals per 10 cm² beyond 3200 m south to a peak of 720 individuals per 10 cm² 200 m southeast.

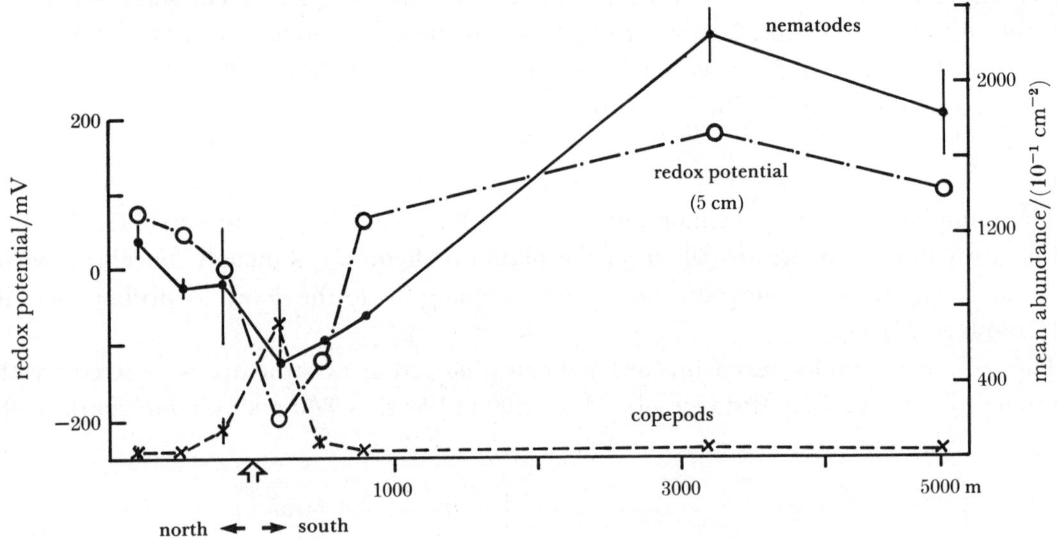

FIGURE 8. Redox potential at a sediment depth of 5 cm and mean densities of meiobenthic nematodes and copepods (with standard error bars) along a transect running north–south through the Beryl A oil platform. (200 m southeast values substituted for missing 200 m south values.)

Because of the lack of replication the 1984 density estimates must be viewed more cautiously. The nematodes displayed an even more drastic reduction in densities, falling from 1918–2439 individuals per 10 cm² at sites beyond 1600 m from the platform to 103–327 individuals per 10 cm² at sites within 500 m of the platform. There was, however, no suggestion of enhanced copepod densities near the platform.

Analysis of meiofaunal species composition has thus far included only a qualitative examination of the copepods from several 1985 sites. At 3200 m south the dominant copepods were *Amphiascus tenuiremis* and *Amphiascoides subdebilis*, whereas at four sites within 800 m of the platform the following species were dominants at one or more sites: *Proameira* aff. *P. simplex*, *Paramphiascella hyperborea*, *Paramphiascopsis longirostris* and *Ancorabolus mirabilis*. *Paramphiascopsis longirostris* copepodites were largely responsible for the enormous copepod density at the most polluted site (200 m southeast).

(iii) *Discussion*

Meiofaunal densities at the Beryl platform may be compared with the data relating to the same time of year from several studies of the nearby Fladen ground, some 140 km to the south. McIntyre (1964) records 1852–2215 nematodes and 33–50 copepods per 10 cm^2 in late April; Heip *et al.* (1983), sampling in May, record 1334–3875 nematodes and 26–66 copepods per 10 cm^2 and Faubel *et al.* (1983) record 1456–2653 nematodes and 28–80 copepods per 10 cm^2 for the period April–June. Thus the nematode and copepod densities recorded at stations beyond 800 m from Beryl seem typical background levels for this part of the North Sea, while nematode densities within 800 m are mostly depressed and copepod densities, at least within 200 m of the platform in 1985, greatly enhanced. Copepod density 200 m southeast of the platform appears to be the greatest recorded for offshore fine sediments.

The alteration in densities may have arisen through changes in the physical condition of the sediment. Complete burial of fauna will occur in the immediate vicinity of the point of discharge of drilling cuttings; even cuttings contaminated with water-based drilling mud are known to adversely affect macrofaunal diversity (Addy *et al.* 1984).

An adverse physical blanketing effect is suggested by the work of Cantelmo *et al.* (1979), who recorded a significant reduction in nematode density when sand maintained in an aquarium was given a 5 mm cover of barite, a major non-toxic constituent of drilling muds. However, the sediment employed was a medium–coarse sand mixture which presumably harboured predominantly interstitial species not adapted for fine sediment. The clean sediment at Beryl is already too fine for the development of an interstitial nematode fauna (Heip *et al.* 1985) and so the resultant change in sediment structure should have less relevance for nematode density here.

Although blanketing is probably a significant factor closer to the platform, it is not the most likely cause of the density changes. The meiofaunal data suggest that some recovery of the fauna was taking place between the surveys of 1984 and 1985 and yet drilling activity was higher before the 1985 survey than before the 1984 survey, the latter taking place after two years of relative inactivity, apart from a short spell of drilling four months beforehand.

The alteration in meiofaunal densities may have arisen through organic enrichment of the sediment, with the infaunal nematodes adversely affected by the reduced sediments, while certain superficially-distributed copepod species can flourish in more oxygenated conditions where food levels are enhanced and predation pressures reduced. Indeed there are now several documented cases (see, for example, Marcotte & Coull 1974; Vidakovic 1983; Gee *et al.* 1985; Moore & Pearson 1986) of harpacticoid copepods flourishing in organically-enriched subtidal sediments. The copepod species composition within 800 m of the platform is also indicative of an organic enrichment effect, the dominant species having been recorded from organically-polluted sediments elsewhere. Thus *Paramphiascella hyperborea*, normally present in clean sediments at low density, has been found to be a very characteristic species of sediments polluted by sewage sludge off the west coast of Scotland, where *Proameira* aff. *P. simplex* was also common (although this appears to be a common and widespread species) (Moore & Pearson 1986). *Paramphiascopsis longirostris*, present in enormous numbers at the most perturbed site at Beryl, is characteristic of sewage-polluted stretches of the river Gota älv in Sweden (Tulkki 1968) and the Firth of Forth (C. G. Moore, unpublished results), while *Ancorabolus mirabilis* was found to increase in density in sediment that had been artificially organically enriched by Gee *et al.* (1985). On the other hand *Amphiascus tenuiremis*, a dominant species on the nearby Fladen ground (McIntyre 1964) and in similar sediments in the Irish Sea (Moore 1979), was only recorded beyond 800 m from the platform.

Although enhanced copepod density is consistent with an organic enrichment effect, this appears unlikely to have depressed nematodes, which are known to flourish in subtidal sediments more strongly reduced, as a result of organic pollution, than the most reduced sediments recorded at Beryl (C. G. Moore, unpublished results).

Explanation of the apparent improvement in meiofaunal density from 1984 to 1985 is complicated by the concomitant increase in drilling activity; however, the change to low-toxicity drilling mud may be implicated. Cessation of the use of diesel-based drilling mud in May 1982 may have initiated a decline in the concentration of the associated toxic fractions in the sediment in the close vicinity of the Beryl platform. Improvement at the sediment–water interface would be expected to precede that in the lower horizons and one might predict an earlier improvement in epifaunal meiofauna. The enhancement of the epifaunal copepods over background levels is perhaps due to enhanced food levels, although reduced predation and competition may also be implicated. The infaunal nematodes, being more closely bound to the sediment, where release and biodegradation of the toxic diesel fractions is perhaps slower, are possibly in an earlier stage of recovery.

A further reason for the slower recovery of nematodes may lie in the relative colonization rates of the two taxa. The superficial distribution and high swimming ability of many harpacticoids facilitates rapid dispersal. It has been shown experimentally that copepods colonize azoic sediments faster than nematodes (Scheibel 1974; Chandler & Fleeger 1983; Alongi *et al.* 1983) and this may contribute to the faster recovery of copepods at Beryl. Because of the reduced nature of the sediment near the platform, successful early nematode colonists might be expected to be species tolerant of hypoxic conditions and hence perhaps typical of subsurface horizons of clean sediments. Such species appear to have relatively slow colonization rates (Alongi *et al.* 1983).

There is also some evidence of a partial recovery in the macrobenthic community following the switch to low-toxicity drilling mud. Moore (1983) recorded a low diversity community within 600 m of the platform in 1982, dominated by the polychaetes, Cirratulidae, *Chaetozone setosa* and *Capitella capitata*. In 1985 recorded values of species richness and evenness close to the platform were higher (D. Moore, personal communication).

Alternative theories could explain the observed meiofaunal density trends. For example, copepod enhancement in 1985 may have been caused by an increase in bacterial food resulting from the renewal of dumping of cuttings before the survey. From the measurement of naphthalene degradation rates in sediment, microbial mineralization of this substance is apparently maintained at an extremely high level in the area of fluctuating meiofaunal densities. However, this heterotrophic activity may not parallel the rate of production of copepod food, which is quite unknown but may track the level of the discharge of cuttings or may be stimulated by constituents in the low-toxicity drilling mud.

(*b*) *Oil spill and experimental studies*

The impact of oiled drilling cuttings on meiofauna has been studied experimentally by Leaver *et al.* (this symposium), who compared the effects of the addition of cuttings, from drilling operations employing diesel-based and low-toxicity drilling muds, on meiofaunal density. They recorded depressed nematode and interstitial harpacticoid copepod populations, with little difference between diesel and low-toxicity treatments, while the non-interstitial harpacticoid copepods exhibited enhanced densities in low-toxicity treatments but differed little from control densities in the diesel treatment. These results are consistent with the hypothesis of enhancement of copepods at Beryl being caused by the switch to low-toxicity mud. The suppression of nematode densities in both diesel and most low-toxicity experimental treatments may suggest that use of these alternatives would produce little difference in field effects on nematodes. However, it is unlikely that the experimental results can be extrapolated directly to field conditions at Beryl. The reduction of nematodes in tanks containing drilling cuttings is likely to have been caused by the lowering in redox potential in the sediments (Leaver *et al.*, this symposium). Unlike the epibenthic copepods, the nematodes and interstitial copepods would be subject to more reduced conditions within the sediment, to which most species were perhaps intolerant, in view of their natural habitat being well-oxygenated sand. Under field conditions faster immigration of species better suited to the altered physico-chemical conditions would be likely. This experimental constraint on the facility for colonization may mask a potentiality for enhanced nematode populations in sediment enriched by low-toxicity drilling mud, which could manifest itself under field conditions. Indeed Leaver *et al.* (this symposium) suggest that fluctuations in nematode density recorded in certain of the low-toxicity treatments may suggest a potential for recovery, full recovery being prevented by lack of suitable colonists.

The work of Grassle *et al.* (1981) suggests that copepod enhancement and nematode depression may represent a typical stage in meiofaunal recovery from oil pollution of subtidal sediments. They recorded depressions in the meiofaunal density of sandy mud retained in tanks subjected to chronic addition of No. 2 fuel oil. The Crustacea (harpacticoid copepods and ostracods) were especially affected. However, two months after cessation of oil addition nematode density was still depressed whereas copepod density had increased fourfold over the density in the control tanks. The reason for the differing responses appears once again to be partly connected with the differing vertical distribution patterns, as oil levels in the overlying water decreased rapidly after cessation of oil addition, whereas sediment oil levels showed no clear reduction.

A recovery stage for oil-polluted subtidal sediments, characterized by enhancement of copepods and depression of nematodes, is also suggested by the work of Alongi *et al.* (1983), whereas the data of Elmgren *et al.* (1983) indicate an enhancement of copepod density after an initial decline after the *Tsesis* spill of fuel-oil. Nematode density did not show a clear trend. The *Amoco Cadiz* spill was found to cause rapid mortality and subsequent recovery of copepods in subtidal sand, but the nematode density suffered a long-term depression (Boucher 1985). Oviatt *et al.* (1982) repeated the experimental work of Grassle *et al.* (1981) (see above) but at a lower level of addition of fuel-oil. They too recorded a significant reduction in harpacticoids and virtual elimination of ostracods. During the recovery period copepods exhibited a slightly greater (although non-significant) density than controls, whereas ostracods remained significantly fewer. Under the lower oil régime nematode density was not significantly affected, although a lesser (but not significant) nematode density after cessation of oil addition may suggest that the oil in the sediment, that had been steadily rising in concentration throughout the period of oil addition, was beginning to exert an effect. Bakke & Johnsen (1979) failed to record any clear impact on meiofaunal density of the experimental addition of crude oil extract to enclosures placed over the sandy bottom of a Norwegian fjord; however, there was no significant accumulation of hydrocarbons in the sediment.

Data on the impact of oil pollution on subtidal meiobenthos at the species level is very sparse. Renaud-Mornant & Gourbault (1980) and Gourbault (1984) could not show that the *Amoco Cadiz* spill clearly affected species composition or diversity in sandy mud in the inner part of Morlaix Bay but Boucher (1983) recorded a long-term depression in nematode diversity and a change in species composition at a fine-sand site in the outer bay. Boucher *et al.* (1982) recorded a sharp fall in diversity and a change in the dominant nematode species in fine sand experimentally subjected to crude oil contamination, but the closed nature of the aquarium system precluded immigration. Alongi *et al.* (1983) only recorded very subtle differences in the nematode fauna colonizing oiled and clean azoic sediments at a shallow estuarine site. Montagna & Spies (1985) recorded a dramatic change in harpacticoid community structure at sites outside and within a natural submarine oil-seep area, with a gross reduction in species richness and change in species composition within the seep area. Diversity values are yet to be determined for the Beryl area but diversity is clearly very low in the immediate vicinity of the platform (200 m southeast), with one sample of 178 copepods consisting of just three species with 76% dominance by *Paramphiascopsis longirostris*.

4. Concluding remarks

It would appear from the studies reported above that North Sea oil developments are not likely to seriously endanger the contribution that meiofauna makes to the functioning of the North Sea ecosystem. Experiments around the Beryl platform suggest that platform pollution causes only a very localized impact on meiofaunal abundance. The nature of the impact, at least in 1985, probably results in an enhancement of that part of the meiofaunal production that is available to predators, the epibenthic copepods being generally regarded as the major meiofaunal group in this respect (Hicks & Coull 1983). However, whether predators such as small fish can take advantage of this production in such an environment is unknown.

Although a marked density effect has been demonstrated in the immediate vicinity of the Beryl platform, the Grangemouth study and many other meiofaunal and macrofaunal studies show that more subtle effects on community structure may be present at lower pollution levels. Thus the real area of perturbation on the meiofaunal community at the Beryl platform is quite unknown but may be expected to extend well beyond 800 m from the platform.

There are certain similarities in the impacts of the oily discharges at Grangemouth and the Beryl platform. Both faunal responses show certain characteristics commonly associated with organic enrichment, such as density enhancement of certain taxa, alteration in species composition and reduction in species richness; however, the patterns of response with distance from the discharges show a number of important differences of detail.

At Grangemouth nematode density recovery precedes that of the copepods, whereas the situation was reversed for the Beryl platform in 1985. Also, the area of apparent toxic impact at the Beryl platform extends beyond that of organic enrichment, nematodes showing no enhancement but only depression before returning to background densities well beyond the area of copepod enhancement.

These differences can be reconciled by appreciation of the distinction between the two-dimensional pollution gradient at Grangemouth and the three-dimensional gradient at the Beryl platform. The meiofauna is confined to the sediment for much of the time on the mudflat and so the response is to a simple horizontal pollution gradient and this results in an impoverished fauna of virtually only nematodes in the most strongly polluted sediments but enhancement of meiofaunal densities under moderate pollution levels. The Beryl platform pattern can be considered as a three-dimensional analogue. Again, in the most strongly polluted habitat, i.e. within the sediment near the platform, the meiofaunal community consists of an impoverished nematode fauna. Moving out of the sediment (along the vertical pollution gradient) we again experience enhancement but this is restricted to those species adapted to live at the sediment–water interface, of which the epibenthic copepods are the predominant group. Such a model suggests that a similar, albeit more gradual, transition to a zone of enhancement of nematodes and burrowing copepods may occur along the horizontal pollution gradient. The resolution of the Beryl surveys was possibly too coarse to identify such a zone. This may explain the apparent discrepancy between the results reported here and the macrofaunal response around the Beatrice platforms, where Addy *et al.* (1984) concluded that the toxic impact of low-toxicity oil-based drilling mud did not extend beyond the area exhibiting the effects of organic enrichment.

There is some evidence that both meiofauna and macrofauna around the Beryl platform are currently in a phase of recovery, which may be associated with the transition to low-toxicity drilling mud. From work (see, for example, Addy *et al.* 1984; Kingston, this symposium) on the impact of drilling mud on macrofauna it is clear that the lower-toxicity muds will continue to exert an impact on the fauna. The time period for the restoration of the meiofaunal *status quo ante* following cessation of all discharges of cuttings cannot be estimated from present data but it is interesting to note that five years after contamination of subtidal fine sand by crude oil from the *Amoco Cadiz*, although no traces of pollution of the sediment could be detected chemically, the nematode population was still showing depressed density and diversity and an altered species composition (Boucher 1983). The sensitivity of nematodes, the most abundant metazoans on the seabed, to certain types of oil pollution might have implications in the field of effects monitoring. Nematode density, a parameter that is very simple to estimate, shows a clear, temporally-consistent, spatial trend at the Beryl platform. It will be very interesting to determine to what extent this pattern is typical of drilling operations and to relate more closely the nature and sensitivity of meiofaunal and macrofaunal responses. Heip *et al.* (1985) review the subject of the role of nematodes in pollution monitoring.

References

Anon. 1976 *Guide to operational procedures for the IGOSS pilot project on marine pollution (petroleum) monitoring.* (IOC/WMO manuals and guides no. 7.) (50 pages.) Paris: UNESCO.

Addy, J. M., Hartley, J. P. & Tibbetts, P. J. C. 1984 Ecological effects of low toxicity oil-based mud drilling in the Beatrice oilfield. *Mar. Pollut. Bull.* **15**, 429–436.

Alongi, D. M., Boesch, D. F. & Diaz, R. J. 1983 Colonization of meiobenthos in oil-contaminated subtidal sands in the lower Chesapeake Bay. *Mar. Biol.* **72**, 325–335.

Bagheri, E. A. & McLusky, D. S. 1982 Population dynamics of oligochaetes and small polychaetes in the polluted Forth Estuary ecosystem. *Neth. J. Sea Res.* **16**, 55–66.

Baker, J. M. 1971 Refinery effluent. In *The ecological effects of oil pollution on littoral communities* (ed. E. B. Cowell), pp. 33–43. London: Institute of Petroleum.

Bakke, T. & Johnsen, T. M. 1979 Response of a subtidal sediment community to low levels of oil hydrocarbons in a Norwegian fjord. In *Proceedings of the 1979 oil spill conference, Los Angeles, California*, pp. 633–639. Washington, D.C.: American Petroleum Institute.

Boucher, G. 1980 Impact of *Amoco Cadiz* oil spill on intertidal and sublittoral meiofauna. *Mar. Pollut. Bull.* **11**, 95–100.

Boucher, G. 1983 Évolution du méiobenthos des sables fins sublittoraux de la baie de Morlaix de 1972 à 1982. *Oceanologia Acta* **1983**, 33–37.

Boucher, G. 1985 Long term monitoring of meiofauna densities after the *Amoco Cadiz* oil spill. *Mar. Pollut. Bull.* **16**, 328–333.

Boucher, G., Chamroux, S., Le Borgne, L. & Mevel, G. 1982 Étude expérimentale d'une pollution par hydrocarbures dans un microécosystème sédimentaire. I. Effet de la contamination du sédiment sur la méiofaune. In *Ecological study of the Amoco Cadiz oil spill* (ed. E. R. Gundlach), pp. 229–243. Report of the NOAA–CNEXO Joint Scientific Commission.

Bouwman, L. A. 1983 A survey of nematodes from the Ems Estuary. Part II: species assemblages and associations. *Zool. Jb. Syst.* **110**, 345–376.

Bouwman, L. A., Romeijn, K. & Admiraal, W. 1984 On the ecology of meiofauna in an organically polluted estuarine mudflat. *Est coast. Shelf Sci.* **19**, 633–653.

Buchanan, J. B. 1984 Sediment analysis. In *Methods for the study of marine benthos* (ed. N. A. Holme & A. D. McIntyre), pp. 41–65. Oxford: Blackwell Scientific Publications.

Cantelmo, F. R., Tagatz, M. E. & Ranga Rao, K. 1979 Effect of barite on meiofauna in a flow-through experimental system. *Mar. environ. Res.* **2**, 301–309.

Chandler, G. T. & Fleeger, J. W. 1983 Meiofaunal colonization of azoic estuarine sediment in Louisiana: mechanisms of dispersal. *J. exp. mar. Biol. Ecol.* **69**, 175–188.

Davies, J. M., Addy, J. M., Blackman, R. A., Blanchard, J. R., Ferbrache, J. E., Moore, D. C., Somerville, H. J., Whitehead, A. & Wilkinson, T. 1984 Environmental effects of the use of oil-based drilling muds in the North Sea. *Mar. Pollut. Bull.* **15**, 363–370.

Decker, C. J. & Fleeger, J. W. 1984 The effect of crude oil on the colonization of meiofauna into salt marsh sediments. *Hydrobiologia* **118**, 49–58.

DeLaune, R. D., Smith, C. J., Patrick, W. H. Jr, Fleeger, J. W. & Tolley, M. D. 1984 Effect of oil on salt marsh biota: methods for restoration. *Environ. Pollut.* **36**, 207–227.

Elmgren, R. 1973 Methods of sampling sublittoral soft bottom meiofauna. *Oikos* **15** (suppl.), 112–120.

Elmgren, R., Hansson, S., Larsson, U., Sundelin, B. & Boehm, P. D. 1983 The 'Tsesis' oil spill. Acute and long-term impact on the benthos. *Mar. Biol.* **73**, 51–65.

Faubel, A., Hartwig, E. & Thiel, H. 1983 On the ecology of benthos in sublittoral sediments, Fladen Ground, North Sea. I. Meiofauna standing stock and estimation of production. '*Meteor*' *ForschErgeb* D **36**, 35–48.

Fleeger, J. W. & Chandler, G. T. 1983 Meiofauna responses to an experimental oil spill in a Louisiana salt marsh. *Mar. Ecol. Prog. Ser.* **11**, 257–264.

Fricke, A. H., Henning, H. F.-H. & Orren, M. J. 1981 Relationship between oil pollution and psammolittoral meiofauna density of two South African beaches. *Mar. environ. Res.* **5**, 59–77.

Gee, J. M., Warwick, R. M., Schaaning, M., Berge, J. A. & Ambrose, W. G. Jr 1985 Effects of organic enrichment on meiofaunal abundance and community structure in sublittoral soft sediments. *J. exp. mar. Biol. Ecol.* **91**, 247–262.

Giere, O. 1979 The impact of oil pollution on intertidal meiofauna. Field studies after the La Coruna-spill, May 1976. *Cah. Biol. mar.* **20**, 231–251.

Gourbault, N. 1984 Fluctuations des peuplements de Nématodes du chenal de la baie de Morlaix. 1. Résultats a moyen terme, après pollution par les hydrocarbures. *Cah. Biol. mar.* **25**, 169–180.

Grassle, J. F., Elmgren, R. & Grassle, J. P. 1981 Response of benthic communities in MERL experimental ecosystems to low level chronic additions of no. 2 fuel oil. *Mar. environ. Res.* **4**, 279–297.

Heip, C. 1980 Meiobenthos as a tool in the assessment of marine environmental quality. *Rapp. P-v. Réun. Cons. int. Explor. Mer* **179**, 182–187.

Heip, C., Herman, R. & Vincx, M. 1983 Subtidal meiofauna of the North Sea: a review. *Biol. Jb. Dodonaea* **51**, 116–170.

Heip, C., Vincx, M. & Vranken, G. 1985 The ecology of marine nematodes. *Oceanogr. mar. Biol.* **23**, 399–489.

Hicks, G. R. F. & Coull, B. C. 1983 The ecology of marine meiobenthic harpacticoid copepods. *Oceanogr. mar. Biol.* **21**, 67–125.

König, D. 1968 Biologische Auswirkungen des Abwassers einer Öl-Raffinerie in einem Vorlandgebeit an der Nordsee. *Helgoländer. wiss. Meeresunters.* **17**, 321–334.

Marcotte, B. M. & Coull, B. C. 1974 Pollution, diversity and meiobenthic communities in the north Adriatic (Bay of Piran; Yugoslavia). *Vie Milieu* **24**, 281–300.

Massie, L. C., Ward, A. P. & Davies, J. M. 1985 The effects of oil exploration and production in the northern North Sea: Part 1. The levels of hydrocarbons on water and sediments in selected area, 1978–1981. *Mar. envron. Res.* **15**, 165–213.

McIntosh, A. D., Massie, L. C. & Mackie, P. R. 1983 *A survey of hydrocarbon levels and some biodegradation rates in water and sediments around North Sea oil platforms, 1981, 1982.* ICES CM 1983/E:42. (Unpublished manuscript.)

McIntyre, A. D. 1964 Meiobenthos of sublittoral muds. *J. mar. biol. Ass. U.K.* **44**, 665–674.

McIntyre, A. D. & Warwick, R. M. 1984 Meiofauna techniques. In *Methods for the study of marine benthos* (ed. N. A. Holme & A. D. McIntyre), pp. 217–245. Oxford: Blackwell Scientific Publications.

McLachlan, A. & Harty, B. 1982 Effects of crude oil on the supralittoral meiofauna of a sandy beach. *Mar. environ. Res.* **7**, 71–79.

McLusky, D. S. 1982 The impact of petrochemical effluent on the fauna of an intertidal estuarine mudflat. *Est. coast. Shelf Sci.* **14**, 489–499.
Montagna, P. A. & Spies, R. B. 1985 Meiofauna and chlorophyll associated with *Beggiatoa* mats of a natural submarine petroleum seep. *Mar. environ. Res.* **16**, 231–242.
Moore, C. G. 1979 Analysis of the associations of meiobenthic Copepoda of the Irish Sea. *J. mar. biol. Ass. U.K.* **59**, 831–849.
Moore, C. G. & Pearson, T. H. 1986 Response of a marine benthic copepod assemblage to organic enrichment. *Syllogeus-Natn. Mus. nat. Sci.* **58** (Proc. 2nd Int. Conf. Copepoda, Ottawa. 1984), 369–373.
Moore, D. C. 1983 Biological effects on benthos around the Beryl oil platform. ICES CM 1983/E:43. (Unpublished manuscript.)
Naidu, A. S., Feder, H. M. & Norrell, S. A. 1978 The effects of Prudoe Bay crude oil on a tidal-flat ecosystem in Port Valdez, Alaska. In *Proceedings of the tenth annual offshore technology conference*, vol. 1, pp. 97–104. Houston, Texas.
Oviatt, C., Frithsen, J., Gearing, J. & Gearing, P. 1982 Low chronic additions of no. 2 fuel oil: chemical behavior, biological impact and recovery in a simulated estuarine environment. *Mar. Ecol. Prog. Ser.* **9**, 121–136.
Pearson, T. H. & Rosenberg, R. 1978 Macrobenthic succession in relation to organic enrichment and pollution of the marine environment. *Oceanogr. mar. Biol. A. Rev.* **16**, 229–311.
Pearson, T. H. & Stanley, S. O. 1979 Comparative measurement of the redox potential of marine sediments as a rapid means of assessing the effect of organic pollution. *Mar. Biol.* **53**, 371–379.
Platt, H. M. & Warwick, R. M. 1980 The significance of free-living nematodes to the littoral ecosystem. In *The shore environment* (ed. J. H. Price, D. E. G. Irvine & W. F. Farnham), vol. 2 (Ecosystems), pp. 729–759. London: Academic Press.
Raffaelli, D. & Mason, C. F. 1981 Pollution monitoring with meiofauna, using the ratio of nematodes to copepods. *Mar. Pollut. Bull.* **12**, 158–163.
Renaud-Mornant, J. & Gourbault, N. 1980 Survie de la méiofaune après l'échouement de l'"Amoco-Cadiz" (chenal de Morlaix, grève de Roscoff). *Bull. Mus. natn. Hist. Nat.* **2**, 759–772.
Rützler, K. & Sterrer, W. 1970 Oil Pollution. Damage observed in tropical communities along the Atlantic seaboard of Panama. *BioScience* **20**, 222–224.
Scheibel, W. 1974 Submarine experiments on benthic colonization of sediments in the western Baltic Sea. II. Meiofauna. *Mar. Biol.* **28**, 165–168.
Tulkki, P. 1968 Effect of pollution on the benthos off Gothenburg. *Helgoländer. wiss. Meeresunters.* **17**, 209–215.
Vidakovic, J. 1983 The influence of raw domestic sewage on density and distribution of meiofauna. *Mar. Pollut. Bull.* **14**, 84–88.
Wormald, A. P. 1976 Effects of a spill of marine diesel oil on the meiofauna of a sandy beach at Picnic Bay, Hong Kong. *Environ. Pollut.* **11**, 117–130.

Discussion

W. A. Hamilton (*Department of Microbiology, University of Aberdeen, U.K.*). Does Dr Moore consider it likely that the effects he has recorded on the meiofauna do not derive directly from the hydrocarbon input to the system, but rather indirectly through other factors in the environment which themselves have been affected by the hydrocarbons? I am thinking particularly of sulphate-reducing bacteria and the production of sulphide.

C. G. Moore. Yes. Although it seems probable that effluent toxicity is important in the immediate vicinity of the effluent channels at the top of the shore, the major impact of the effluents on the meiofauna of the area appears to be largely a consequence of the enhancement of activity by reducing bacteria.

J. K. Rudd (*Amoco Europe and West Africa, London, U.K.*). Did Dr Moore consider the effects of variations in salinity of the refinery effluent and what happened as the effluent water became purer?

C. G. Moore. By comparing the impact of the Grangemouth petrochemical effluents with those of freshwater discharges to the mudflats, it would appear that the low salinity of the effluents is of considerably less importance than the organic loading. This is not surprising because we are dealing with an estuary, where many of the species are tolerant of wide fluctuations of salinity. The paper describes the effect of the effluents on the meiobenthic community with distance from the effluent channels, and hence under different levels of effluent contamination. The purity of the effluents is unlikely to diminish significantly with distance along the effluent channels because of the high velocities and short distances involved.

P. K. Probert (*Wimbol Limited, Swindon, Wiltshire, U.K.*). Does Dr Moore think that biodeposition from fouling organisms on the platform could contribute significantly to the organic enrichment of the sediment in the immediate vicinity of the platform?

C. G. Moore. I think this is improbable. Unlike drilling cuttings, waste material from fouling organisms is of low specific gravity and might be expected to be widely dispersed by currents in such an exposed situation.

Phil. Trans. R. Soc. Lond. B **316**, 545–565 (1987)
Printed in Great Britain

Field effects of platform discharges on benthic macrofauna

By P. F. Kingston

Institute of Offshore Engineering, Heriot-Watt University, Research Park, Riccarton, Edinburgh EH14 4AS, U.K.

Although there were originally no statutory obligations for North Sea oilfield developers to monitor the environmental impact of their activities, many companies undertook such studies voluntarily.

The central and northern North Sea is principally a level-bottom habitat and can be broadly considered as being dominated by variations of the classical *Amphiura* benthic community. Early approaches to monitoring studies involved the use of grids of sampling stations extending several kilometres in every direction from the proposed site of the installation. More recently it has been found that the major impact on the environment is from the discharge of oil-based drilling cuttings at the platforms and drilling rigs. Efforts are now concentrated closer to the installations, using transects starting as near to the source of the discharge as possible.

By using community parameters such as diversity and equitability, it has been shown that the fauna responds with a dramatic drop in values of these measures close to the platform. However, in most surveys, background values are regained between 500 and 1000 m from the installation. This seems to be the case regardless of whether diesel or low toxicity oil-based drilling fluids are used. Numbers of individuals and biomass responded in a similar way at some installations using 'low toxicity' oil-based drilling muds but increased at others using diesel oil-base. The latter response is similar to that of areas of great organic enrichment while the drop in numbers is more indicative of disturbed or toxic conditions. The markedly patchy distribution of drilling cuttings around the production platforms calls into question the sampling strategies that have been adopted for offshore surveys in the past. The extreme variation of figures, particularly oil levels in sediments, makes it almost impossible to establish firm connections between cause and effect.

The effects of the discharge of cuttings on the benthic environment has been shown to be very severe, but only in a very localized area around the installations. It is suggested that attention is now focused on the persistence of the oil in the cuttings and that future monitoring strategies should include this in their scope.

Introduction

The North Sea has been a focus of study for European marine biologists since the last century. This relatively small sea area is still one of the most productive fishing areas in the world and as such holds considerable public attention as a major resource for the countries that border it. Considerable effort has been spent studying and monitoring North Sea fisheries and there is a wealth of data available mainly concerned with fishing statistics and the effects of natural and man-induced factors on the long-term changes in fish stocks. Until the discovery of oil in the North Sea, the man-induced factors that concerned fishing interests and governments were principally overfishing and pollution (mainly coastal). Because a large proportion of the North Sea fish catch comes from the central and northern North Sea, the development of the oilfields in these areas naturally caused concern (figure 1). Nevertheless it was not until 1984, over ten years after the first production platforms were placed in position, that any statutory obligation was placed on operators to monitor the environmental impact of their activities. Although British authorities did not enforce environmental monitoring in the vicinity of platforms, many operators initiated their own programmes of pre-operational surveys and monitoring.

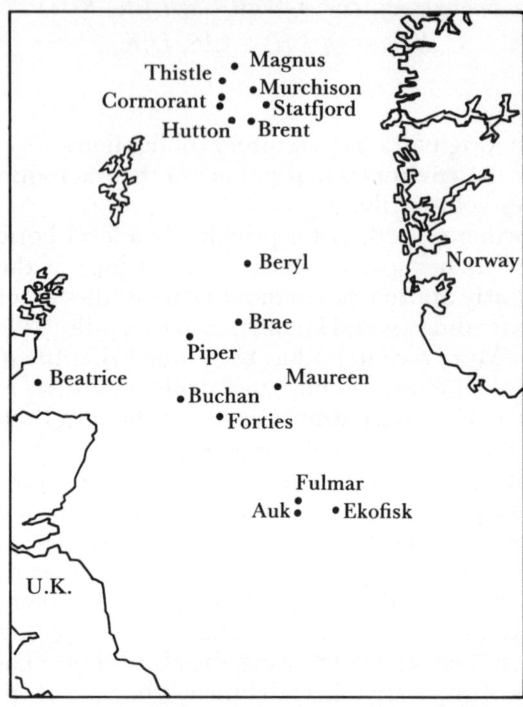

FIGURE 1. Major North Sea oilfields.

With no statutory requirement for an operator to monitor the impact of its offshore activity, it is not surprising that the approach to the problem varied considerably between companies. Two main lines of investigation finally emerged: (1) physical and chemical analysis of the seabed sediments and (2) monitoring of benthic macrofauna. The concentration of the studies around the major oilfields has led to an unprecedented wealth of information about the benthos of localized areas of the central and northern North Sea being accumulated. This contrasts with the paucity of information available for the rest of the area (Kingston & Rachor 1982). The aim of this paper is to consider the effects of oilfield development (particularly discharges) on the benthic environment. However, before doing so, it is worth considering the distribution of benthic communities in the area as a whole.

Benthic communities

For the most part, the central and northern North Sea supports benthic faunal assemblages that can be equated broadly to the classical *Amphiura* community, first described by Petersen (1914) for Danish inshore waters, and later by Jones (1950) (as Boreal offshore muddy sand and mud associations). In the broadest sense, these benthic associations include the foraminiferan-dominated communities of the soft muds of the Fladen Ground (McIntyre 1961), and the Forties area (Hartley 1979) and many that appear to conform with the time-honoured definitions of *Echinocardium–filiformis* and *Brissopsis – chiajei* communities (Buchan (Oil Pollution Research Unit (O.P.R.U.) 1981, 1983), and Brae (Institute of Offshore Engineering (I.O.E.) 1982*b*), (unpublished reports). The complex and variable sediments that are found in the area result in a host of intermediate communities, all of which are basically similar and, from the persistence of *Amphiura*, can be included in Petersen's, (1914) broad label.

The coarser sediments of the area around Magnus (British Petroleum) and Murchison (Conoco) support communities that may be described as variants of the 'deep *Venus*' communities of Petersen (1914), with the muddier areas to the south again tending towards *Amphiura*-type associations. Although these are very broad generalizations they serve to illustrate the dominance of mud and muddy sand associations over most of this area.

Relationship of communities with the sediment

Level-bottom communities, such as occur all over the North Sea, are composed predominantly of infaunal animals, that is, they live within the sediment, either moving through it or using it as a support for their burrows. The sediment not only offers support and protection, but is also a source of food. Benthic communities occurring in mud or muddy sand are characteristically dominated by deposit feeding species and small predatory forms feeding on meiofauna. This is reflected in the distribution of faunal groups within the community which are usually heavily dominated by the Polychaeta, the majority of which are sedentary forms feeding either directly or indirectly on the sediment. Offshore communities such as these are therefore heavily reliant on the sediment as a principal source of food and are thus very vulnerable to the consequences of sediment contamination.

History of North Sea benthic monitoring

Although originally there was no legal requirement for United Kingdom sector offshore operators to perform environmental studies around their installations, many undertook to do so to establish the status of the seabed environment before and during drilling operations and oil production. In the early days of North Sea oil development, attempts were made to predict the area of potential environmental damage to establish suitable sampling programmes. At that time it was thought that the major input of oil contaminants would be in the form of oily water resulting from platform drainage, production water and formation water (Department of the Environment 1976). No emphasis was placed on the dumping of drilling cuttings since at that time the use of oil-based muds was not anticipated. Indeed few studies had sampling stations within the 500 m prohibited zone that had been set up to protect the installations.

Predictions of a total area of up to 11 km^2 d^{-1} being affected by platform discharges with oil levels in excess of 10 $\mu g\ l^{-1}$ in the 500 m zone (Department of the Environment 1976) prompted operators to opt for surveys which endeavoured to cover as wide an area as possible with extensive grids of 20–30 sample stations. This resulted in excellent background information about various areas in the North Sea that had never been studied with such intensity before.

With the development of more and more oil-based mud drilling operations it soon became clear that any effects on the environment were going to be very localized and so sampling strategy switched from the wide grid approach to transects which originated very close to the platforms and emphasized the line of prevailing bottom currents.

Sampling methods

Obtaining reliable quantitative samples of offshore sediments with its fauna is surprisingly difficult. The method employed in almost all North Sea surveys has involved the use of some sort of bottom grab sampler.

The majority of surveys have used either the van Veen, Day or Smith McIntyre grabs. Each takes a sample of 0.1 m^2, the depth of penetration depending upon the firmness of the substratum. The usual practice has been to take five grab samples from each site for faunal analysis and one or two for physical and chemical analysis of the sediments. Recently there has been a trend towards reducing the number of faunal samples to two per station. The fauna are separated from the sediments by screening, by using either a 0.5 mm or 1.0 mm aperture mesh. The material retained is then fixed and preserved for analysis in the laboratory. Such methods have been used almost universally in North Sea studies and were employed in all the surveys that are to be discussed here.

Changes in community structure associated with platform activity

Methods of measurement

In most benthic studies undertaken in the North Sea oilfields changes in community structure have been measured principally by looking for changes in species abundance distribution, that is, the way in which individuals are allocated to species in a given community. The most widely used approach has been to attempt to measure diversity. In an ecological context, diversity refers to two community components; the first is species richness, that is, the total number of species in the community and the second their equitability, that is, the evenness with which the individuals in the community are distributed among species. Many indices of diversity have been suggested. For comparison, one of the most commonly used indices, the Shannon–Wiener Index, ($H_{(s)}$) has been used here. This is a measure of how difficult it would be to predict correctly the species identity of the next individual collected from the community under study and is based on information theory (Shannon & Weaver 1963). The formula for its calculation is

$$H_{(s)} = -\Sigma p_i \log_2 p_i,$$

where p_i = the proportion of the ith species.

Diversity indices such as this have been widely used as a means of integrating the complexity of a community into a single measure that can be used to monitor pollution-induced change. The theory behind their use is based on the premise that communities with a high diversity result from less environmental stress than those with a low diversity. The validity of this assumption has often been questioned, however, where gross pollution is involved, such measures appear to be quite useful for comparative purposes.

The equitability component of diversity has also been frequently used in North Sea oil related survey work to indicate the degree to which a species abundance distribution may be dominated by a proportion of its members. One of the most commonly used indices is J (Pielou 1966). This gives the ratio of the measured value of $H_{(s)}$ to the theoretical value if all species were represented by equal numbers of individuals and is expressed as

$$J = \frac{H_{(s)}}{H_{(max)}} \quad \text{or} \quad \frac{H_{(s)}}{\log_2 S},$$

where S = number of species.

The final diversity measure that will be referred to here, that of Hurlbert (1971), is based on an algorithm of Sanders (1968). This generates a curve by interpolating the number of species that would have been recorded had a series of progressively smaller samples been taken instead of the original. Because the diversity of the sample is represented by the shape of the curve produced (the steeper the curve the greater the diversity), this 'rarefaction' technique allows the comparison of samples of different sizes.

Changes in diversity

Figure 2 shows the values of $H_{(s)}$ with respect to distance from five North Sea oil production platforms, Brent (Shell), Murchison (Conoco), Statfjord (Mobil, Norway), and Beatrice (Britoil, data from Addy *et al.* (1984)). The graphs show a striking similarity in the way in which the diversity of the benthos is affected. The maximum and minimum $H_{(s)}$ values obtained from a range of pre-operational benthic surveys are also indicated (I.O.E. 1978, 1981, 1982*b*).

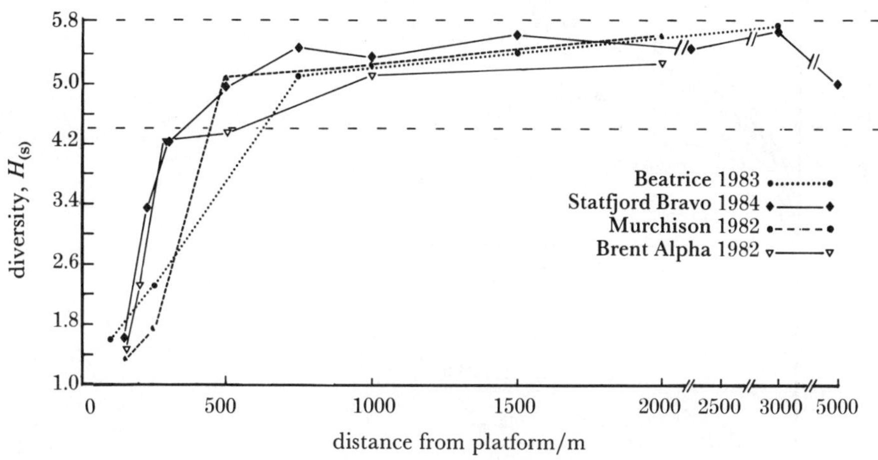

FIGURE 2. Values of Shannon–Wiener diversity index ($H_{(s)}$) with respect to distance from four North Sea oil production platforms. Broken lines indicate maximum and minimum values obtained from pre-operational surveys.

It can be seen that 'pre-operational' values of $H_{(s)}$ are reached for all platforms between 500 and 750 m from the installation suggesting that the extent to which community structure is disrupted, as defined by $H_{(s)}$, is confined to a relatively small area around each platform. A similar picture emerges when the rarefaction curves are examined (figure 3). As with the $H_{(s)}$ figures, the upper and lower limits of curves obtained on pre-operational surveys at the same or nearby sites are shown as dotted lines. The extent of the effect, as indicated by the upward progression of each curve with increasing distance from the platform, supports the figures obtained from the Shannon–Wiener diversity measurement showing containment within 750 m for these locations. However, within this zone the drop in diversity as the platform is approached is dramatic, indicating a gross change in community structure and suggesting gross disturbance of the seabed environment at the proximal stations.

Changes in equitability

The marked drop in diversity in the vicinity of the production platforms is echoed by a similar pattern of response for the equitability measure J used here (figure 4). Values of J, like $H_{(s)}$, increase with distance from the platforms and appear to stabilize at a distance between 500

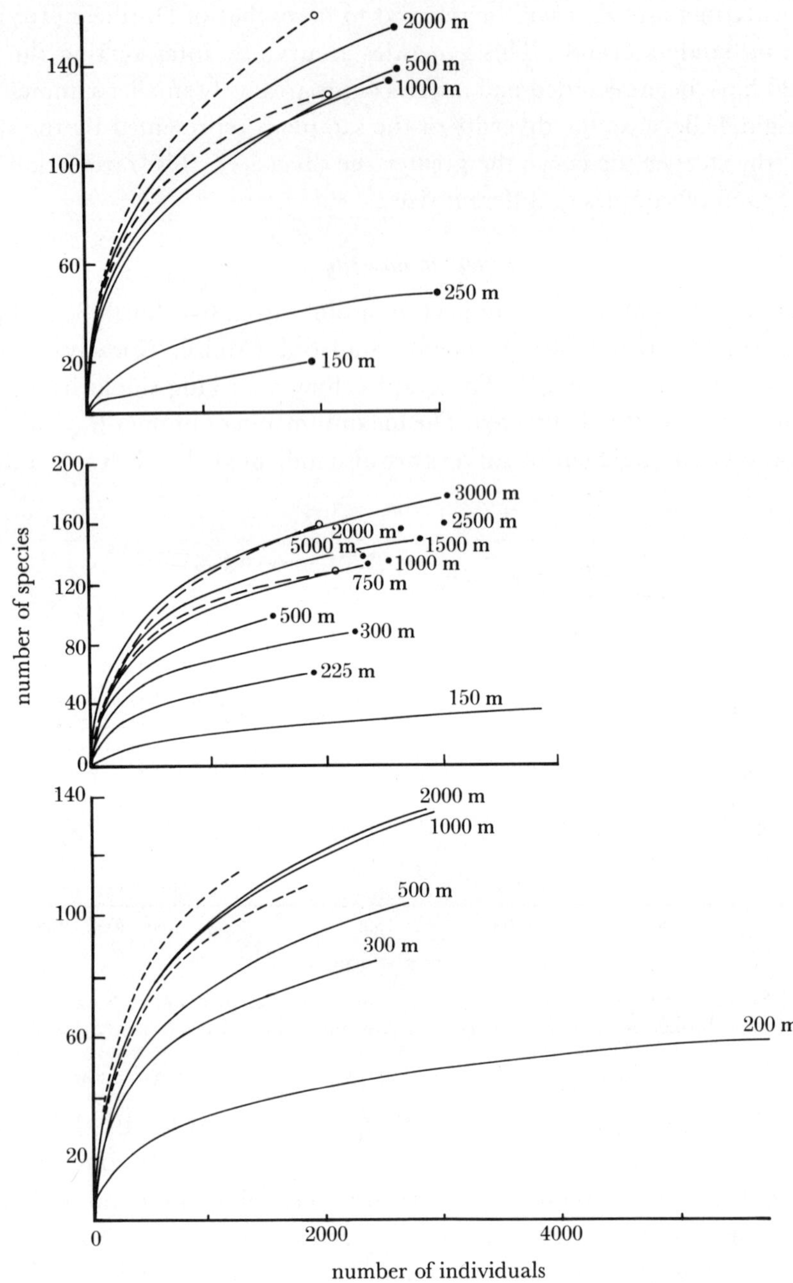

FIGURE 3. Hurlbert rarefaction curves for Murchison (top graph), Statfjord B (middle graph) and Brent Alpha (bottom graph). Dotted lines give limits for pre-operational surveys either at the same location or nearby fields.

and 750 m. The very low values of J recorded at the inner stations are indicative of a species abundance distribution that is greatly dominated by a few individuals and is a classical response to environmental stress (Gray 1976).

Changes in numbers of species and individuals

Diversity, as has already been said, incorporates two properties of a faunal community. One of these, equitability, has already been shown to reflect the trends in diversity; the other, species richness, as might be expected also follows the same general trend. To enable a comparison

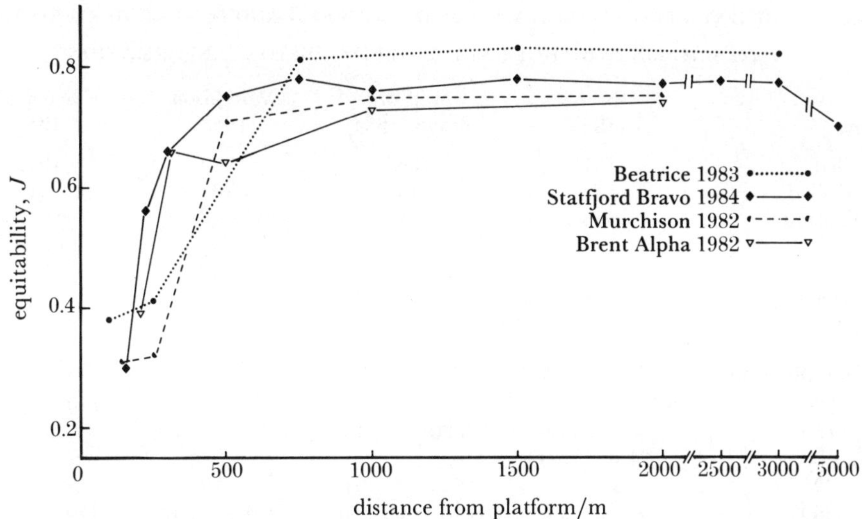

FIGURE 4. Values of Pielou's equitability index (J) with respect to distance from four North Sea oil production platforms.

of data obtained with the variety of sampling methods employed in the studies under consideration here, figures relating to species numbers have been standardized by expressing the number of species for each platform as a percentage of the number recorded at the most distant station. These are presented in figure 5. A full list of the data from which these values were derived, together with information regarding sampling frequency and size of sieve mesh used to separate the fauna from the sediment sample, is given in table 1.

The fact that species richness mirrors diversity is not surprising, although it is clear that it is a more variable parameter. The most important feature of figure 5, however, is the very low numbers of species found at the stations closest to the platforms, particularly at Beatrice and Murchison, where the representation was 16% or less of the value for the most distant station of the transects.

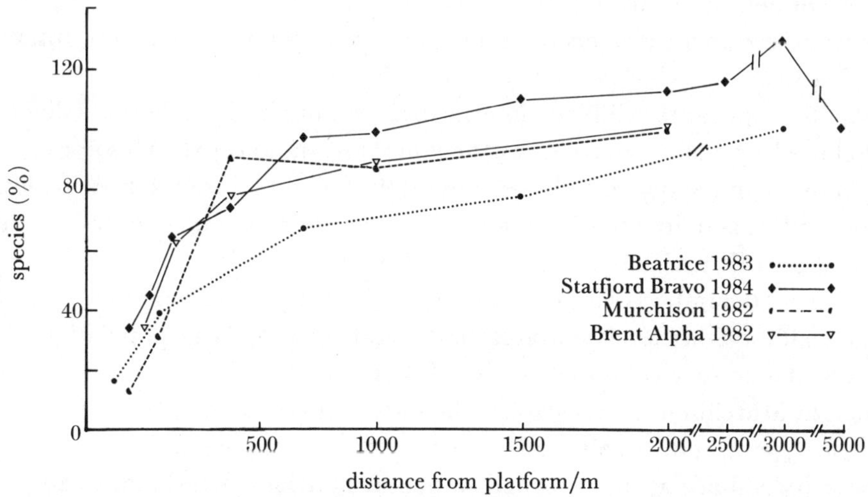

FIGURE 5. Number of species with respect to distance from four North Sea oil production platforms expressed as the percentage of number at the most distance sampling site.

Table 1. Summary of data from which values in figures 5 and 6 were derived together with sampling methods and drilling history information

Platform	Brent (1982)		Statfjord Bravo (1984)		Murchison (1982)		Beatrice† (1983)	
first well drilled	1977		1982		1980		1978	
no. of wells drilled	17		13		20		21	
type of drilling mud	diesel		diesel		WBM (3)‡ diesel (17)		WBM (13)‡ LTM (8)§	
sieve size used/mm	0.5		0.5		0.5		1.0	
sampling frequency (per 0.1 m²)	5		5		5		2	
distance from platform/m	N	S	N	S	N	S	N	S
100	—	—	—	—	—	—	140	19
150	—	—	7590	46	1911	20	—	—
200	5745	60	—	—	—	—	—	—
225	—	—	1923	61	—	—	—	—
250	—	—	—	—	2975	48	416	45
300	2406	86	2274	87	—	—	—	—
500	2679	105	1576	100	2695	138	—	—
750	—	—	2402	132	—	—	661	78
800	—	—	—	—	—	—	—	—
1000	2894	135	2580	134	2621	134	—	—
1200	—	—	—	—	—	—	—	—
1500	—	—	2855	149	—	—	591	91
2000	2852	136	2695	154	2675	154	—	—
2500	—	—	3064	158	—	—	—	—
3000	—	—	3110	177	—	—	—	—
5000	—	—	2347	136	—	—	618	117

† From Addy *et al.* (1984), data for NW transect only.
‡ WBM, water-based mud.
§ LTM, low-toxicity oil-based mud.

If the numbers of individuals are plotted in a similar way a different picture emerges (figure 6). The responses of individual abundances for each of the platforms appear to group into three categories:

(1) platforms where there is an increase in numbers of individuals as the installation is approached (Brent and Statfjord B);

(2) platforms where there is a decrease in numbers of individuals as the installation is approached (Beatrice);

(3) platforms that appear to exhibit an intermediate condition where there is a slight increase in numbers followed by a slight decrease at the innermost stations (Murchison).

In the case of Brent and Statfjord B the response of the individual abundances to the presence of the platform and its activity appears to follow a sequence very similar to that elicited from a point source of organic enrichment (Pearson & Rosenberg 1978).

At Beatrice, the total numbers of individuals are considerably depressed near to the platforms suggesting that conditions are less favourable for the support of the large populations of benthic animals that are able to survive in the vicinity of Brent and Statfjord B.

The situation at Murchison is less clear. At the station closest to the platform (150 m) there were less than 30% fewer individuals than at the most distant stations and at 250 m 10% more (cf. 320% more individuals at the proximal station at Statfjord B and almost 80% fewer at some of the inner stations at Beatrice). This suggests either an intermediate effect or perhaps a transitional condition from one extreme to the other.

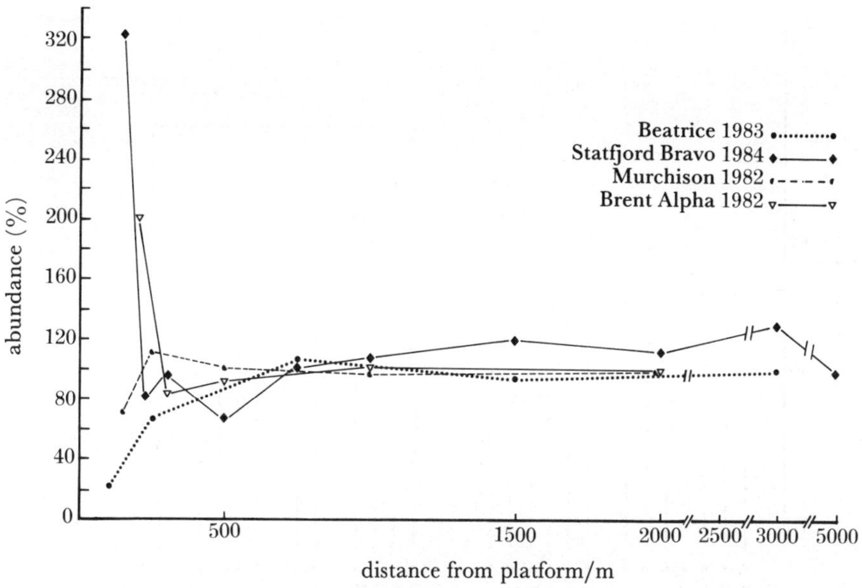

FIGURE 6. Individual abundances with respect to distance from four North Sea oil production platforms expressed as the percentage of the abundance at the most distance station.

Patterns in spatial distribution

The benthic faunal parameters considered so far have dealt in statistical terms exclusively with no regard to individual species identity. Clearly there have been considerable changes in community structure in the vicinity of the production platforms and the question arises whether these responses can be attributed to the same species in each case. In a recent study done at the Statfjord oilfield (Matheson *et al.* 1986) three detailed surveys were done around each of the production platforms Statfjord Alpha, Bravo and Charlie. At the time of the survey, which was conducted in 1984, Statfjord Alpha had a six-year drilling history, Statfjord Bravo had a two-year history and Statfjord Charlie, a concrete gravity platform, had been in place for less than two weeks.

The data obtained were subjected to presence–absence classification analysis using Jaccard's coefficient of similarity:

$$S_J = a/(a+b+c),$$

where a = number of species occurring in both samples; b = number of species occurring in b only, and c = number of species occurring in c only.

The results of the analysis are shown as a dendrogram in figure 7. The stations cluster into five main groups which appear to correspond to two possible environmental régimes:

(1) groups 1 and 2, stations within an area of effect responding to discharges at Statfjord Alpha and Bravo;

(2) groups 3–5, stations unaffected by drilling activity, representing undisturbed communities, each group approximately corresponding to the area around a particular platform.

The validation of groupings 1 and 2 in terms of the community parameters previously discussed is given by Matheson *et al.* (1986). The important point here is that these groups refer to the species identity of the faunal communities at each station and suggest a strong element of similarity between the stations under the influence of platform discharge. Similar

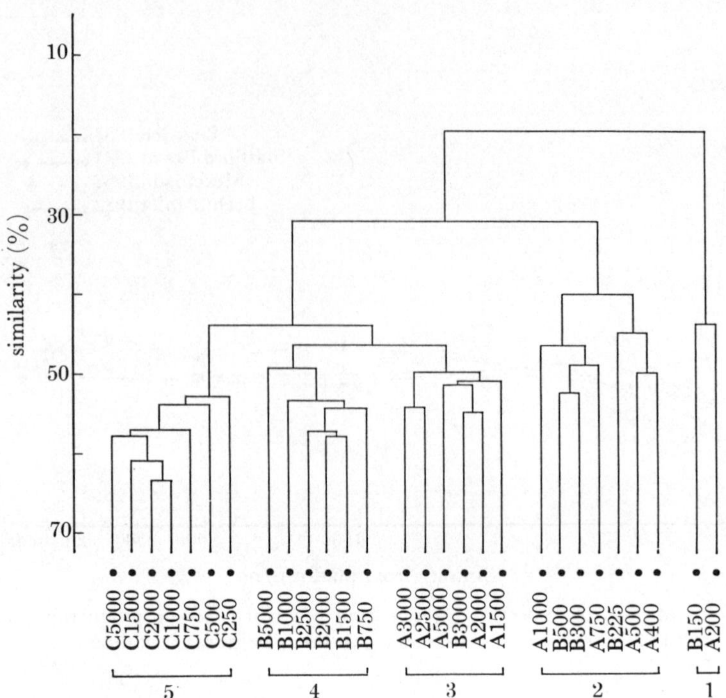

FIGURE 7. Cluster dendrogram using Jaccard's coefficient of similarity on faunal presence–absence data; Statfjord (1984).

patterns of spatial distribution have been demonstrated for many other North Sea installations, for example, Murchison (I.O.E. 1981), Magnus, (I.O.E. 1984*a*), Hutton (I.O.E. 1981), Brent (I.O.E. 1982*a*).

Faunal responses

Table 2 shows the top faunal ranked species from a similar range of stations sampled at Brent Alpha, Murchison and Statfjord Bravo. The data from the proximal station for each platform display two striking features:

(1) the remarkable similarity of the species making up the lists;

(2) the fact that in each case these five species account for between 89–99% of the total numbers of individuals from the station.

It is also apparent from the table that some species are more abundant at the 500 m stations than at either of the others, whereas others reach their maximal numbers away from the platforms. This is illustrated for the Brent Alpha fauna in figure 8.

The three types of faunal distributions more or less conform with the *r*, *T* and *K* strategists suggested by Gray (1979) as adaptive strategies to pollution: *r* strategists are able to mature and reproduce very rapidly and therefore quickly colonize and dominate areas subject to disturbance; *T* strategists are less fecund but are able to tolerate stressful conditions and therefore out-compete less tolerant species: *K* strategists are controlled by the normal forces of interspecific competition.

In this case the *r* strategists are represented by species such as *Capitella capitata*, which is able to switch reproductive modes from pelagic to benthic larvae (Grassle & Grassle 1977), the

TABLE 2. TOP FIVE RANKED SPECIES FROM THREE STATIONS AT COMPARABLE DISTANCES FROM THE BRENT (1982), MURCHISON (1982) AND STATFJORD BRAVO (1984) PLATFORMS

Brent			Murchison			Statfjord Bravo		
species	no. per 0.5 m²	cum (%)	species	no. per 0.5 m²	cum (%)	species	no. per 0.5 m²	cum (%)
200 m			150 m			150 m		
1. Ctenodrilidae	3460	60.2	*Ophryotrocha puerilis*	1438	75.2	*Ophryotrocha puerilis*	5176	68.2
2. *Capitella capitata*	760	73.5	*Ophryotrocha* sp.	185	84.9	Ctenodrilidae	948	80.7
3. *Pholoe minuta*	398	80.4	*Spiophanes* sp.	147	92.6	*Capitella capitata*	922	92.8
4. *Ophryotrocha puerilis*	273	85.1	*Pseudopolydora paucibranchiata*	70	96.3	*Pseudopolydora paucibranchiata*	218	95.7
5. *Thyasira* sp.	241	89.3	*Capitella capitata*	54	99.1	*Pholoe minuta*	93	96.9
500 m			500 m			500 m		
1. *Pholoe minuta*	525	19.8	*Exogone verugera*	338	12.5	*Pholoe minuta*	250	15.9
2. *Prionospio cirrifera*	359	33.4	*Caulleriella killariensis*	305	23.9	*Pseudopolydora paucibranchiata*	175	27.0
3. *Thyasira croulinensis*	345	46.4	*Pholoe minuta*	206	31.5	*Ophiura affinis* (juvenile)	114	34.2
4. *T. pygmaea*	299	57.7	*Spiophanes* sp.	204	39.1	*Exogone verugera*	104	40.8
5. *T. succisa*	217	65.9	*Pseudopolydora paucibranchiata*	111	43.2	*Glycera capitata*	71	45.3
1000 m			1000 m			1000 m		
1. *Thyasira croulinensis*	367	12.8	*Exogone verugera*	241	9.2	*Exogone verugera*	247	9.6
2. *T. pygmaea*	279	22.5	*Glyphanostomum macroglossum*	184	16.2	*Pholoe minuta*	224	18.3
3. *T. succisa*	233	31.0	*Protodorvillea kefersteini*	171	22.7	*Pseudopolydora paucibranchiata*	179	25.2
4. *Myriochele* sp.	188	37.6	*Spiophanes* sp.	164	29.0	*Ophiura affinis* (juvenile)	170	31.8
5. *Pholoe minuta*	179	43.8	*Caulleriella killariensis*	146	34.6	*Glycera capitata*	142	37.3

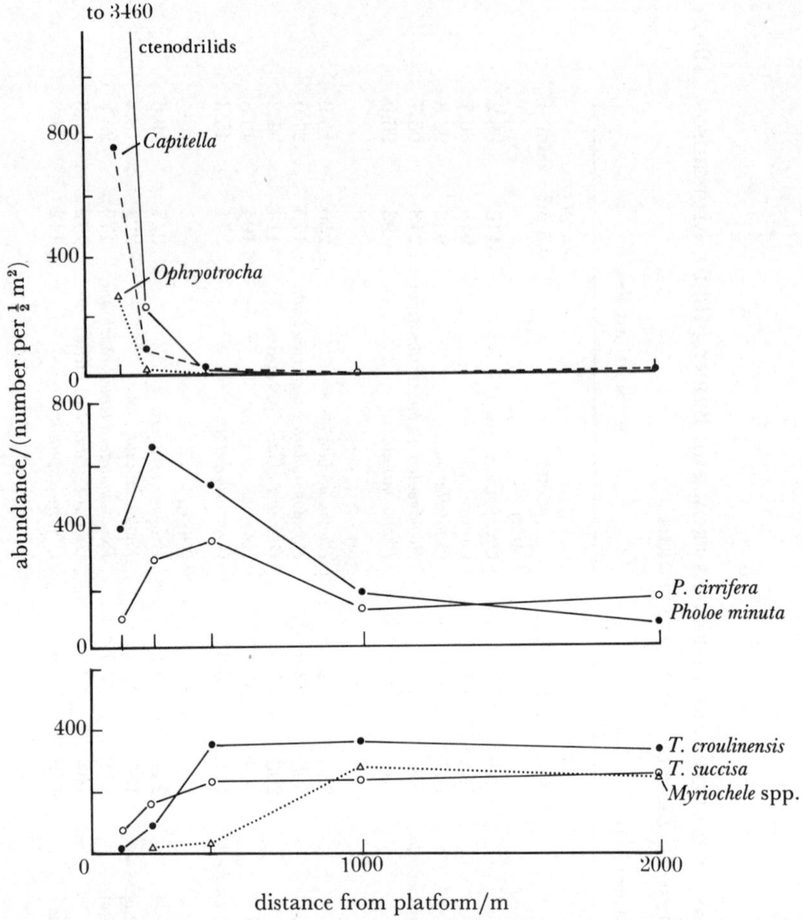

FIGURE 8. Selected examples of three types of abundance variations of benthic species under the effects of disturbance (from Kingston 1983).

Ctenodrilidae, which are able to reproduce vegetatively by fragmentation (Fauvel 1927), or *Ophryotrocha* which is able to change its mode of reproduction to suit local conditions (Åkesson 1973). The *T* strategists are possibly represented by species such as *Pholoe minuta*, *Prionospio cirrifera* (Brent, Alpha), *Pseudopolydora paucibranchiata* (Murchison and Statfjord Bravo) and other species not represented in the top five rankings. These species may be interpreted as being tolerant of slight pollution and are able to increase in numbers by filling the niche vacated by the more sensitive species. The *K* strategists include most of the remaining species which are unable to compete where there is environmental stress and so decrease in numbers as the platform and its point source of contamination are approached.

DISCUSSION

It cannot be denied that the installation and operation of an oil production platform will have a profound effect on the benthic environment in its immediate vicinity. The mere act of placing a structure on the seabed will disturb its inhabitants. The dumping of material whether inert or toxic is bound to result in the physical destruction of the communities underneath. The important questions here concern the spread, toxicities and persistence of the discharge.

Zones of effect

There is plenty of evidence that the zones of effect are quite limited in their extent and that disruption of the biological system may extend no further than 500–1000 m from drilling platforms where there is a point source of contamination (see figure 7; Davies *et al.* 1984; Addy *et al.* 1984). Whether these zones have stabilized or not is open to question. It could be argued that the observations thus far reported represent only a transitory phase on a route to more widespread environmental damage. It is certainly true that the longer a drilling platform has been in operation, the wider the affected area. Matheson *et al.* (1986) demonstrated this for the Statfjord Field. Here the zone of biological effect, confined to 500–750 m at the Bravo platform with its two-year drilling history, was significantly greater (750–1000 m) at the Alpha platform, at which drilling had been taking place for six years.

In the case of the Statfjord installations comparisons of zones of effect may be made with a reasonable amount of confidence because the platforms are geographically very near to one another and are subjected to broadly similar hydrographic conditions. However, platforms in different localities and run by different operators will, in addition to a range of drilling histories, have widely differing spread characteristics of cuttings based upon local tidal flow, height of discharge chute and nature of discharge. To attempt to establish generalized zones of effect based upon data subject to such a range of variables inevitably must lead to zones of such wide overlap as to be of little practical use. However, in none of the studies cited here has there been any detectable deleterious effects on the benthic communities outside 1000 m from the platform. In most cases community parameters return to background values soon after 500 m.

Toxicity of discharges

The toxicity of drilling cuttings arises principally from the use of oil-based drilling muds and the considerable quantities of base-oil that adhere to the cuttings after they are dumped. Diesel-oil was used as the original base-oil in these muds because it was readily available and cheap. The final washed cuttings that were discharged to the seabed may have contained 6–17% by mass diesel-oil (Blackman & Law 1981). Diesel contains a relatively high concentration (20–30%) of low molecular mass aromatic substances and is known to be toxic to marine organisms. Recent concern in the United Kingdom over the effect these diesel oil-based drilling muds might be having on the seabed environment has culminated in United Kingdom Government legislation to control the use of oil-based drilling muds and to encourage the use of low aromatic content base-oils (the so called 'low toxicity' base-oils). The United Kingdom legislation prohibits the discharge of whole muds, requires the use of efficient solids-control equipment for low-toxicity muds and specifies additional treatment equipment where diesel-based muds are used. Similarly in Norway the discharge of whole oil-based muds is forbidden. The State Pollution Control Authority (S.P.C.A.) must approve the installation's treatment systems for oil-contaminated cuttings. The S.P.C.A. classify drilling muds as toxic (e.g. diesel-base) moderately toxic (e.g. low aromatic oil-base) and low toxic (e.g. water-base) using acute toxicity tests. In Denmark diesel-based mud has been used on only one occasion and cuttings were brought ashore for treatment and disposal. Drilling with low aromatic oil-base mud has been permitted on an experimental basis for the development of the Dan and Tyra fields. In Denmark, permits for drilling using a specific mud system are granted only on a case by case basis. In the Netherlands provisional approval has been given to the use of oil-based muds containing less than 5% aromatics by mass providing they meet specified toxicity-test requirements. These, and controls over discharges of contaminated cuttings, are provisional and under review. The effect has been an almost total switch from diesel-base to the use of low toxicity oil-base muds in the North Sea.

Community parameters such as diversity, equitability and species richness have been shown to respond quite dramatically to the influence of discharge of drill cuttings both for the oilfields considered here (figures 3–5) and at other parts of the North Sea (Davies *et al.* 1984).

If this response is then related to total levels of hydrocarbon in the sediment a very strong negative correlation can be demonstrated. An example is shown in table 3. This is not surprising when the concentrations of hydrocarbons in the sediment relative to distance from the platforms is considered. Figure 9 shows such values from several North Sea oilfields, following a trend that is the inverse of that found for species richness and diversity (Davies *et al.* 1984). However, the use of such community parameters tell us little about the nature of the cause of effect. A drop in diversity, for example, could equally well be signalling toxicity or organic enrichment because responses of both conditions could lead to a drop in number of species with an increase

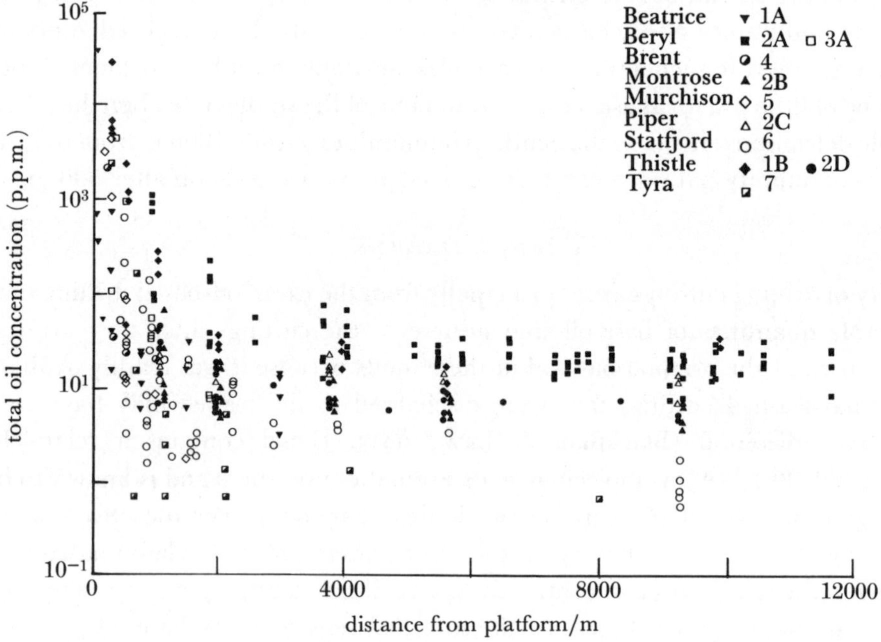

FIGURE 9. The concentration of total oil related to distance from a production platform (from Davies *et al.* 1984).

TABLE 3. VALUES OF PEARSON'S PRODUCT MOMENT CORRELATIONS FOR DIVERSITY ($H_{(S)}$), NUMBER OF SPECIES (S) AND NUMBERS OF INDIVIDUALS (N) WITH RESPECT TO THE PERCENTAGE ORGANIC AND TOTAL OIL CONTENT OF SEDIMENTS AROUND STATFJORD ALPHA AND BRAVO PRODUCTION PLATFORMS

	Statfjord Alpha			Statfjord Bravo		
	% org.	total oil IR	total oil UV	% org.	total oil IR	total oil UV
$H_{(S)}$	−0.852	−0.76	−0.916	−0.346	−0.928	−0.948
S	−0.756	−0.755	−0.880	−0.183	−0.827	−0.817
N	0.761	0.982	0.884	0.327	0.763	0.827

in dominance of opportunistic or pollution tolerant forms. Change in diversity is thus a good measure of environmental disturbance and little more. Pearson & Rosenberg's (1978) classical response to organic enrichment will not apply if the levels of toxic substances prevent the opportunistic species from thriving. Rygg (1986) has shown that in Orkdalsfjorden, where heavy metal contamination from effluents has a toxic effect, populations, as one might expect, are diminished. Thus, one might expect an increase in the total number of individuals where organic pollution dominates environmental insult and a reduction where toxicity is the more important factor.

Figure 6 shows that both types of response can be recognized around North Sea production platforms. The interesting feature of the figures, however, is that the organic enrichment effect appears here to be associated with the use of diesel oil-based muds (Brent A and Statfjord B) and the 'toxic effect' with the use of 'low toxicity' oil-based muds (Beatrice).

Blackman *et al.* (1983) have shown that most 'low toxicity' base-oils are at least an order of magnitude less toxic than the diesel equivalent when appraised using the brown shrimp, *Crangon crangon*, in 96 h LC_{50} tests. However, the results they obtained were very variable, the measured differences in their toxicity being attributed to normal experimental variability. Not only did Blackman *et al.* (1983) find that the alternative base-oils were less toxic, but also that the toxicity of the base-oils varied much more than that of the muds which were formulated from them. In addition, the toxicity of the drilling mud with the base-oil added was lower than that of the base-oil alone. They concluded that the acute toxicity of the base-oil and the drilling mud formulated from it bore no strict relation, even when the same manufacturer's mud-solids formula is used with a range of different oils.

This suggests a certain element of unpredictability in the performance of an oil-based mud in the field regardless of the base-oil used and might go some way to explaining why the field observations on faunal response are at variance with laboratory evidence based on *Crangon crangon* 96 h LC_{50} toxicity tests.

Persistence of contaminating oil

It is only relatively recently that the question of persistence of the toxic effects of discharged drilling cuttings, in the context of North Sea operations, has been addressed with any seriousness. The rate at which degradation of the toxic components of the discharge takes place will affect, not only how quickly the area returns to normal after drilling stops, but also how far the toxic effects will spread. Most of the evidence presently available comes from exprimental work. This is hardly surprising because there are few developments in the North Sea where drilling activity has ceased altogether.

The large numbers of individuals recorded close to the production platforms at some oilfields investigated indicate that there is considerable biological activity regardless of the potentially toxic nature of the discharges. The enhanced number of individuals of macrofauna at the innermost stations at Brent A (see figure 6) is also reflected in the values for standing crop biomass which increase dramatically at this station (figure 10). While this reinforces the notion that the benthic fauna is responding to organic pollution at this site, it can also be inferred that considerable biodegradation of the contaminating oil might be taking place because it is difficult to see where else the energy required to sustain the dense populations would be coming from. Brent A is one of the few studies in which biomass determination has been undertaken (I.O.E. 1982*a*).

The standing crop biomass value of just under 5.0 g m^{-2} ash free dry weight (AFDW) is not

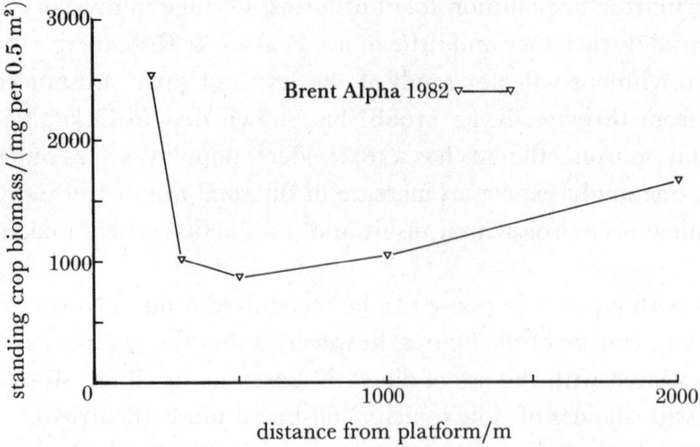

FIGURE 10. Changes in standing crop biomass with respect to distance from Brent A platform (totals for up to ten species ranked by biomass (I.O.E. 1982*a*)).

particularly high compared with other North Sea muddy bottom habitats (Buchanan & Warwick 1974; Rachor 1982; Creutzberg *et al.* 1984). However, it must be remembered that, unlike these unpolluted areas, the members of the fauna immediately around the production platforms are composed almost exclusively of small species the majority of which have individual AFDWs of less than 0.1 mg. Most of the species were earlier identified as the opportunistic species associated with organic pollution and as such are characterized by having very rapid population turnovers. Small species with short life cycles may have very great production/biomass (P/B) ratios. Values of 14 have been reported for some species of *Ophryotrocha* (Tennant 1984).

It is unfortunate that so few data exist regarding the distribution of biomass around the production platforms because the relevance of this to the eventual fate of the discharged material cannot be overemphasized. For approximate comparison, biomass values for the 200 m station at Statfjord Alpha have been prepared from individual abundance data and applying the biomass values for the principal species obtained from the 1982 Brent A study. Values for the Beatrice platform have also been prepared, this time by simply multiplying the individual abundance figures by the mean AFDW of *Capitella capitata*, because this species accounted for up to 98% of the individuals at the inner stations (Addy *et al.* 1984). The results are given in table 4. If the P/B ratio values for *Ophryotrocha* are applied to the biomass values obtained at Statfjord Alpha (it was the most abundant species at 200 m at over 16000 per 0.5 m²) then possible annual production rates of over 40 g m⁻² per year can be expected for

TABLE 4. BIOMASS OF TOP FIVE SPECIES (RANKED BY ABUNDANCE) FROM BRENT A AND BIOMASS ESTIMATES† FOR STATFJORD ALPHA AND BEATRICE DERIVED FROM BRENT BIOMASS VALUES FOR INDIVIDUAL SPECIES

platform	distance/m	biomass/(mg m^{-2})	% organic	sieve size/mm
Brent A	200	2379	1.94	0.5
Statfjord A	200	3270†	1.22	0.5
Statfjord A	200	516†	1.22	1.0
Beatrice‡	65–100	66–221†	*ca.* 10–5	1.0
Beatrice‡	250	50–166†	*ca.* 1	1.0

‡ Original figures from Addy *et al.* (1984), values for SSW and NNE transects given.

the top three opportunists alone. This is over 20 times the production rate reported by Buchanan & Warwick (1974) for a muddy bottom community off the Northumberland Coast.

Regardless of the fact that the only oil-based muds used at Beatrice have been of the 'low toxicity' type (Addy *et al.* 1984), derived values indicate that the standing crop biomass of macrofauna around this platform is less than that around the installations where diesel-based muds have dominated drilling activity. Such a response could be explained if the relative toxicity of the two types of drilling mud is not the main factor in determining the biodegradability of the hydrocarbon content of the cuttings. There is an increasing amount of evidence to support this contention.

Blackman *et al.* (1983) reported 'the formation of heavy algal and some fungal mats on the sediment surface' of some of the test tanks used for settlement trials. One of these was a tank containing diesel. Gillam *et al.* (1986) have shown that one of the organisms responsible for these surface mats, *Streptomyces* sp., is twice as efficient as utilizing diesel than it is at utilizing 'low toxicity' base-oil at diesel concentrations of less than 3 % by volume. They go on to suggest that, although potentially more toxic than the more refined base-oils, diesel-oils may possibly be broken down at a greater rate in the marine environment owing to the greater diversity of potential carbon sources provided by the more complex diesel-oil.

The evidence presented here has suggested that, at least in the case of Brent A and Statfjord Alpha, macrofaunal activity of a few opportunistic species has been greatly enhanced by the use of diesel-based drilling muds when compared with the results obtained by Addy *et al.* (1984) for the Beatrice oilfield at which 'low toxicity' drilling muds were used.

There is simply not enough evidence available yet to enable firm conclusions to be reached regarding the cause of the drop in number of individuals near this installation. The high levels of organic matter in the sediments immediately around the platform suggest that not only is there very little physical spread of cuttings but also that the rate of dispersion–breakdown of the oil is too slow to prevent accumulation taking place. In addition, the less stringent washing treatment of cuttings, required when dealing with 'low toxicity' oil-based muds, may lead to greater amounts of oil finding its way to the seabed. Similar decreases in individual abundance have been observed around other North Sea platforms using 'low toxicity' drilling muds exclusively (I.O.E. 1986), but without the enhanced organic matter levels reported by Addy *et al.* (1984) for the Beatrice platform. Similarly, decreases in individual abundance (although not so severe) were found at Murchison (see figure 6) where, up to the time of the survey diesel muds had been used. Again organic matter levels in the sediment here, at approximately 1 %, were a fraction of that at the Beatrice platform.

It is likely that there are a variety of factors determining the size of populations that a cuttings pile and the immediate environs can support, not least in the total formulation of the drilling mud. The use of low-toxicity base-oils may avoid the aromatics found in the diesel, but the increased amounts of mud, as opposed to cuttings, that are likely to be deposited as a result of less stringent cleaning treatment requirements may mean that other mud additions become limiting.

The markedly patchy distribution of drilling cuttings on the seabed around the production platforms calls into question the sampling strategies that have been adopted for offshore surveys in the past, and indeed at the present. The extreme variation of figures, particularly hydrocarbon concentrations in sediments close to the platforms (see table 5) makes it very difficult to establish firm correlations between cause and effect.

The presently accepted standard of taking single samples to estimate levels of hydrocarbons

TABLE 5. CONCENTRATION OF TOTAL OIL (IR) IN RELATION TO DISTANCE FROM FIVE NORTH SEA OIL PRODUCTION PLATFORMS (MICROGRAMS PER KILOGRAM)

sampling	Brent		Murchison		Stafjord Alpha		Statfjord Beta	
distance/m	1	2	1	2	1	2	1	2
100–200	344	1993	2993	4387	52777	1314	2838	832
250–300	84	33	1050	535	4541	2182	1439	1344
500	23	22	12	11	2221	2412	80	22
800–1000	13	19	8	11	164	209	227	42
2000–2500	19	19	6	6	19	24	19	27

in the sediments in the vicinity of the cuttings pile is very unsatisfactory. The use of duplicate samples for estimating macrofauna in such variable conditions is equally dubious. The effects of drilling cuttings have clearly been shown to be highly disruptive to the seabed environment but at the same time very localized. The need to design monitoring studies to detect environmental disturbance in the very inner areas is really no longer necessary, if indeed such studies ever were. However, two aspects do remain of concern:

(1) the possible continued spread of effects;

(2) the rate of hydrocarbon degradation.

Both are interconnected in that a sustained high level of hydrocarbon degradation will restrict the area eventually affected.

Future approaches to monitoring the seabed around production platforms should now be concentrated on the persistence of the oil in the cuttings as well as maintaining a watchful eye over spread effects. Sampling strategies should focus on determining the degree of biological activity close to the platform with the aim of monitoring rates of biodegradation of the base-oils of the various mud formulations.

I am very grateful to the members of the benthic section at the Institute of Offshore Engineering who contributed to the studies described here. In this respect I particularly thank Dr Hamish Mair and Mrs Susan Hamilton. I am also grateful to Professor Cliff Johnston, Jonathan Side and Iain Matheson for helpful discussions and advice during the course of the work and preparation of this paper. Finally my thanks to Conoco (North Sea) Inc., Shell Exploration Ltd and Mobil (Norway) (on behalf of Statfjord Owners) for commissioning the studies described here and allowing the publication of the data.

REFERENCES

Addy, J. M., Hartley, J. P. & Tibbetts, P. J. C. 1984 Ecological effects of low toxicity oil-based mud drilling in the Beatrice oilfield. *Mar. Pollut. Bull.* **15** (12), 429–436.

Åkesson, B. 1973 Morphology and life history of *Ophryotrocha maculata* sp.n. (Polychaeta, Dorvilleidae). *Zool. Scr.* **2**, 141–144.

Blackman, R. A. A. & Law, R. J. 1981 The oil content of discharged drill-cuttings and its availability to benthos. ICES CM: 1981/E:23. (7 pages.) Copenhagen.

Blackman, R. A. A., Fileman, T. W. & Law, R. J. 1983 The toxicity of alternative base oils and drill muds for use in the North Sea ICES CM: 1983/E:11. (7 pages.) Copenhagen.

Buchanan, J. B. & Warwick, R. M. 1974 An estimate of benthic macrofaunal production in the offshore mud off the Northumberland coast. *J. mar. biol. Ass. U.K.* **54**, 197–222.

Creutzberg, F., Wapenaar, P., Duinveld, G. & Lopez Lopez, N. 1984 Distribution and density of the benthic fauna in the southern North Sea in relation to bottom characteristics and hydrographic conditions. *Rapp. P.-v. Réun. Cons. int. Explor. Mer.* **183**, 101–110.

Davies, J. M., Addy, J. M., Blackman, R. A., Blanchard, J. R., Ferbrache, J. E., Moore, D. C., Somerville, H. J., Whitehead, A. & Wilkinson, T. 1984 Environmental effects of the use of oil-based drilling muds in the North Sea. *Mar. Pollut. Bull.* **15**, (10), 363–370.

Department of the Environment 1976 The separation of oil from water for North Sea Oil Operations. *Pollution Paper No. 6*. London: HMSO.

Fauvel, P. 1927 Polychètes sédentaires. *Faune Fr.* **16** (494 pages).

Gillam, A. H., O'Carrol, K. & Wardell, J. N. 1986 Biodegradation of oil adhering to drill cuttings. In *Proceedings of Conference on Oil-based Drilling Fluids – Cleaning and Environmental Effects of Oil Contaminated Drill Cuttings, Trondheim, Norway, Feb. 1986*, pp. 123–136.

Grassle, J. F. & Grassle, J. P. 1977 Temporal adaptations in sibling species of *Capitella*. In *Ecology of marine benthos* (ed. B. C. Coull), pp. 177–190. Columbia: University of South Carolina Press.

Gray, J. S. 1976 Are baseline surveys worthwhile? *New Scient.* **70**, 219–221.

Gray, J. S. 1979 Pollution-induced changes in populations. *Phil. Trans. R. Soc. Lond.* B**286**, 545–561.

Hartley, J. P. 1979 Biological monitoring of the seabed in the Forties Field. In *Proceedings of Conference on Ecological Damage Assessment, Arlington, Virginia*, pp. 215–253.

Hurlbert, S. H. 1971 The non-concept of species diversity: a critique and alternative parameters. *Ecology* **52**, 578–586.

I.O.E. 1978 Murchison field environmental baseline study: August 1978. Unpublished report of the Institute of Offshore Engineering to Conoco North Sea Inc. (257 pages.)

I.O.E. 1981 Hutton–Murchison Field. Environmental baseline study volume 2. Macrofaunal assessment: August 1980 survey. Unpublished report of the Institute of Offshore Engineering to Conoco North Sea Inc. (248 pages.)

I.O.E. 1982*a* Drilling cuttings study. Benthic sediment survey. Brent A 1982. Unpublished report of the Institute of Offshore Engineering to Shell U.K. Exploration & Production Ltd. (139 pages.)

I.O.E. 1982*b* Brae Field environmental baseline study. Volume 1, August 1981 survey. Unpublished report of the Institute of Offshore Engineering to Marathon Oil U.K. Ltd. (174 pages.)

I.O.E. 1984*a* Environmental assessment. BP Magnus Field: August 1983 survey. Unpublished report of the Institute of Offshore Engineering to BP Development Ltd. (189 pages.)

I.O.E. 1984*b* Statfjord environmental survey, June 1984. Final report. Unpublished report of the Institute of Offshore Engineering to Mobil Exploration Norway Inc. on behalf of Statfjord Unit Owners. (203 pages.)

I.O.E. 1986 Environmental assessment. Southeast Forties: July 1985 survey. Unpublished report of the Institute of Offshore Engineering to BP Petroleum Development Ltd. (129 pages.)

Jones, N. S. 1950 Marine Bottom Communities. *Biol. Rev.* **25**, 283–313.

Kingston, P. F. 1983 Effects of toxicants on benthic ecosystems. In *Environmental Toxicology: Proceedings of a course held in Edinburgh, August 1982* (ed. J. H. Duffus & J. I. Waddington), pp. 228–248. World Health Organisation Interim Document 13.

Kingston, P. F. & Rachor, E. 1982 North Sea level bottom communities. ICES CM: 1982/L:41. (16 pages.) Copenhagen.

McIntyre, A. D. 1961 Quantitative differences in the fauna of boreal mud associations. *J. mar. biol. Ass. U.K.* **41**, 599–616.

Matheson, I., Kingston, P. F., Johnston, C. S. & Gibson, M. J. 1986 Statfjord field environmental study. In *Proceedings of Conference on Oil-based Drilling Fluids – Cleaning and Environmental Effects of Oil Contaminated Drill Cuttings, Trondheim, Norway, February 1986*, pp. 3–16.

O.P.R.U. 1981 Biological survey of the Buchan oilfield. April 1980. Unpublished report of the Oil Pollution Research Unit to BP Petroleum Ltd. (17 pages plus figures and tables.)

O.P.R.U. 1983 Environmental monitoring of the Buchan oilfield. April 1982. Unpublished report of work carried out by the Oil Pollution Research Unit and Mass Spec. Analytical for BP Petroleum Ltd. (13 pages plus figures and tables.)

Pearson, T. H. & Rosenberg, R. 1978 Macrobenthic succession in relation to organic enrichment and pollution of the marine environment. *Oceanogr. mar. Biol.* **16**, 229–311.

Petersen, C. G. J. 1914 Valuation of the Sea. II. the animal communities of the sea bottom and their importance for marine zoogeography. *Rep. Dan. biol. Stn* **21**, 1–46.

Pielou, E. C. 1966 The measurement of diversity in different types of biological collection. *J. theor. Biol.*, **13**, 131–144.

Rachor, E. 1982 Biomass distribution and production estimates of macro-endofauna in the North Sea. ICES CM: 1982/L:2. (10 pages.) Copenhagen.

Rygg, B. 1986 Heavy-metal pollution and log-normal distribution of individuals among species in benthic communities. *Mar. Pollut. Bull.* **17** (1), pp. 31–36.

Sanders, H. L. 1968 Marine benthic diversity; a comparative study. *Am. Nat.* **102**, 243–282.

Shannon, C. E. & Weaver, W. 1963 *The mathematical theory of communication.* (117 pages.) Urbana, Illinois: University of Illinois Press.

Tennant, V. 1984 Energy partitioning and reproductive strategies in four species of the meiofaunal polychaetes of the genus *Ophryotrocha*. Ph.D. thesis, University of Exeter.

Discussion

LYNDA M. WARREN (*Department of Life Sciences, Goldsmiths' College, London. U.K.*). At least two of the genera of polychaetes recorded from areas close to production platforms form species complexes of morphologically very similar species distinguished mainly on ecological and physiological characteristics. Are there any indications that the species of *Capitella* and *Ophryotrocha* found vary either with distance from the platform, and hence degree of organic enrichment, or with length of time since the original disturbance? Both genera are commonly associated with unstable environments and it would be surprising if the population structures remained constant for very long. An analysis of the species composition at various sites and over a period of time might provide far more useful information concerning small changes in the environment than is available from data referring to genera only.

P. F. KINGSTON. Although we are aware of the existence of species complexes in *Capitella* and *Ophryotrocha*, we have not been able to distinguish between species by using ecological and physiological characteristics at the level of taxonomic resolution used here. To undertake the analysis Dr Warren proposes would be most interesting but presently is not practical given the sampling frequency of current monitoring studies.

R. G. HUGHES (*School of Biological Sciences, Queen Mary College, University of London, U.K.*). Dr Kingston has related the decline in diversity and number of species only to the oil-based discharge close to the platforms. Is there any effect on these statistics of the sometimes considerable 'rain' of fouling organisms that are periodically scraped off the supporting structures and accumulate underneath the platforms?

P. F. KINGSTON. Enhanced productivity resulting from debris from fouling organisms has been demonstrated around offshore platforms. The best evidence for this comes from the Gulf of Mexico where productivity is very high and the water shallow (mostly less then 40 m depth). It is unlikely that in the central North Sea, where the average depth around production platforms is 100 m, that a significant amount of material from this source accumulates around the base of the structure. There is no evidence that organic material accumulating from fouling organisms has any measurable effect on the gross perturbation of benthic community structure by drilling cuttings.

R. EARLL (*Marine Conservation Society, Ross on Wye, Herefordshire, U.K.*). I have two questions for Dr Kingston. Firstly, I was interested to hear Dr Kingston's comments about the bacterial mats, methanogenesis and fish farms. Could he say whether he finds animals living under or in the bacterial mats. Secondly, most people will be aware of the recent observations which have been made concerning the anoxic areas of seabed off the German coast. How would he compare the effects of oil installations on the benthos (in terms of areas covered or affected or both) with what we know about the areas affected by these anoxic events?

P. F. KINGSTON. To answer Dr Earll's first question, in our experience the areas directly under fish cages used on fish farms have been devoid of macrofauna. Secondly, the anoxic areas of the seabed off the German coast are caused by a variety of factors which include the high amount or organic loading originating from the Rhine, Elbe and other sources and the frontal areas occurring in the region which result in poor mixing and stratification of the water column. The anoxic conditions around oil production platforms are confined to the sediments and the area of effect is insignificant by comparison.

R. Earll. I am interested to hear Dr Kingston confirm the impression obtained from survey work on Western Isles fish farms.

J. G. Parker (*Shell Exploration and Production, Aberdeen, U.K.*). Firstly, what was the purpose of replicate sampling in the macrobenthos programme, given that samples were pooled for the subsequent numerical analysis? Secondly, does Dr Kingston have any sympathy for the view that when examining variability in benthos over a gradient it is better to take single closely spaced samples rather than replicated samples from sites located further apart?

P. F. Kingston. The sampling strategy adopted in the studies described here reflect what is practical when working offshore rather than what is ideal. Replicate samples are taken to enable estimates of individual abundance of benthic species to be obtained with an acceptable degree of precision. Benthic macrofauna is characteristically contagiously distributed and theoretically the size and number of the unit samples should reflect the degree and scale of contagion. However, offshore conditions place a considerable restraint on the number of hauls that may be made per site and the paucity of suitable benthic samplers place a further restraint on the size of the unit sample.

Experience has shown that five samples from each site using a 0.1 m^2 grab will generally yield species abundance curves for benthic data that begin to approach their asymptote. In addition, values for commonly used diversity measures also begin to approach their maximal values for North Sea benthos at this frequency per unit area of sampling. It is not usual to pool replicate benthic samples prior to enumeration of the fauna. The data here have been pooled after sample analysis where sample area is important in the stabilization of the community parameters used.

In answer to the second part of the question, the use of single, closely spaced samples in gradient analysis may be appropriate when monitoring a homogeneous benthic environment and a point source of contamination. However, sample variability resulting from the contagious distribution of benthos around production platforms and the patchy nature of the drill cuttings piles favour the approach of using replicate samples from discrete stations along a transect.

J. S. Gray (*Department of Marine Biology, University of Oslo, Norway*). Given the spatial variability that Dr Kingston records, and the clearly different benthic communities along the gradient from platforms, I would have thought that one could eliminate the need for grab sampling, sorting, identification and counting by using remote camera systems. There is a commercial system available and tested which gives rapid analysis (in minutes) and should give similar precision at less cost (see Rhoads & Germano 1982).

Reference

Rhoads, D. C. & Germano, J. D. 1982 Characterization of organism-sediment relations using sediment profile imaging: an efficient nethod of remote ecological monitoring of the seafloor (Remots™ system). *Mar. Ecol. Prog. Ser.* **8**, 115–128.

Phil. Trans. R. Soc. Lond. B **316**, 567–585 (1987)
Printed in Great Britain

Effects of discarded drill muds on microbial populations

By P. F. Sanders[1] and P. J. C. Tibbetts[2]
[1]*Micran Ltd, Berryden Business Centre, Berryden Road, Aberdeen AB2 3SA, U.K.*
[2]*M-Scan Ltd, Silwood Park, Sunninghill, Ascot, Berkshire SL5 7PZ, U.K.*

Drilling operations from platforms in the North Sea result in the production of large quantities of drill cuttings. These are a variable mixture of rock chippings, clays and original drilling fluids. Drilling mud is cleaned on the platform to remove rock chips before re-use of the mud. The rejected fraction from the clean-up plant (the cuttings) contains some of the base drilling fluid, and this can lead to an organically rich input to the sea-bed. Cuttings are discarded immediately underneath the platform jacket and thus build-up over the natural seabed sediment. In many cases this cuttings pile may cover considerable areas of seabed, leading to seabed biological effects and potential corrosion problems.

Different types of cuttings have different environmental impacts, this being partly dependent upon their hydrocarbon component. Diesel-oil based cuttings contain significant amounts of toxic aromatic hydrocarbons, whereas low-toxicity, kerosene-based cuttings contain less. Both types of cuttings support an active microbiological flora, initiated by hydrocarbon oxidation. This paper presents a study of microbiological degradation of hydrocarbons in cuttings piles around two North Sea platforms. Results indicate that there is a close correlation between microbiological activity and hydrocarbon breakdown in the surface of cuttings piles and that both of these parameters reach their maximum values closer to the platform when low-toxicity muds are in use.

Introduction

Drilling operations from North Sea platforms have a number of ecological impacts. By far the most important of these is the deposition of drilling-mud cuttings onto the seabed. This deposition has a number of effects on life in the sediment, primarily physical smothering and organic enrichment but also a direct toxic effect from the hydrocarbons, especially the aromatic fractions (Heildelberger 1975).

Drilling muds are used to lubricate the drill bit, to remove rock chips to the surface and to control reservoir pressure (figure 1).

Drilling mud is pumped down the inside of the drill string and forced up the well casing, carrying the drilled rock chips with it. The used mud is cleaned up by means of shale shakers, mud cleaners, to separate the rock chips (cuttings) from the mud. Mud is then re-used for drilling; cuttings are generally deposited down a cuttings caisson onto the seabed immediately beneath the platform structure, in most cases a tubular steel crossed braced jacket. In the northern North Sea, cuttings can build up into a deep pile on the seabed because of low residual seabed currents. The cuttings pile may be more than 3 cm deep at a distance of 250 m from the platform jacket. In spite of the clean-up and separation process, the cuttings contain significant amounts of mud associated with the rock chips. It is this residual mud, and its associated hydrocarbons, that causes some of the observed environmental impacts (Davies *et al.* 1984; Addy *et al.* 1984).

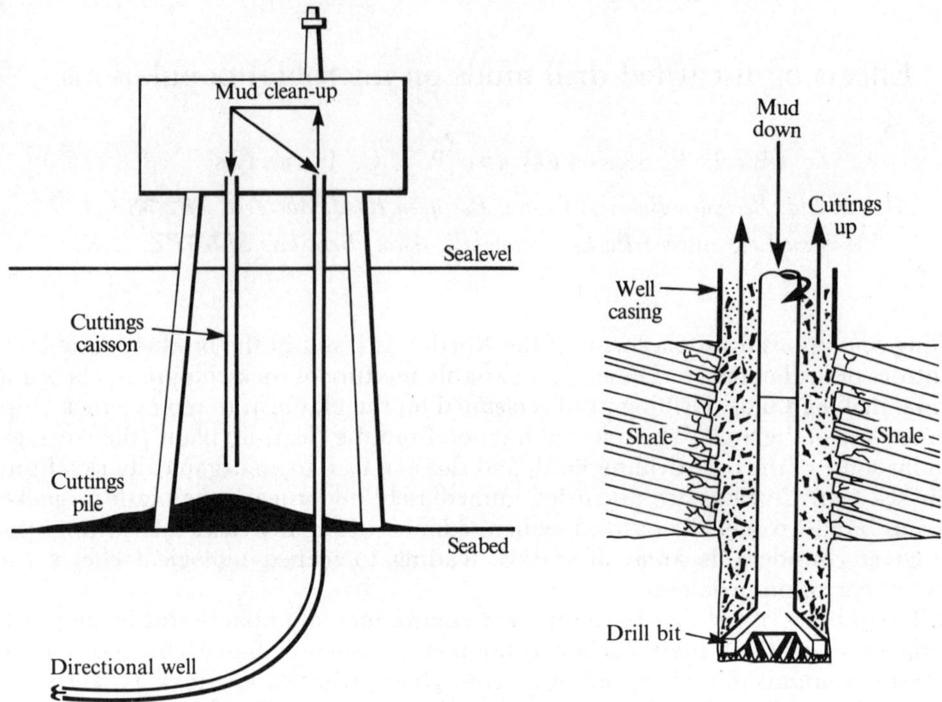

FIGURE 1. Diagrammatic representation of typical cuttings discharges and drilling operations in the North Sea.

There are many types of drilling mud, but the three main types are: water-based, diesel-based and low-toxicity muds. The composition of these three fluids is given in table 1.

Water-based muds are the most simple of the three types and have been used extensively in the North Sea. Their organic content is usually low and their environmental impact from hydrocarbon enrichment is not as significant as the other types. In addition, because the continuous phase is water, rather than hydrocarbons, they are more easily dispersed by seabed currents.

Water-based muds suffer from a number of limitations especially when used in deep, directionally drilled, wells. To reduce these limitations, oil-based muds have been increasingly used in the North Sea since 1978, either for drilling the entire well or for drilling the final sections. Owing to the large amount of diesel-oil in such fluids, there was concern about the

TABLE 1. TYPICAL COMPOSITION OF THREE TYPES OF DRILLING FLUIDS USED IN THE NORTH SEA

water-based muds	diesel-based muds	low-toxicity muds
water	water in oil emulsion	water in oil emulsion
polymers	diesel oil (70–80 %)	mineral oil–kerosine (70–80 %)
clays	water	water
barite	polymers	polymers
drilled solids (cuttings)	clays	clays
	barite	barite
	drilled solids (cuttings)	drilled solids (cuttings)

immediate and long-term environmental impact of dumping diesel-based cuttings because fresh cuttings may contain up to 17% residual hydrocarbons (Wilkinson 1982; Addy *et al.* 1984). In response to pressure to minimize these effects, low-toxicity drilling muds were developed. These are based upon mineral-oils or kerosenes, are low in aromatic components and have largely replaced diesel-based muds since 1982.

Diesel is the petroleum distillate fraction immediately above (higher boiling point than) kerosene. Gas–liquid chromatography (GLC) analysis of diesel oil generally displays a carbon number range of approximately C_{12} to C_{26} with the greatest concentration at C_{15} to C_{16}. This 'cut' contains a high relative abundance of aromatic hydrocarbons such as naphthalene, fluorene and phenanthrene and their alkylated homologues, which are thought to contribute greatly to the overall toxicity of diesel. Most of the low-toxicity base-oils, however, are based on the kerosene fraction of a crude-oil, and are further refined by solvent washing to remove the majority of the aromatic hydrocarbons. GLC analysis of a low-toxicity base-oil of this type generally shows a carbon number range of approximately C_9 to C_{16}, with the greatest concentration at C_{12}.

Oil discharged with the cuttings undergoes both biological and chemical weathering and physical loss by dispersion. The major factor involved in the degradation of petroleum hydrocarbon in the marine environment is likely to be microbiological, with numbers of hydrocarbon degrading microbes increasing because of the elevated hydrocarbon concentration (Atlas 1981). Active hydrocarbon degradation has been found in seabed sediment even in cold-water systems (Haines & Atlas 1983). Microbiological hydrocarbon oxidation is exclusively an aerobic phenomenon, mainly mediated by aerobic bacteria (eg. *Pseudomonas* spp., *Micrococcus* spp. and *Acinetobacter* spp.) although other aerobic organisms, such as yeasts and fungi, may also be involved. These microbes are able to utilize a very wide range of hydrocarbon fractions as sources of both carbon and energy. Oxidation liberates breakdown products such as fatty acids, aldehydes and alcohols (McKenna & Kallio 1965). There is usually a gradual succession in the fractions of oil which are degraded; *n*-alkanes followed by branched alkanes, cycloalkanes and then aromatic compounds. Although branched alkanes, such as the isoprenoids pristane (Pr) and phytane (Ph) are biodegraded, the process is slower than for the *n*-alkanes because branching, in general, increases the resistance of hydrocarbons to microbial attack (Pirnik *et al.* 1974). On analysis by GLC this can be observed as a decrease in the value of n-C_{17}/Pr and n-C_{18}/Ph.

Because of the similar physical properties of n-C_{17} and pristane, and n-C_{18} and phytane, these ratios are not significantly altered by other physical weathering processes. GLC analysis of an oil generally produces a complex series of resolved components (peaks) superimposed on a 'hump', which is known as the unresolved complex mixture (UCM). The UCM comprises thousands of structurally complex (multi-branched and cyclic) compounds, which, as stated above, are resistant to biodegradation (Farrington & Tripp 1977). Biodegradation therefore results in an increase in the relative concentration of UCM (Davies & Tibbetts 1987). This effect can be quantified by calculating the value of *n*-alkanes/UCM %.

Co-oxidation is also known to occur (Raymond *et al.* 1967; Perry 1979). This enables a compound, which cannot be degraded in isolation, to be enzymatically attacked by a consortium of micro-organisms growing primarily on other hydrocarbons within the oil. Aromatic compounds, the most toxic fraction to macrobenthic species, may thus be utilized by a mixed microbiological population. Such microbiological biodegradation increases the organic status of the cuttings and, although this may reduce the direct toxic effect of the aromatic fractions, partial biodegradation may also produce further toxic or carcinogenic compounds (Cerniglia *et al.* 1982). Microbiological hydrocarbon degradation and the effects of hydrocarbon pollution on microbiological processes are well reviewed by Atlas (1984).

Microbiological processes in sediment are complex, with different microbiological populations being spatially separated due to physical and chemical stratification (Jørgensen 1977*a*, *b*, 1982). The surface of the sediment is aerobic; the depth of this aerobic layer will be dependent upon the organic status and physical properties of the sediment. An organically enriched sediment, such as a drill-cuttings pile, will have a very active aerobic microflora in the surface layer. The consequences of this growth are twofold: low molecular mass substrates are produced from incomplete hydrocarbon oxidation and oxygen is rapidly consumed from the pore water. This sets up an anaerobic zone in the subsurface sediment and also provides a wide range of carbon sources, stimulating further microbial populations.

In oil-based cuttings piles, the aerobic layer may be less than 2 cm thick, the remainder of the sediment being anaerobic and highly reduced. Fermentation reactions by anaerobic bacteria liberate yet more carbon compounds and produce reduced conditions suitable for the growth of sulphate-reducing bacteria (SRB). These bacteria are unable to utilize hydrocarbons directly (Postgate 1984) but are able to grow on the products of aerobic hydrocarbon degradation (Laanbroek & Pfennig 1981). SRB produce sulphide as a by-product of their metabolism and, under favourable conditions, maintain high concentrations of toxic sulphide (both soluble and insoluble) and very low redox potentials. These effects add to the direct toxic effects of the discarded cuttings. In addition, such sulphide generation and low redox potential produces a corrosive environment which poses a risk to structures, pipelines and associated steel covered by the mud (Fischer 1983; Sanders 1984; Sanders & Hamilton 1986). This simplified microbiological consortium is shown in figure 2.

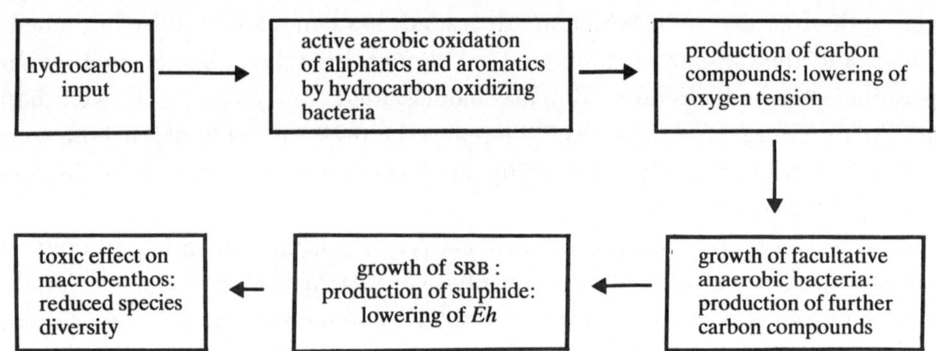

FIGURE 2. Microbiological processes involved in the degradation of hydrocarbons in the seabed sediment.

In sediment with a low organic content (e.g. unpolluted sandy beaches), the aerobic layer is thicker; nutrients for SRB are in low concentration, sulphide production is low and therefore toxic effects are minimal. A high species diversity of the sediment macrobenthos is commonly found in such environments, although biomass and productivity may be low.

The aim of this paper is to highlight the effects of drill-cuttings piles on seabed microbiological populations by means of hydrocarbon, chemical and microbiological analysis around two comparable North Sea installations. A brief drilling history of the two locations is shown in table 2.

TABLE 2. SIMPLIFIED DRILLING HISTORIES OF THE TWO FIELDS STUDIED

Field A	Field B
16 water-based mud wells to 1979 28 diesel-based mud wells to 1983 3 low-toxicity mud wells to 1985 depth: 3700–4300 m drilling ceased end 1985	16 wells to Jan. 1986 depth: 3400–4300 m All low-toxicity mud, little diesel

Field A used predominantly diesel-based muds (28 wells), with 3 final low-toxic mud wells before cessation of drilling in 1985. The field is situated in 160 m of water in the northern North Sea, with a small residual seabed current in a south easterly direction. The natural seabed sediment is a fine sand (mean particle size *ca.* 0.2 mm diameter) with a diverse macrobenthic community.

Field B, south of Field A, by comparison, used low-toxicity muds exclusively, 16 wells being drilled up to January 1986, with the drilling programme still in progress at the time of the survey. This field is situated in 110 m of water, with a residual seabed current of 2 cm s^{-1} in a southerly direction. The natural seabed sediment is of a fine clay consistency (6–15% silt), again with a diverse macrobenthic community.

METHODS

Samples were mostly collected from the seabed by using Perspex hand-held coring devices, shown in figure 3. Some samples were collected by taking Day grab samples, with cores being taken immediately after recovery to surface. The cores of sediment were immediately subsampled for microbiological and hydrocarbon analysis. The samples for hydrocarbon analysis were kept frozen (−20 °C) in pre-cleaned and de-greased aluminium cans.

Hydrocarbon analysis

Cores for hydrocarbon analysis were transported to the laboratory in a frozen condition to minimize hydrocarbon breakdown by the microbiota. Frozen cores were allowed to thaw for 2 h before being extruded onto pre-cleaned aluminium foil using a wooden ram covered with pre-cleaned aluminium foil. The sediment core was marked with a clean stainless steel spatula, at intervals of 2 cm from the top down. The core in its semi-frozen state was then easily sectioned with the spatula, which was wiped clean and rinsed with isopropanol between each sectioning. Each section was placed in a Kilner jar and mixed thoroughly with isopropanol (50 cm^3), which acts as a biocide.

Hexane (20 cm^3) and isopropanol (80 cm^3) were added to the sediment and the sample extracted by using ultrasonication (2 × 5 min with stirring in-between). The solvent was then decanted and partitioned between water and hexane (2:3 by volume; 120 cm^3). The organic layer was then collected in a pre-cleaned round-bottomed flask. A further 100 cm^3 of isopropanol and hexane (4:1) was added to the sediment, the extraction procedure repeated, and the two organic extracts combined and evaporated to dryness at ambient temperature under vacuum (Buchi Rotavap R).

The resulting total organic extract (TOE) was weighed to estimate total hydrocarbon

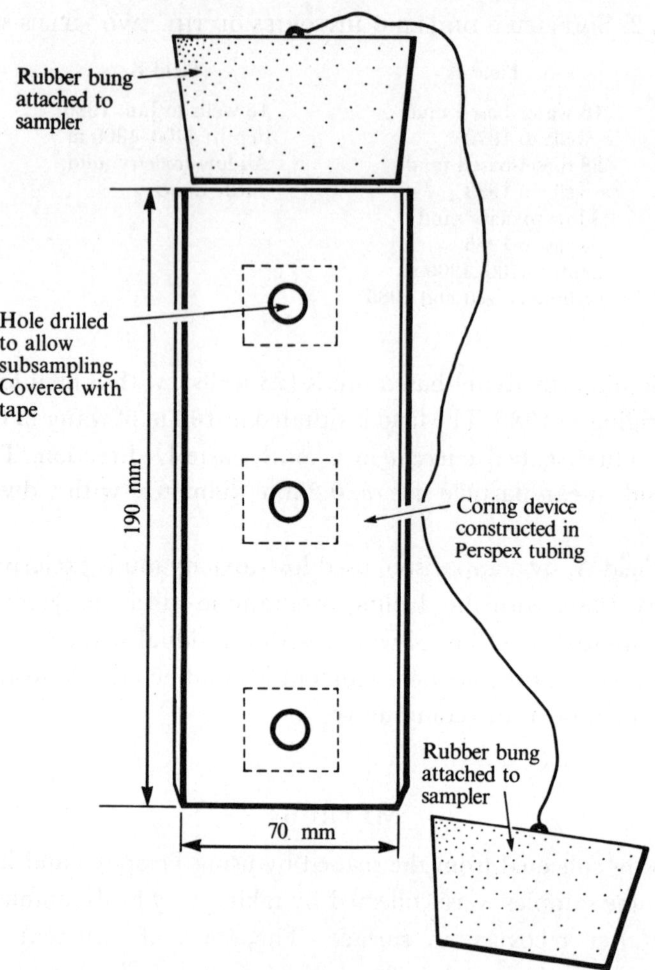

FIGURE 3. Diagram of Perspex hand-held corer used for collecting undisturbed seabed sediment.

content of the sample. A known aliquot of the TOE was taken and 2,21-dimethyldocosane (25 μg) was added as internal standard. The aliquot was dissolved in hexane (2 cm^3) and 0.5 μl was injected for analysis by GLC under the following conditions: instrument, Carlo Erba 4160; detector, flame ionization at 320 °C; column (aliphatic), 25 m × 0.32 mm internal diameter MEGA OV-1 WCOT fused silica; temperature programme, 40 °C–320 °C at 8 °C min^{-1}; carrier gas, helium.

To estimate the concentrations of *n*-alkanes actually removed from the oil in the sediment by biodegradation it was assumed that the UCM in the oil remains unbiodegraded (Van Der Linden 1978). Thus, by calculating the concentration of *n*-alkanes remaining in the sediment and comparing this with the concentration of *n*-alkanes relative to UCM in the oil used in the drilling-mud formulation on the platform (reference oil), it is possible to estimate the amount of *n*-alkanes lost by biodegradation as follows:

$$n\text{-alkanes biodegraded} = \frac{(n\text{-alkanes (reference oil)} \times \text{UCM (sample)})}{(\text{UCM (reference oil)})} - n\text{-alkanes (sample)}.$$

Microbiological analysis

All the samples for microbiological analysis were stored in an insulated coolbox, with frozen ice-packs to keep them cool. They were analysed immediately after collection on board the diving vessel. This minimized any changes in the microbial population which begin immediately after sample collection (the so-called 'bottle effect'). Samples were processed and analysed within a maximum of 12 h of collection. Samples were shipped to Aberdeen in a cold, insulated icebox for final analysis.

(*a*) *Microbial numbers*

Sediment samples were sub-sampled by placing a known mass of the material (approximately 1 g) into a tube of sterile, anaerobic, reduced diluent (10 cm^3) (medium 1, table 3). In some cases, sub-samples were taken at various depths in the core to obtain a sediment profile. An even suspension was made by shaking, stirring and mixing, followed by ultra-sonic dispersion to break up clumps of cells.

The first dilution tube, set up offshore, was used to prepare a 1:9 dilution series in Aberdeen. A fresh, sterile syringe was used to add 1 cm^3 of this suspension to a fresh tube containing 9 cm^3 of anaerobic dilution fluid and mixed. Similar 1:9 dilutions were made to achieve a final dilution of $1:10^{-8}$ (eight tubes).

The dilution series were tested for a number of bacterial groups by inoculation onto selective nutrient media. Results for SRB and hydrocarbon oxidizing bacteria are presented here.

SRB were enumerated by injecting 1 cm^3 amounts of each dilution tube into 9 cm^3 of Postgate broth medium (medium 2, table 3) in triplicate by using standard procedures modified from

TABLE 3. MICROBIOLOGICAL MEDIA USED

(All amounts are percentages, mass/volume or volume/volume.)

medium 1: dilution fluid (anaerobic) (pH 7.6)	
peptone	0.5
yeast extract	0.3
ascorbic acid	0.01
sodium thioglycollate	0.01
seawater	75
medium 2: seawater Postgate medium (pH 7.6) for SRB	
Na_2SO_4	0.1
$CaCl_2 . 6H_2O$	0.01
$MgSO_4 . 7H_2O$	0.2
yeast extract	0.2
$FeSO_4 . 7H_2O$	0.05
sodium thioglycollate	0.01
ascorbic acid	0.01
sodium lactate (70% solution)	0.35
NaCl	2.5
medium 3: Bushnell–Haas broth (pH 7.0) for hydrocarbon oxidizers	
$MgSO_4 . 7H_2O$	0.02
$CaCl . 6H_2O$	0.002
KH_2PO_4	0.1
K_2HPO_4	0.1
$FeCl_3 . 6H_2O$	0.005
seawater	75
carbon source (hydrocarbon)	1

the American Petroleum Institute (API) recommended practices (1975). Growth of SRB was indicated by blackening of the tube; SRB numbers were estimated by most probable number (MPN) tables from the highest dilution giving blackening after 28 days growth at 30 °C.

Aerobic diesel-oil oxidizing bacteria were enumerated in Bushnell–Haas broth, by using diesel-oil or mineral-oil as the carbon source (medium 3, table 3). Samples (100 μl) of each dilution were added to 10 cm^3 of the broth in triplicate, with 100 μl of sterile hydrocarbon added afterwards. Growth is indicated by an increase in turbidity and emulsification of the oil after 10 d growth with gentle shaking at 30 °C. Numbers were estimated from MPN tables as above.

(*b*) *SRB activity measurement*

SRB activity was measured by using methods similar to those previously described for coastal sediments (Jørgensen 1978; Rosser & Hamilton 1983). A known mass of the sediment was placed in the tube, as shown in figure 4. The tube contained 4 cm^3 of vacuum de-gassed sulphate-free artificial seawater, to which 1 μCi† of [^{35}S] sodium sulphate was added. The tube

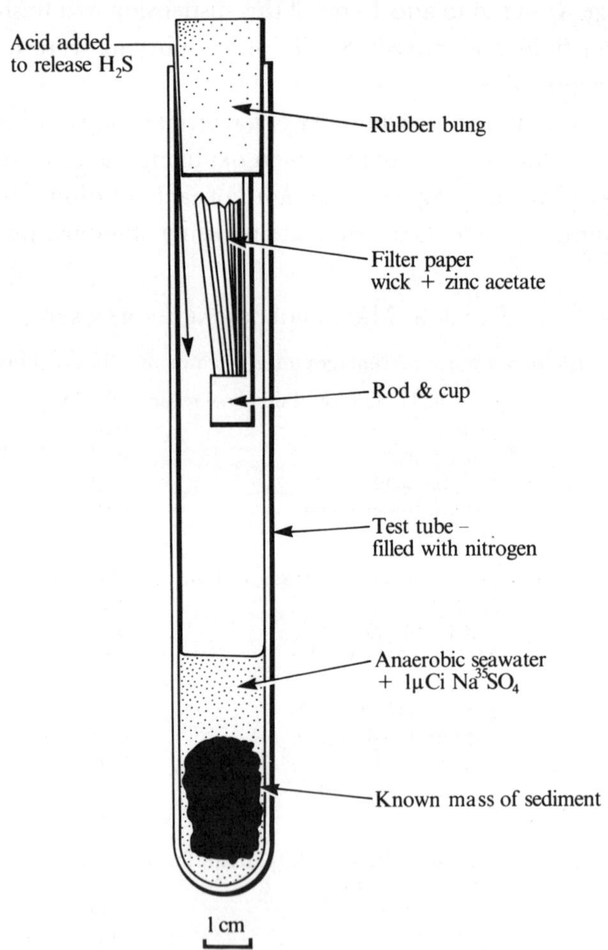

FIGURE 4. Diagram of tube used to assess SRB activity in sediment.

† 1 Ci = 3.7×10^{10} Bq.

was gassed with oxygen-free nitrogen before sealing with an H_2S impermeable butyl rubber bung, from which a rod, cup and wick assembly was suspended. Zinc acetate was added to the filter paper wick at the start of the 24 h incubation period (20 °C) to trap any H_2S evolved. Five tubes were set up from each sample. Four of these were used as 'live' tubes, incubated for 24 h to allow the $^{35}SO_4^{2-}$ to be reduced by SRB to $^{35}S^{2-}$. The fifth tube was killed with acid to liberate H_2S and kill bacteria at the start of the incubation, thus acting as a background, control, tube. The four live tubes were killed with hydrochloric acid after incubation, the H_2S being released by gentle shaking in a shaking water-bath.

The rods, cups and wicks were removed after shaking, placed in scintillation vials and the amount of ^{35}S quantified by liquid scintillation counting. The sulphate reduction activity was calculated from the amount of sulphate in the tube, the amount of $^{35}SO_4^{2-}$ added, the amount of $^{35}S^{2-}$ produced, the mass of deposit and the incubation time, by subtracting the background from the live values.

Results and discussion

Field A, 1982

Microbiology

The depth profile for SRB activity shown in figure 5 is similar to that found in natural sediments, with sulphate reduction being localized within depths of 2–5 cm below the sediment surface. There were also more aerobic bacteria in the surface layers, but they were distributed throughout the sediment profile. Aerobic bacterial activity (including microbial hydrocarbon

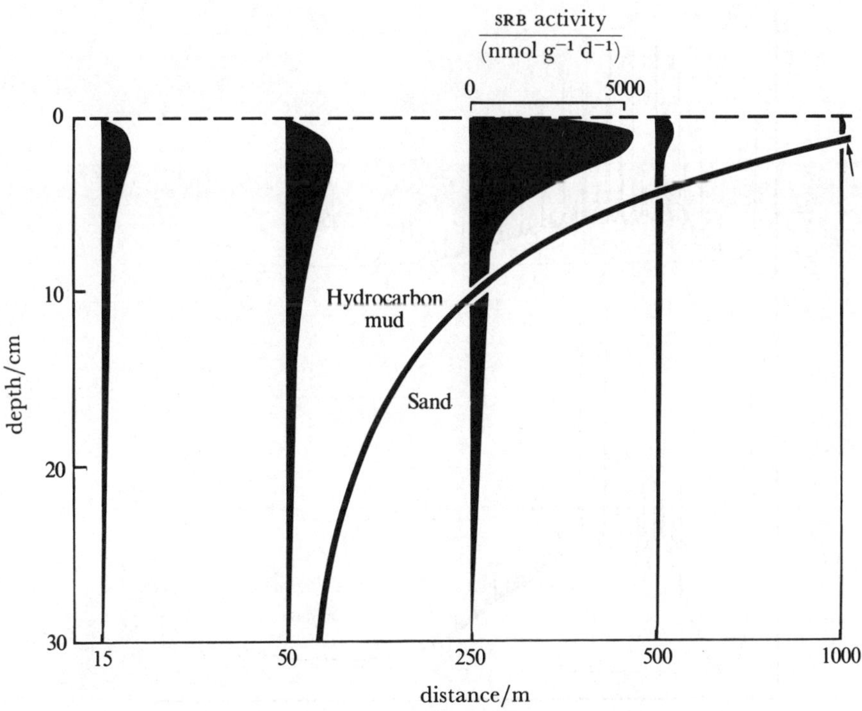

FIGURE 5. Depth profile of SRB activity along a transect at Field A in 1982. Maximum SRB activity was found in the surface or subsurface core at a distance of 250 m from the jacket. The arrow denotes the limit of hydrocarbon-affected sediment.

oxidation) would be localized in the more oxidized surface layers, because high sulphide concentrations and low redox potentials were found at all depths below 2 cm. The stations closest to the jacket were composed entirely of drill cuttings, whereas the top 10 cm at 250 m was affected in this way. Even at 500 m, there was a thin layer of the fine fractions of the mud, overlying the natural seabed sand, with elevated hydrocarbon concentrations in the top 4 cm. The highest sulphate reduction rates were found at the 250 m station (5300 nmol g^{-1} d^{-1}), with lower rates being found close to the jacket (200–1300 nmol g^{-1} d^{-1}). At stations unaffected by the hydrocarbon input, sulphate reduction rates were very low (less than 50 nmol g^{-1} d^{-1}). It thus appears that the microbiological profiles in these mud deposits are similar to those in natural, undisturbed sediment, but that the activity and population size of SRB is enhanced in the subsurface layers of the sediment close to the platform jacket. This is likely to be because of the organic loading from the cuttings.

Hydrocarbons

Examples of the GLC traces obtained from sediment extracts from Field A in 1982 are shown in figure 6. The uppermost trace (50 m from the platform) shows a range of alkanes from C_{12}

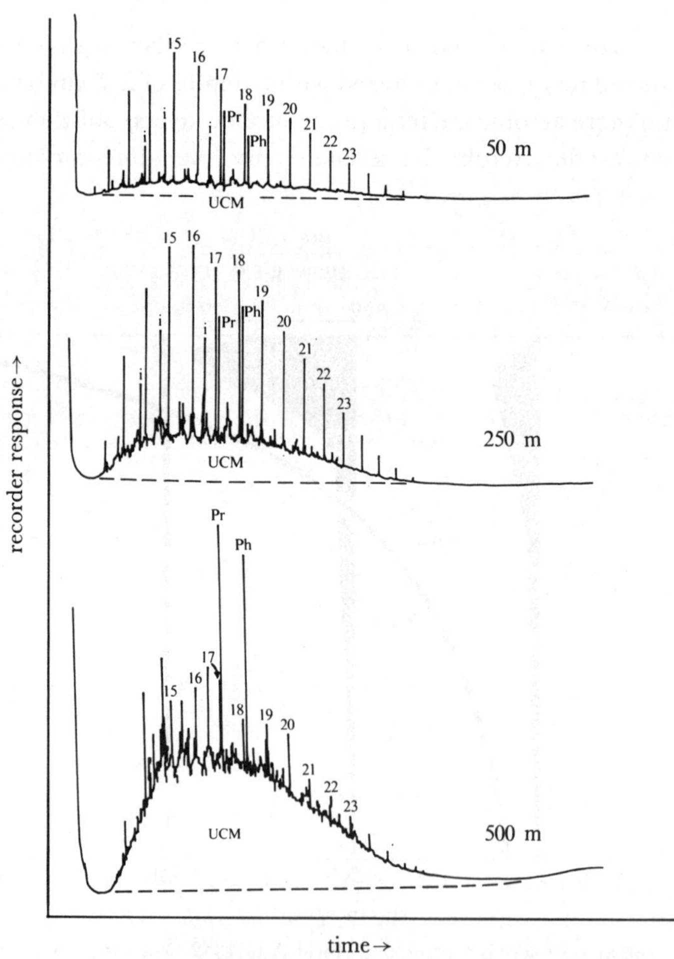

FIGURE 6. GLC traces of TOE from samples along a transect at Field A in 1982. The traces are typical of a biodegraded diesel-oil, especially at the outer stations. Key: i, isoprenoids; Pr, pristane; Ph, phytane; numbers refer to *n*-alkane carbon number; UCM, unresolved complex mixture.

to C_{26}, which is typical of a diesel-oil. As expected, all of the sediments in this survey contained a major contribution from diesel, which had been weathered to a greater or lesser extent.

The degree of biodegradation, estimated from the value *n*-alkanes/UCM (table 4), indicates that the degree of biodegradation of the oil in the sediment increases with increasing distance from the platform. This effect can be seen in figure 6 where the relative concentration of the

TABLE 4. RESULTS OF GLC ANALYSIS (BY MASS) OF DRY SEDIMENT, FIELD A, 1982 AND 1986

	distance /m	*n*-alkanes/UCM (%)	*n*-alkanes biodegraded (p.p.m.)	*n*-alkanes biodegraded (%)
1982	15	18	0	0
	50	11	280	61
	250	3	520	79
	500	2	180	89
	1000	1	30	91
1986	50	40	12100	33
	250	0	4900	100
	500	0	340	100
	750	0	100	100
	1000	0	70	100

UCM in the oil appears to increase away from the platform and the n-C_{17}/Pr value falls from approximately 1.4 to 0.3. However, if the sedimentary concentration of the diesel is taken into consideration, the actual concentration of *n*-alkanes removed by biodegradation is at a maximum 250 m from the platform (table 4).

This effect has been observed previously around other North Sea platforms (Tibbetts & Large 1986). At the innermost stations it is thought that the constant rain of drill cuttings causes rapid burial under oily sediment, through which there can be little water (and therefore oxygen) percolation. Aerobic biodegradation therefore has a very limited time in which to take place. In the outermost stations, the amount of *n*-alkanes removed by biodegradation is limited by the lower total concentration of *n*-alkanes deposited in the sediment.

General

The above results show a close correlation between the profiles of SRB activity and the concentration of *n*-alkanes biodegraded (figure 7), with maximum values of both parameters being observed at a distance of 250 m from the platform. SRB activity is the final microbiological stage but is dependent upon initial hydrocarbon oxidation by aerobic bacteria. These results therefore suggest that SRB activity can be used as an indirect indicator of the extent of hydrocarbon biodegradation.

Field A, 1986

Sediment profiles were examined for microbiology in 1986 and SRB were again found to be most active in the subsurface zone. Results from this stratum are presented in figures 8 and 9. Although depth profiles for hydrocarbon were not carried out in 1986 a physical examination of the cores suggested a similar depth profile to that seen in 1982.

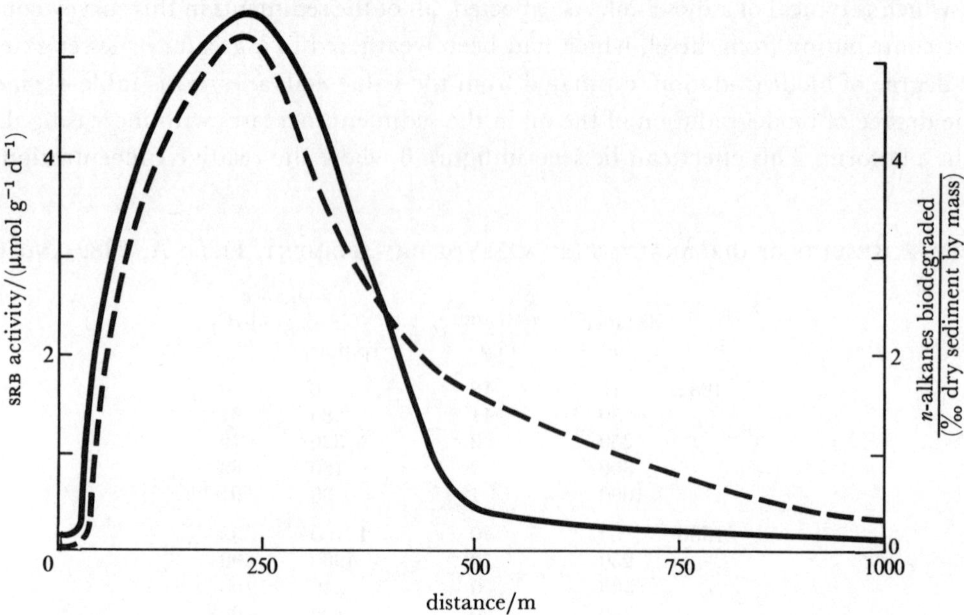

FIGURE 7. Comparison of SRB activity and n-alkane biodegradation around Field A in 1982, showing a close correlation.

Hydrocarbons

The GLC traces of the TOE from the sediments around Field A in 1986 are shown in figure 8. Because of the utilization of low-toxicity base-oil on the platform since 1983, the 50 m station shows a kerosene-type input (alkanes from n-C_{12} to n-C_{16}), which represents the major source of contamination in the sediment. The other stations, however, all show weathered diesel as the major contaminant.

Although biodegradation of the n-alkanes appears complete (table 4) from 250–1000 m, when the hydrocarbon loading of the sediment is taken into consideration these are not the stations from which the most n-alkanes have been removed. On this survey the maximum concentration of n-alkanes biodegraded was only 50 m from the platform (table 4 and figure 9), at which point over 1200 p.p.m. dry sediment by mass of n-alkanes have been removed. This represents a ten- to twentyfold increase in the concentration of n-alkanes removed compared with that found in 1982.

Microbiology

Figure 9 summarizes the microbiological results from the survey in 1986. SRB activity was found to be maximal at the 50 m station, compared with the 250 m station in 1982. Maximum sulphate reduction rates were lower than in 1982 (maximum 2.2 μmol g^{-1} d^{-1}); this apparent decrease in overall SRB activity may be due to a number of factors, one of which is the change in the nature and level of the platform discharges. SRB numbers were also highest at the 50 m and 125 m stations (10^5 g^{-1}). Immediately beneath the jacket (0 m), numbers were lower by an order of magnitude, and low numbers (10^1–10^2 g^{-1}) of SRB were found between 500 m and 1000 m. Hydrocarbon oxidizing bacteria were also found in highest numbers close to the platform (10^3 g^{-1} at 0–125 m), reaching low background values (0–10) between 500 m

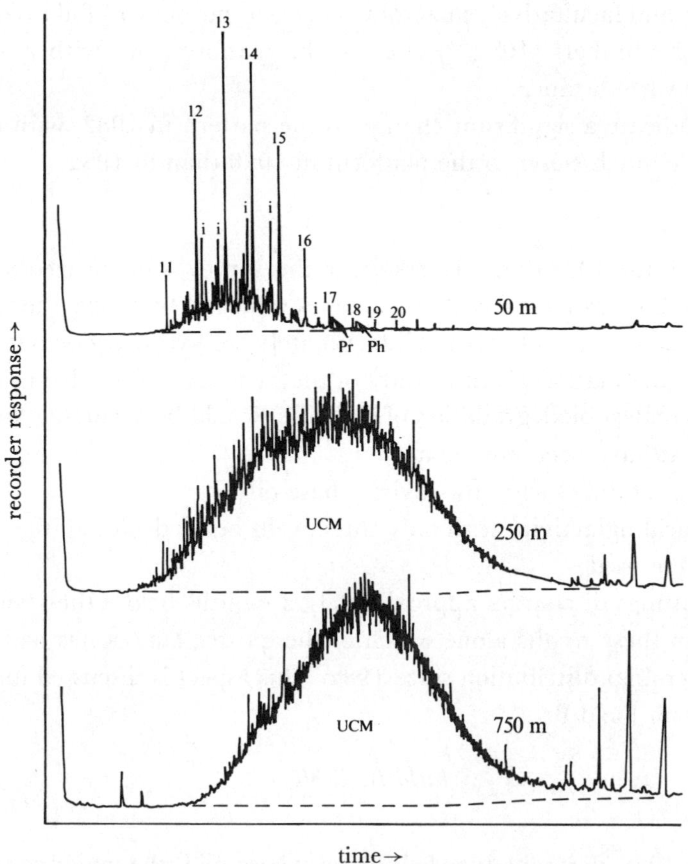

FIGURE 8. GLC traces of TOE from samples around Field A in 1986. Diesel-oil is completely biodegraded at the outer stations whereas the trace at 50 m shows partly biodegraded low-toxicity mud. Key: i, isoprenoids; Pr, pristane; Ph, phytane; numbers refer to *n*-alkane carbon number; UCM, unresolved complex mixture.

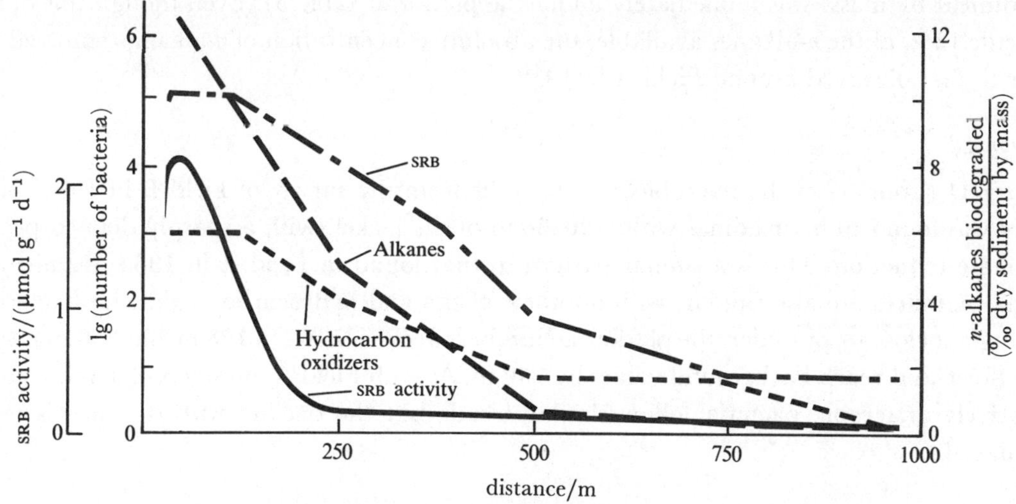

FIGURE 9. Comparison of hydrocarbon and microbiological analyses of Field A, 1986, showing a slightly different zonation when compared with 1982 (figure 7).

and 1000 m. Aerobic and facultatively anaerobic bacteria (not shown) followed similar trends, being present in high numbers (10^6 g^{-1}) close to the platform, but with a less pronounced decrease in numbers with distance.

All these results indicate a significant change to the pattern in 1982, with microbiological processes taking place much closer to the platform in 1986 than in 1982.

General

By comparison with the 1982 data, the results of this survey, summarized in figure 9, show the greatest values of both SRB activity and *n*-alkanes removed by biodegradation to occur at a distance of only 50 m from the platform. Unfortunately, no hydrocarbon data were obtained immediately beneath the platform, but a study of the hydrocarbon oxidizing bacteria (figure 9) suggests that the greatest biodegradation of *n*-alkanes would be occurring there. Since 1982, major parameters that have been altered are:

(1) the replacement of diesel with low-toxicity base-oil;

(2) a decrease in drilling activity with only three wells being drilled in the two years when low-toxicity muds were used;

(3) cessation of cuttings discharges approximately 2 months before the 1986 survey.

It is not clear from these results alone which of these three factors has had most influence on the change in microbial distribution since 1982. This aspect is discussed further, below, in the light of results from Field B.

Field B, 1986

Hydrocarbons

As with Field A in 1986, kerosene-type low-toxicity base-oil is the major contaminant in the sediment within 40 m of the platform (figure 10). At 250 m 96% of *n*-alkanes have been removed (table 5) but isoprenoid hydrocarbons (i) are still present. At 500 m and beyond, however, all of the *n*-alkanes and isoprenoids have been removed by biodegradation.

In this case, the greatest concentration of *n*-alkanes removed by biodegradation (1900 p.p.m. dry sediment by mass) was immediately under the platform (table 5). Even though this only represents 12% of the *n*-alkanes available, the absolute concentration of *n*-alkanes removed is similar to that observed around Field A in 1982.

Microbiology

Figure 11 summarizes the microbiological results from the survey of Field B in 1986. SRB activity was found to be maximal within 20–30 m of the jacket, with a sharply defined peak of sulphate reduction. This is a similar pattern to that found in Field A in 1986. Results of bacterial numbers are also similar, with numbers of SRB and hydrocarbon-oxidizing bacteria being highest close to, or under, the platform. Numbers decline between 125 m and 250 m from the platform and reach background values by 750 m. As with Field A, numbers of aerobic and facultatively anaerobic bacteria follow similar trends but the decline with distance is less pronounced.

General

These results again show close correlation between the profiles of SRB activity and the concentration of *n*-alkanes removed by biodegradation (figure 11), except immediately under

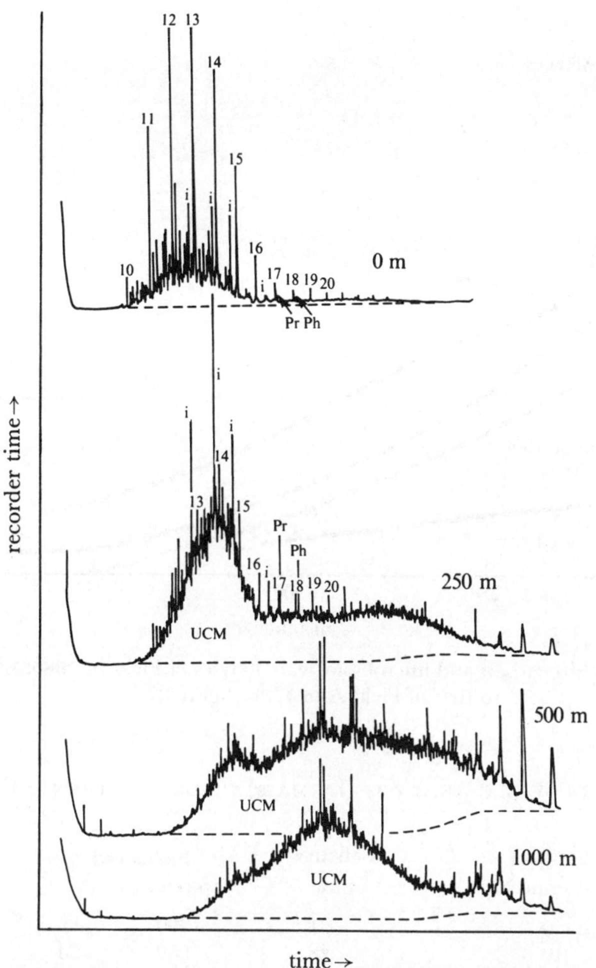

FIGURE 10. GLC traces of TOE from samples around Field B in 1986. Traces indicate biodegradation of the low-toxicity oil, together with some diesel-oil at the outer stations. Key: i, isoprenoids; Pr, pristane; Ph, phytane; numbers refer to *n*-alkane carbon number; UCM, unresolved complex mixture.

the platform, where little SRB activity was observed but where the greatest concentration of *n*-alkanes is estimated to have been biodegraded.

Inter-field comparison

From a comparison of the data obtained for Field A in 1982 and 1986 (table 4), it appears that there has been a general increase in the quantity of *n*-alkanes removed from the sediment by biodegradation since 1982. This could be accounted for by any of the three changes in drilling activity between 1982 and 1986 outlined earlier. However, when the results from Field B are considered (table 5) it can be seen that the absolute concentrations of *n*-alkanes removed by biodegradation are similar to those in Field A in 1982. In both of these cases drilling activity was in progress. This suggests that the increase in the concentrations of *n*-alkanes removed by biodegradation around Field A in 1986 is mainly due to the reduction and subsequent cessation of drilling activity at Field A rather than the change to low-toxicity drilling mud. It is unknown at this stage whether this is because of an increase in the rate of biodegradation or merely reflects

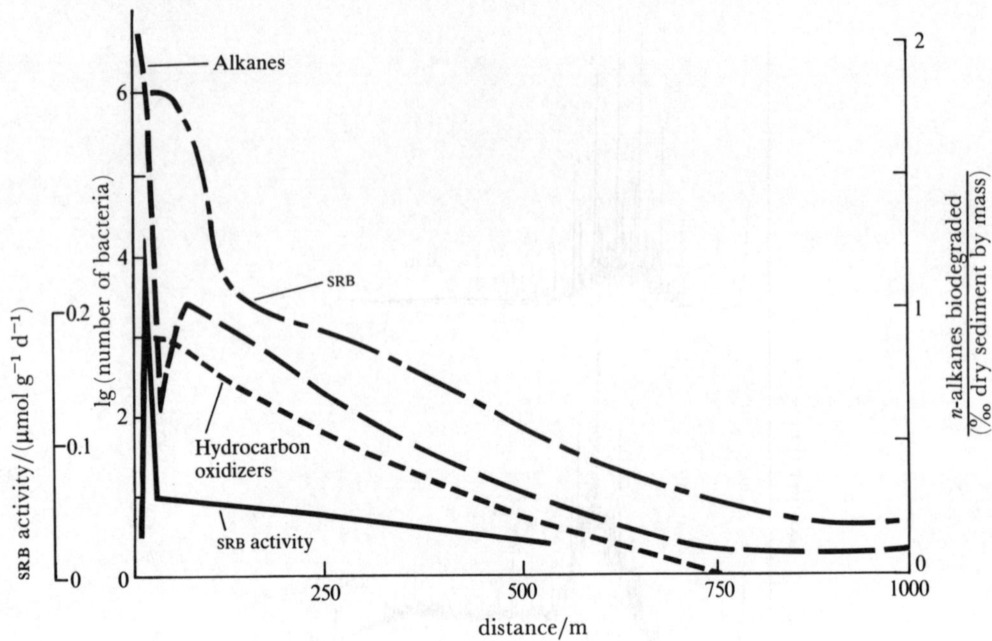

FIGURE 11. Comparison of hydrocarbon and microbiological analyses of Field B, 1986, showing a pattern similar to that of Field A in 1986 (figure 9).

TABLE 5. RESULTS OF GLC ANALYSIS (BY MASS) OF DRY SEDIMENT, FIELD B, 1986

distance/m	*n*-alkanes / UCM (%)	*n*-alkanes biodegraded (p.p.m.)	*n*-alkanes biodegraded (%)
0	31	1900	12
40	28	550	23
100	22	1070	40
250	2	520	96
500	0	120	100
1000	0	64	100

the greater length of time for biodegradation to occur as a result of the reduction in drilling activity.

It is evident from a comparison of the three surveys (figures 7, 9 and 11) that both SRB activity and *n*-alkane biodegradation peak closer to the platforms when low-toxicity drilling muds were in use (Field A, 1986 and Field B, 1985) than at Field A in 1982 when diesel-based mud was in use. The similarity of the profiles in the two low-toxicity mud studies and their marked difference to the profiles seen when diesel-based mud was used, suggests that the spatial distribution of *n*-alkane biodegradation and SRB activity is governed by the base-oil type. This effect may occur because at the same concentration of total oil in the sediment, diesel contains higher relative concentrations of aromatic hydrocarbons than low-toxicity oil. Because it is generally accepted that the aromatic hydrocarbons are the toxic fraction of an oil it is likely that diesel could suppress microbial activity to a greater distance from a platform.

References

Addy, J. M., Hartley, J. P. & Tibbetts, P. J. C. 1984 Ecological effects of low toxicity oil based mud drilling in the Beatrice oilfield. *Mar. Pollut. Bull.* **15**, 429–436.

American Petroleum Institute 1975 *API recommended practice for biological analysis of subsurface injection water*, 3rd edn. Division of production, Dallas, U.S.A.

Atlas, R. M. 1981 Microbiological degradation of petroleum hydrocarbons; an environmental perspective. *Microbiol. Rev.* **45**, 180–209.

Atlas, R. M. 1984 *Petroleum microbiology*. New York: MacMillan Pub. Co.

Cerniglia, C. E., Dodge, R. H. & Gibison, D. T. 1982 Fungal oxidation of 3-methyl-cholanthrene: formation of proximate carcinogenic metabolites of 3-methyl-cholanthrene. *Chem. Biol. Interact.* **38**, 161–173.

Davies, J., Addy, J., Blackman, R., Ferbrache, J., Moore, D., Somerville, H., Whitehead, A. & Wilkinson, T. 1984 Environmental effects of oil-based mud cuttings. *Mar. Pollut. Bull.* **15**, 363–370.

Davies, J. M. & Tibbetts, P. J. C. 1987 The use of *in-situ* benthic chambers to study the fate of oil in sublittoral sediments. *Estuar. coast. Shelf Sci.* **24**, 205.

Farrington, J. W. & Tripp, B. W. 1977 Hydrocarbons in Western North Atlantic surface sediments. *Geochim. cosmochim. Acta* **41**, 1627–1641.

Fisher, K. P. 1983 Cathodic protection in saline mud containing sulphate-reducing bacteria. In *Microbial corrosion* (ed. A. D. Mercer, A. K. Tiller and R. W. Wilson), pp. 110–116. London: Metals Society.

Haines, J. R. & Atlas, R. M. 1983 Biodegradation of petroleum hydrocarbons in continental shelf region of the Bergen sea. *Oil Petrol. Poll.* **1**, 85–96.

Heidelberger, C. 1975 Chemical carcinogenesis. *A. Rev. Biochem.* **44**, 79–121.

Jørgensen, B. B. 1977*a* The sulphur cycle of a coastal marine sediment (Limfjorden, Denmark). *Limnol. Oceanogr.* **22**, 814–837.

Jørgensen, B. B. 1977*b* Bacterial sulphate reduction within reduced microniches of oxidised marine sediments. *Mar. Biol.* **41**, 7–17.

Jørgensen, B. B. 1978 A comparison of methods for the quantification of bacterial sulphate reduction in coastal marine sediments. I. Measurement with radiotracer techniques. *Geomicromicrobiol. J.* **1**, 11–27.

Jørgensen, B. B. 1982 Ecology of the bacteria of the sulphur cycle with special reference to anoxic–oxic interface environments. *Phil. Trans. R. Soc. Lond.* B **298**, 543–561.

Laanbroek, H. J. & Pfennig, N. 1981 Oxidation of short chain fatty acids by sulphate-reducing bacteria in freshwater and marine sediments. *Arch. Microbiol.* **128**, 330–335.

McKenna, E. J. & Kallio, R. E. 1965 Biology of hydrocarbons. *A. Rev. Microbiol.* **19**, 182–206.

Perry, J. J. 1979 Microbiological co-oxidation involving hydrocarbons. *Microbiol. Rev.* **43**, 59–72.

Pirnik, M. P., Atlas, R. M. & Bartha, R. 1974 Hydrocarbon metabolism by *Brevibacterium erythrogenes*: normal and branched alkanes. *J. Bact.* **119**, 868–878.

Postgate, J. R. 1984 *The sulphate-reducing bacteria*, 2nd edn. Cambridge University Press.

Raymond, R. L., Jamison, V. W. & Hudson, J. O. 1967 Microbial hydrocarbon co-oxidation. 1. Oxidation of mono- and dicyclic hydrocarbons by soil isolates of the genus Nocardia. *Appl. Microbiol.* **15**, 857 865.

Rosser, H. R. & Hamilton, W. A. 1983 Simple assay for accurate determination of (^{35}S) sulphate reduction activity. *Appl. environ. Microbiol.* **45**, 1956–1959.

Sanders, P. F. 1984 Assessment of bacterial activity and corrosion in offshore environments. *Proceedings IRM Offshore Inspection Repair and Maintainance Conference, November 1984, Aberdeen.*

Sanders, P. F. & Hamilton, W. A. 1986 Biological and corrosion activities of sulphate-reducing bacteria in industrial process plant. In *Biologically induced corrosion* (ed. S. C. Dexter) pp. 47–68. Houston: NACE.

Tibbetts, P. J. C. & Large, R. 1986 Degradation of low-toxicity oil-based drilling mud in benthic sediments around the Beatrice Oilfield. *Proc. R. Soc. Edinb.* B **91**, 349–356.

Van Der Linden, A. C. 1978 Degradation of oil in the marine environment. In *Developments in the biodegradation of hydrocarbons – 1* (ed. R. J. Watkinson) pp. 165–200. London: Applied Science.

Wilkinson, T. G. 1982 An environmental programme for offshore oil operations. *Chemy Ind, Lond.*, Feb., pp. 115–123.

Discussion

R. E. Jones (*School of Ocean Sciences, University College of North Wales, U.K.*). Does the nature of drilling cuttings change from field to field depending on the strata being drilled and the techniques used? Would such changes account for some of the variability by altering the permeability of the sediments and the transport of oxygen into and hydrocarbons out of the sediments?

P. F. Sanders. Drilling cuttings will vary from field to field, since the rock cuttings derive from the strata drilled. The residual mud in which the rock chips are embedded, however, will be dependent on the type of drilling mud in use. It is quite likely that such localized changes in mud chemical and physical characteristics could account for differences in biodegradation between the sites. We would, however, expect such differences to be slight because the predominant effect is from the mud itself.

R. J. Watkinson (*Shell Research Ltd., Sittingbourne, Kent, U.K.*). The analytical data on hydrocarbon distribution between the two sites are used to make statements on relative degradability. Oxygen will have a large influence on rates of degradation of the two mud-type hydrocarbons. Could the differences between the two sites be explained in terms of rates of oxygen diffusion rates between the two sites?

P. F. Sanders. This study was purely concerned with the relative degree of biodegradation of *n*-alkanes in the base-oil and no attempt was made to establish the rate of biodegradation. The same degree of biodegradation is possible with either a high rate for a short time or a low rate for a long time. Similar considerations apply when trying to extrapolate these data to assess rates of seabed recovery.

R. A. A. Blackman (*MAFF, Burnham-on-Crouch, Essex, U.K.*). I would like to ask Dr Sanders for clarification on the relative degradability of diesel and low-toxicity alternative base-oils. He has suggested that the zone of maximum degradation of the latter lies much closer to the discharge point than for diesel-oil but Dr Kingston suggested (this symposium) that at sediment concentrations of 3% (still very high concentrations only to be expected very close to the platform) diesel-oil was more readily degradable than low-toxicity oils.

P. F. Sanders. There are many examples in the literature where studies conducted under controlled conditions in the laboratory with single pure cultures provide conflicting data to those obtained from field measurements.

P. F. Kingston (*Institute of Offshore Engineering, Heriot-Watt University, Edinburgh, U.K.*). In answer to Dr. Blackman's question I should like to make it clear that the experimental results that I quoted from Gillam *et al.* (1986) referred to one species of filamentous bacterium isolated from diesel-contaminated cuttings (*Streptomyces* sp.). The fact that such species exist and that at diesel concentrations below 3% may utilize diesel as a substrate twice as fast as low-toxicity base-oil suggests that diesel may be biodegraded at a much greater rate than was previously thought when compared with 'low-toxicity' base-oils.

Reference

Gillam, A. H., O'Carrol, K. & Wardell, J. N. 1986 Biodegradation of oil adhering to drill cuttings. In *Proceedings of Conference on Oil-based Drilling Fluids – Cleaning and Environmental Effects of Oil Contaminated Drill Cuttings, Trondheim, Norway, Feb. 1986*, pp. 123–136.

P. F. Sanders. Undoubtedly, some organisms use diesel-oil equally rapidly, if not more rapidly, than other oils, including low-toxicity base-oils. The reverse is also true, however, and the microbial population in nature will be adapted to the oil types in the environment. The rate of breakdown of the oil will ultimately depend upon the nature of the consortium of microbes so that results from single pure cultures may not reflect the situation in the natural environment. It is also possible that concentrations of diesel greater than 3 % occur close to the discharge. This would explain our findings of depressed bacteria activity close to the platforms using diesel oil-based muds.

W. A. Hamilton (*Department of Microbiology, University of Aberdeen, U.K.*). The last few questions have discussed various factors concerned with breakdown of hydrocarbon fractions. I should like to shift our attention back to what is the main point being made by Dr Sanders; that is to say, the direct relation between hydrocarbon breakdown and the sulphate-reducing bacteria.

Earlier in this meeting C. G. Moore discussed the effects of meiofauna in both a mudflat environment and also under production platforms. In our own early work we were looking at a polluted estuary where the organic input was in the form of cellulose. Each of these three environments is therefore quite different in terms of physical and chemical components, and of the microorganism responsible for the primary biodegradation. What they all have in common however is the consequences of organic overloading: reduced conditions, growth and activity of sulphate-reducing bacteria, production of sulphide. What Dr Sanders has shown is this direct relation between these indices of sulphate-reducing bacteria and of hydrocarbon loading and its biodegradation in discarded drill cuttings.

The point we would make is that low redox conditions and sulphide production are major direct causes of the effects that can be observed in the meiobenthos and macarobenthos, and on corrosion.

Phil. Trans. R. Soc. Lond. B **316**, 587–602 (1987)
Printed in Great Britain

The importance of the planktonic ecosystem of the North Sea in the context of oil and gas development

By P. C. Reid

Natural Environment Research Council, Institute for Marine Environmental Research, Prospect Place, The Hoe, Plymouth PL1 3DH, U.K.

Planktonic organisms are the primary source of food for the top level of the marine food chain, the fish. Yet only part of the plankton is ingested by fish, the remainder sediments to the bottom to provide food for benthic organisms (which may in turn be grazed by demersal fish) and to contribute to a detrital sink. Although the relative proportions of the plankton entering each of these compartments is still a matter of debate, some indication of its importance as a resource can be gauged from the North Sea fishery that has yielded 2–3 Mt per year since the mid-1960s. Calculations for the North Sea of the annual production of phytoplankton, zooplankton, fish and benthos as energy equivalents are contrasted with the annual energy yields of oil and gas.

Since 1948 the plankton of the North Sea has been monitored at a depth of 10 m by the Continuous Plankton Recorder (CPR) survey. Results for two large areas which encompass the Brent, Beryl and Forties oilfields are presented. Between 1948 and 1982 the plankton of these areas showed similar large changes in population structure in both phytoplankton and zooplankton with almost tenfold changes in levels of biomass during this 35 year period. The development of oil-related activities is discussed in relation to the plankton time-series with comment on possible causes of the changes which are believed to be the result of natural variability.

Introduction

Planktonic organisms are small, not readily visible to the naked eye, with patchy vertical and horizontal distributions in the water column and large seasonal variability. They are thus difficult to study in relation to pollution events and any potential pollution effect raises little public concern because of a lack of a visible impact. Their importance is underrated however, as the plankton includes the primary energy source and subsequent food for almost all other components of the ecosystem. Their size and rapid rates of production may make them particularly susceptible to small concentrations of pollutants and any impact is likely to be reflected higher up the food chain. To demonstrate the importance of the plankton the simplified energy budget for the North Sea compiled by Steele (1974) has been revised with the inclusion of an estimate of the input of solar energy. Using these results, and to place a value on the plankton, a comparison is made between yields of North Sea oil and gas with energy yields from components of the ecosystem.

Oil production activities off the British coast are concentrated in the central and northern North Sea. In these areas there is evidence of localized effects of oil pollution on the benthos and fish (Addy *et al.* 1984; Davies *et al.* 1984*a*, *b*). Time-series of plankton measurements are presented for these same areas, together with data on their physical environment as a backdrop against which the development of petroleum extraction can be compared and possible pollution impacts assessed.

Energy budget calculations

The North Sea as defined here is equivalent to the fishery statistical division IV of the International Council for the Exploration of the Seas (ICES). This area (figure 1), which covers 0.5872×10^{12} m^2, includes all currently producing oil and gas reserves in the North Sea. A

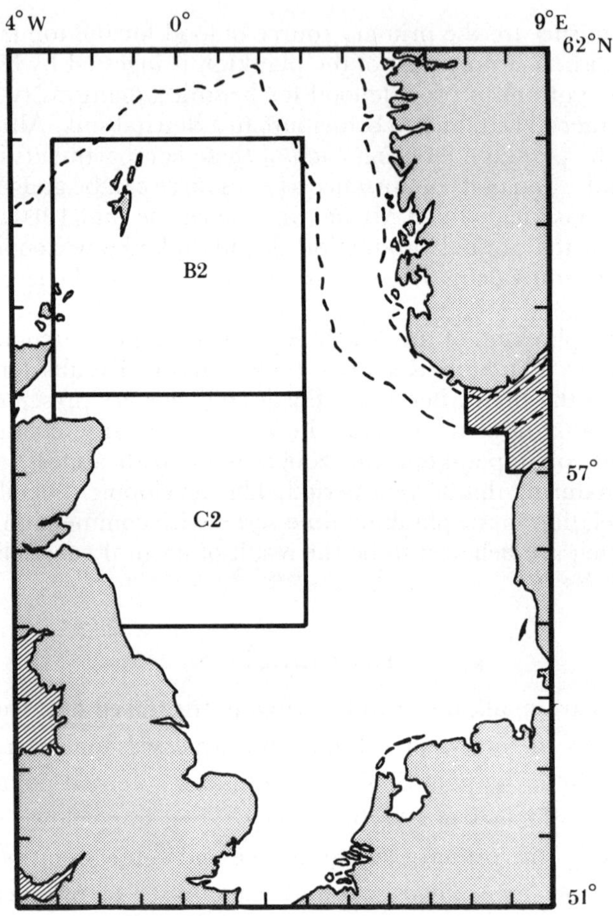

Figure 1. ICES fishery statistical division IV which extends from 51° N to 62° N and from 4° W to approximately 8° E. Sea areas outside the division boundary are hatched. The 200 m contour (broken line) and areas B2 and C2 for which data from the CPR survey have been processed are superimposed.

considerable variation in depth and hydrography is evident from north to south with a boundary between shallower mixed and deeper stratified waters at approximately 54°N in the summer months of the year (Pingree *et al.* 1978).

Solar energy input to the North Sea

The mean radiation penetrating the sea surface (R) was calculated by using tables and equations from Goldsmith & Bunker (1979). Cloud cover observations taken by merchant ships from 1961 to 1976 were provided by the Meteorological Office, Bracknell, U.K. The data used, in Watts per square metre, were $R = 96.6$, Q (the radiation received at the surface after penetrating cloud) = 104.3 and Q_0 (the radiation at the bottom of the atmosphere without cloud) = 195.7. Confirmation of these estimates was made by calculating an annual global irradiation on a horizontal surface of 95.7 W m^{-2} for Aberdeen from table 51 in Anon. (1980). Only 50% of R is available to plants (Monteith 1973) to use as 'Photosynthetically Active Radiation' (PAR) which I have taken as 48.3 W m^{-2}. Of this value an unknown proportion is utilized by plants and the remainder dissipates as heat.

Primary production

Until recently an annual net primary production of approximately 100 g C m^{-2} per year (the value accepted by Steele (1974)) was consistently quoted for the North Sea. This estimate was raised to 130 g C by Jones (1984) based upon data up to 1980. Since 1980 a growing number of studies (Fransz & Gieskes 1984; Gieskes & Kraay 1984; Joiris *et al.* 1982; Weichart 1985) have recorded primary production rates which are more than double the early estimates. (These estimates do not take into account production by benthic micro- and macroalgae and contributions from river inflow.) A thorough seasonal and spatial coverage of production measurements has not yet been made to evaluate true phytoplankton production, but it does not seem unrealistic to take an estimate of 200 g C m^{-2} for annual net primary production by all plants in the North Sea.

Zooplankton herbivore production

Traditionally, the copepods were considered to be the predominant herbivore grazers and they are the only group for which there are any accurate estimates of production. These values range between 2.5 g C m^{-2} per year (Evans (1977), but they are now considered to be an underestimate (F. Evans, personal communication)) and 12 g C m^{-2} per year (Fransz *et al.* 1984; Fransz & Gieskes 1984). As copepods are not the only component of the herbivores, and herbivorous planktonic larvae may at times dominate the plankton, an average total herbivorous production of 15 g C m^{-2} per year is taken as a fair estimate (see Baars & Fransz 1984).

There is at present no information on rates of production of the microzooplankton in the North Sea although they are known to occur in large numbers at certain times of the year (Fransz & Gieskes 1984). These small organisms may at times have an important grazing impact on nano- and picophytoplankton which appear to be the most productive size fractions of the phytoplankton (Joint *et al.* 1986). A microzooplankton production of approximately 12 g C m^{-2} per year calculated by Burkill (1982) for Southampton Water is used here as an estimate.

Benthic production

Rachor (1982) estimated a mean macrofauna biomass of 1.28 g C m^{-2} for the North Sea. Only 36% of the bottom of the North Sea has been sampled quantitatively for benthos, and most of these samples were taken with Van Veen or McIntyre grabs which only sample the top 10 cm or so of sediment. There are great contrasts between the macrofauna biomass of the deep northern North Sea and the more productive south. Recent benthic studies in the southern North Sea and especially those using box corers (De Wilde *et al.* 1984) give much higher biomass estimates than were previously determined. Almost 50% of the biomass was in sediment below 10 cm which would have been missed by grabs. In the northern North Sea in contrast, macrofauna levels may be very low (Hartwig & Thiel 1983) so Rachor's average estimate is accepted here. The productivity/biomass (P/B) ratio of the macrobenthos is dependent on the lifespan of individual organisms, a mean value for the North Sea of 1.5 is used to give a production of approximately 2 g C m^{-2} per year using Rachor's biomass estimate and De Wilde's conversion factor of 2.5 g ash free dry weight (AFDW) = 1 g C.

There are few production estimates for the meiofauna in the North Sea; they range from 1.5–2.0 g C m^{-2} per year in Belgian coastal waters (Heip *et al.* 1984) to 2.4 g C m^{-2} at the Oyster Ground (De Wilde *et al.* 1984) to 2.42 g C m^{-2} per year in the Fladen Ground (Faubel & Hartwig 1983). For the budget a mean value of 2.0 is used.

A proportion of the benthic epifauna is harvested for human consumption. Statistics on these catches have been kept by ICES since 1966. Adopting the conversion used by Steele (1974) for fish, of 1 g wet mass = 0.1 g C, the mean catches for 1966–69 and 1978–82 are 0.09 and 0.07 g C m^{-2} per year respectively. Total production is calculated by assuming the production of the commercial stock is ten times the fishing yield, multiplied by two for the epifauna which does not contribute to the commercial fishery, to give a total production of approximately 0.2 g C m^{-2} per year.

Fish production

In his energy budget, Steele (1974) took a mean of the years 1965 to 1969 for his estimates of fish production. This sequence of years was atypical as it represented a transition period which coincided with the introduction of Purse Seines, the development of a major industrial fishery on the mackerel and a rapid decline in the contribution of herring to catches. The high yields of the late 1960s represented an almost doubling of the catch over the immediate postwar years to 1964 which had an average yield of 1.75 Mt. Catches were again reduced to a lower level of 2.44 Mt from 1978 to 1982. Yang (1982) categorized the fish of the North Sea into three groups (planktophagic, benthophagic, and ichthyophagic) dependent on their dominant feeding habits. Production estimates for three periods 1948–52, 1965–69 and 1978–82 have been calculated with a subdivision according to Yang's (1982) feeding groups. A natural mortality estimate of 20% for the benthophagic and 50% for planktophagic species was used as by Steele (1974), an intermediate value of 35% was taken for the ichthyophagic species.

Detrital pool, bacteria and Protozoa

All organisms in the North Sea ecosystem contribute to a pool of soluble and particulate carbon through excretion. A further contribution to this pool comes from dead and decaying products. Soluble carbon becomes immediately available for bacterial production. The particulate fraction was divided into 'labile' and 'inert' by Postma & Rommets (1984). The former was estimated to turnover every five days and the latter remains at a background level of 40 mg m^{-3} over a longer period. If the average depth of the North Sea is taken as 87 m this gives a constant background detrital pool of 3.4 g C m^{-2}, which is approximately equal to the average biomass of the phytoplankton. The labile fraction provides an important substrate for bacteria and Protozoa and is an additional food source for higher trophic levels. Rates of carbon flow through the detrital pool are not known.

Ecosystem energy budget

An energy budget for the North Sea was calculated by Steele (1974) with fish yields from ICES statistics as the terminal product of a simplified food web (figure 2*a*). For comparison with Steele (1974) all units used in this section are in kilocalories† per square metre per year. A conversion factor of 10 $kcal_{th}$ = 1 g C as per Steele (1974) has been used. From an estimated primary production of 900 $kcal_{th}$ m^{-2} per year a yield of 8 pelagic and 2.6 $kcal_{th}$ m^{-2} per year for demersal species was obtained. Steele (1974) concluded that transfer efficiencies of at least 20% were necessary to obtain such high yields. If only a 10% transfer efficiency was used, all the planktonic production would have been needed to satisfy the food requirements of the fish. By determining a more accurate estimate of fish food-requirements, Jones (1984) showed that transfer efficiencies of 10% would be possible in the unreal situation of a lack of predation by fish on other small fish. In reality, small-bodied fish such as sprats, Norway Pout, sand eels and juvenile specimens provide the main source of food for ichthyophagic species. Jones (1984) overcame this dilemma by raising primary production to 1300 $kcal_{th}$ m^{-2} per year and confirming the necessity for transfer efficiencies of 15–20%.

Figure 2*b* plots a revised energy budget for the North Sea from 1965–69 by using the data given in the previous sections. Almost all of the estimates of production are lower than those given by Steele (1974). This is partly because of his underestimation of the size of the ICES

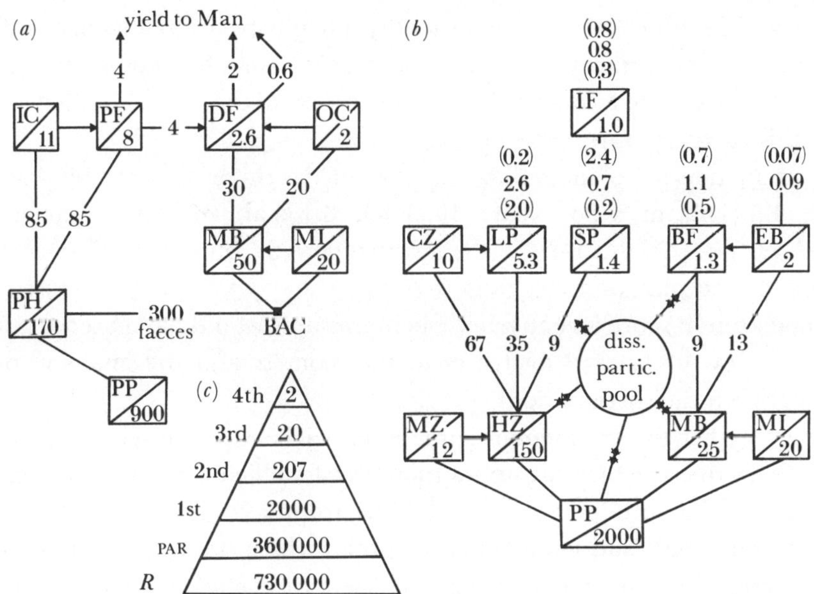

Figure 2. (*a*) A simplified energy budget for the North Sea redrawn after Steele (1974). IC, invertebrate carnivores; PF, pelagic fish; DF, demersal fish; OC, other carnivores; MB, macrobenthos; MI, meiobenthos; PH, pelagic herbivores; BAC, bacteria; PP, primary production. (*b*) A revised energy budget for the North Sea. IF, icthyophagic fish; CZ, carnivorous zooplankton; LP, large pelagophagic fish; SP, small pelagophagic fish; BF, benthophagic fish; EB, epibenthos; MZ, microzooplankton; HZ, large herbivorous zooplankton; MB, macrobenthos; MI, meiobenthos; PP, primary production. The circle represents a detrital pool of dissolved and particulate material. Superimposed above the fish and epibenthos production boxes are yields to Man, from top to bottom, for the years 1978 to 1982, 1965 to 1969 and 1948 to 1952. (*c*) A pyramid of energy production by each trophic level in the North Sea with, at the base, estimated PAR, photosynthetically active radiation and *R*, radiation penetrating the sea surface. 1st, 2nd, 3rd and 4th, trophic levels.

† 1 cal_{th} = 4.184 J.

fishery area (0.5×10^{12} m^2) compared with my calculation of 0.5872×10^{12} m^2. Mean fish and crustacean yields from the commercial fisheries are included above the production boxes for the periods 1948–52 (fish only), 1965–69 and 1978–82 with the oldest period at the bottom and the most recent at the top. A complete reversal in the contribution of the small and large pelagic fish between the first and last period is evident with a major contribution from benthophagic species (predominantly haddock) in the middle period. Yields of ichthyophagic species as a proportion of catches increased substantially in the two most recent periods. An unknown, but possibly significant, portion of the detrital pool may be lost to the system by sedimentation or export from the North Sea.

If net primary production is 2000 kcal$_{th}$ m^{-2} per year, transfer efficiencies must still be at least 15–20% through secondary producers to allow for a contribution to herbivorous microzooplankton, the detrital pool and recycling by bacteria and Protozoa. Transfer efficiencies of at least 15% are also necessary at the next feeding level for a flexible account of the energy flow. Food requirements based upon a 15% transfer efficiency are plotted on the diagram, giving an excess for predators on the herbivores and benthic macrofauna of 39 and 3 respectively. The diet of ichthyophagic species depends upon their size and age, and although they predominantly feed on small pelagic species and juvenile fish of other groups they also feed on plankton and benthic organisms, especially when young. The total production of all fish and crustaceans at the third trophic level was estimated to be 10 kcal$_{th}$ m^{-2} per year from 1965–69 which would not balance the estimated food requirement (6.7, rising to 7.5 kcal$_{th}$ m^{-2} per year from 1978–82) of the ichthyophagic species at a 15% transfer efficiency. Direct feeding of the ichthyophagic species on the plankton and benthos is necessary to balance the budget. Such a change in feeding habits is also implied in the marked reduction in the mean size of fish in this category now seen in catches.

The total production for planktophagic species in the three different periods is estimated as: 1948–52, 4.5 kcal$_{th}$ m^{-2} per year; 1965–69, 6.7 kcal$_{th}$ m^{-2} per year and 1978–82, 5.4 kcal$_{th}$ m^{-2} per year. High values in the last two periods, plus an additional demand from the fish eaters on the plankton, suggests that there may have been increased pressure on herbivorous plankton in more recent years. Such pressures would have been reinforced by the evidence given below of marked decreases in the biomass and, by analogy, production of zooplankton between the three periods.

By summing together the production estimates for each trophic level a 'pyramid of energy' (Odum 1953) is produced which suggests a total transfer efficiency between each trophic level of 10% (figure 2*c*). The production of the highest trophic level was doubled to include the impact of birds, mammals and ichthyophagic species which do not contribute to the fishery. Input of light energy as PAR with 0.5% transfer efficiency is included at the base of the pyramid. No account was taken of the detrital pool in these calculations; it is possible that fluxes through this compartment, which appears to remain at a relatively constant value, may have little effect on average trophic interactions.

Comparison of petrogas and ecosystem yields

The North Sea basin has existed since at least the Permian although source rocks for oil and gas are believed to be of Late Jurassic age. Fossil sedimented planktonic organisms are the most likely source of this petroleum energy. By assuming that the palaeo-sea in which the fossil hydrocarbons were formed was a similar size to the present North Sea, an attempt to place a value on the plankton is made by comparing it with oil and gas yields. Figure 3 presents a time series of annual gas and oil yields for the North Sea as million tonnes of oil equivalent (MTOE) from estimates compiled by the British Petroleum Company. As an energy rate, 1 MTOE per year = 1.37 10^9 W, mean gas production from 1968 to 1984 and oil from 1971 to 1984

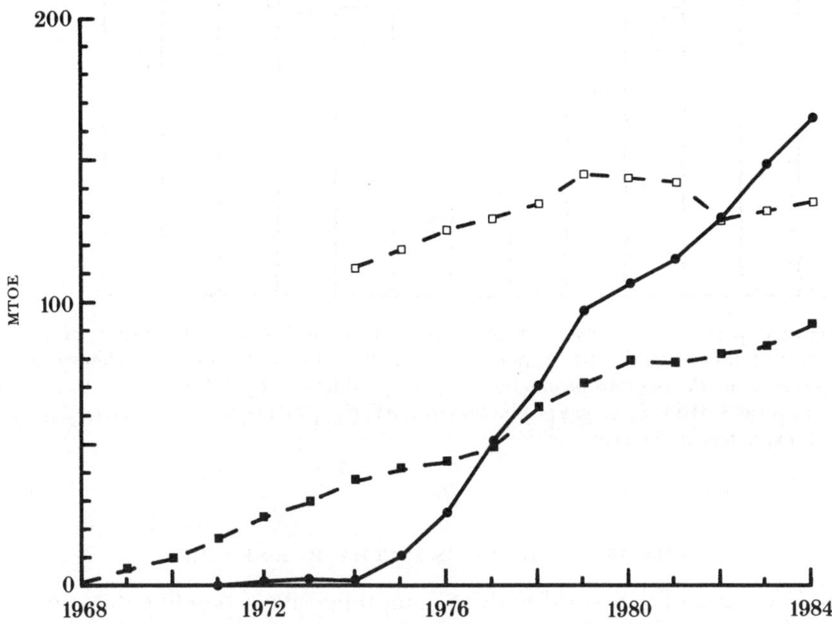

FIGURE 3. Total North Sea oil and gas production from first yields to 1984. (Sources Mr G. Pyne, British Petroleum Company, and Anon. 1985*a*). Oil is represented by the continuous line, gas is represented by the broken line; the top plot represents gas production from 1974 and includes output from onshore fields in the Netherlands. MTOE, Million Tonnes Oil Equivalent.

for the North Sea equals 0.11 and 0.15 W m^{-2} respectively to give a total for oil and gas of 0.26 W m^{-2}. A higher rate of extraction was calculated if the onshore gasfields of the Netherlands, which can be considered as within the same geological basin, were included. A conservative estimate of proved North Sea reserves for gas and oil (sources as for figure 3) is respectively 5.37×10^9 MTOE and 3.96×10^9 MTOE which gives a life of 39 years for gas extraction and 19 years for oil extraction at 1984 rates.

The average oil and gas energy yield per year from the North Sea over the past fifteen years and yields for gas and oil in 1984 are approximately equal to the annual primary production (figure 4). This figure also gives a comparison between yields and reserves of oil and gas as energy rates in the unreal situation of total extraction of the reserves in one year. As an energy equivalent, the annual mean fish yield to man for the three periods studied here is 0.2 % of mean oil and gas yields i.e. 500 years of fish harvest from the North Sea represents the mean oil and gas harvested in the last fifteen years. Extraction at 1984 rates is equivalent to almost 1600 years of fish harvest.

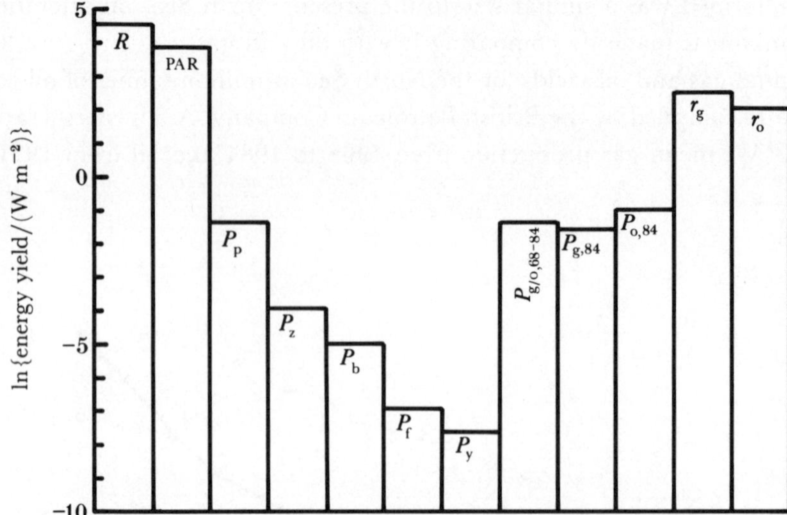

FIGURE 4. A comparative plot of the energy yields of components of the North Sea ecosystem, and oil and gas. R, radiation penetrating the sea surface; PAR, photosynthetically active radiation; P_p, primary production; P_z, zooplankton production; P_b, benthos production; P_f, fish production; P_y, fish yield; $P_{g/o,68-84}$, mean gas and oil production from 1968–1984; $P_{g,84}$, gas production in 1984; $P_{o,84}$, oil production in 1984. r_g, gas reserves, and r_o, oil reserves if exploited in one year.

TEMPORAL CHANGES IN THE PLANKTON

There can be few areas of the world with as comprehensive a baseline data set of plankton measurements before the development of an industrial resource as the North Sea. From 1948 Continuous Plankton Recorders (CPRs) have surveyed the North Sea on a number of standard routes at approximately monthly intervals. Details of the survey and methods are given in Edinburgh Oceanographic Laboratory (1973). Unfortunately, this survey only gives information on standing stocks; there are no comparable data sets of production for the plankton. However, gross changes in annual production can be inferred when marked reductions in biomass occur if P/B ratios are assumed to remain approximately equal. Figure 5 presents the mean monthly variability of phytoplankton and zooplankton sampled by the CPR in two areas (B2 and C2) of the North Sea (figure 1) between January and December from 1948 to 1982. Counts of the major zooplankton groups (copepods, thecosomes, euphausiids, large chaetognaths and hyperiids) were transformed to biomass by using mean wet mass values. The biomass estimates must only be considered as an index of biomass since the CPR only samples at 10 m depth and does not adequately sample young stages and soft-bodied plankton. Phytoplankton colour is a visual estimate of chlorophyll on the filtering silks; individual samples are allocated to one of four numerical categories according to their greenness. Counts of phytoplankton species are also made on CPR samples but can only be used for coarse estimates of abundance because of the size of the 270 μm mesh filtering silk which lets most phytoplankton pass through.

In areas B2 and C2, there was a marked shortening of the growing season and a progressive decline in levels of the zooplankton biomass index from 1948 to 1982; between the two sequences of years 1948–52 and 1978–82 this represents a reduction to 34% and 20% of the earlier period in areas B2 and C2 respectively. For area B2 phytoplankton colour showed the

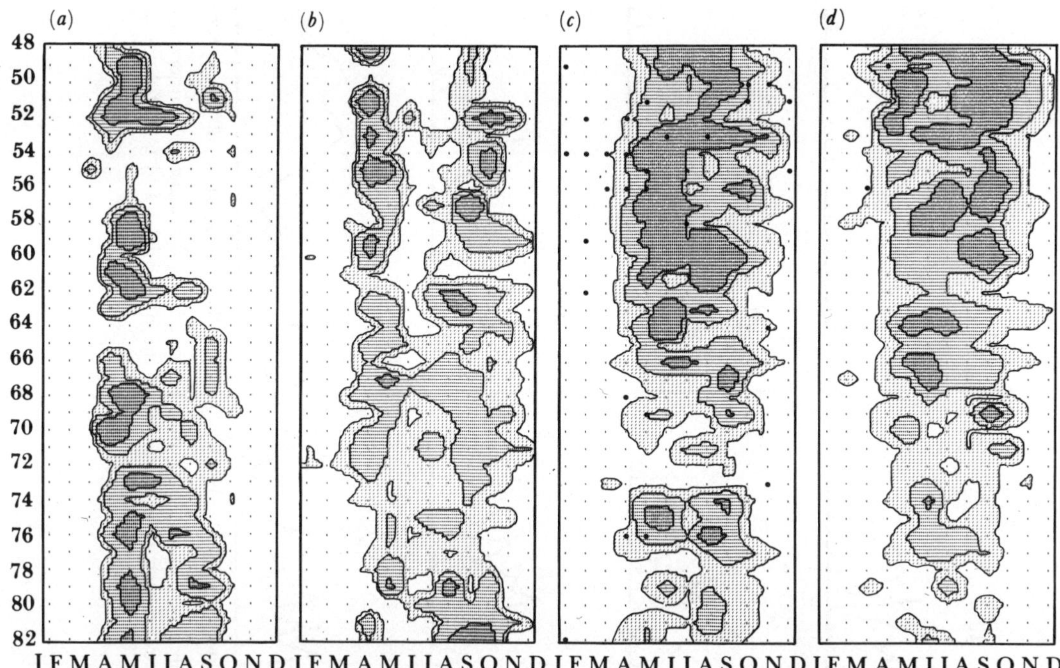

FIGURE 5. Monthly fluctuations in areas (*a*, *c*) B2 and (*b*, *d*) C2 of an index of (*a*, *b*) phytoplankton colour and (*c*, *d*) zooplankton biomass with values for every month (horizontally) in years (vertically) from January 1948 to December 1982. In (*c*) and (*d*) filled circles represent months that were not sampled but which have been interpolated in the contouring procedure. Contour levels for the phytoplankton are 0.07:0.12:0.23 units and for the zooplankton are 15:40:72 units.

reverse trend to the zooplankton (correlation −0.63*** (a correlation significant at the 0.1 % level is indicated by ***)). Colour for area C2 also showed an increasing trend until approximately 1973 when a period of low levels occurred, but was not significantly related to the zooplankton trend. A marked seasonal and interannual variability is superimposed on these general trends. Large changes in the composition of the plankton have also occurred; for example, thecosomes which are small planktonic molluscs which once formed a major food source for the herring in the northern North Sea declined from 38 % to 2 % of the herbivorous zooplankton biomass averaged for B2 and C2 between 1948–52 and 1978–82.

A number of studies using data from the CPR survey have noted empirical relations between physical environmental variables and long-term changes in the plankton; Colebrook (1978, 1985) found relations with temperature and Lamb's westerly weather index, Dickson & Reid (1978) found correlations with wind strength, direction and turbidity, Reid & Budd (1979) with salinity and Robinson (1983) with wind and Lamb's westerly weather index. We do not know the mechanisms behind these apparent relations, changes in the plankton cannot be forecast and any impact from pollution is difficult to assess.

Annual means for zooplankton biomass, phytoplankton colour, diatoms and fish yield are plotted in figure 6 with similar graphs for temperature, salinity, scalar wind speed, cloud and Lamb's westerly weather index. Some of the relations mentioned above are confirmed with correlations between Lamb's westerly weather index and zooplankton for both area B2 and C2 of respectively 0.53*** and 0.62*** and between zooplankton and salinity in area C2 of

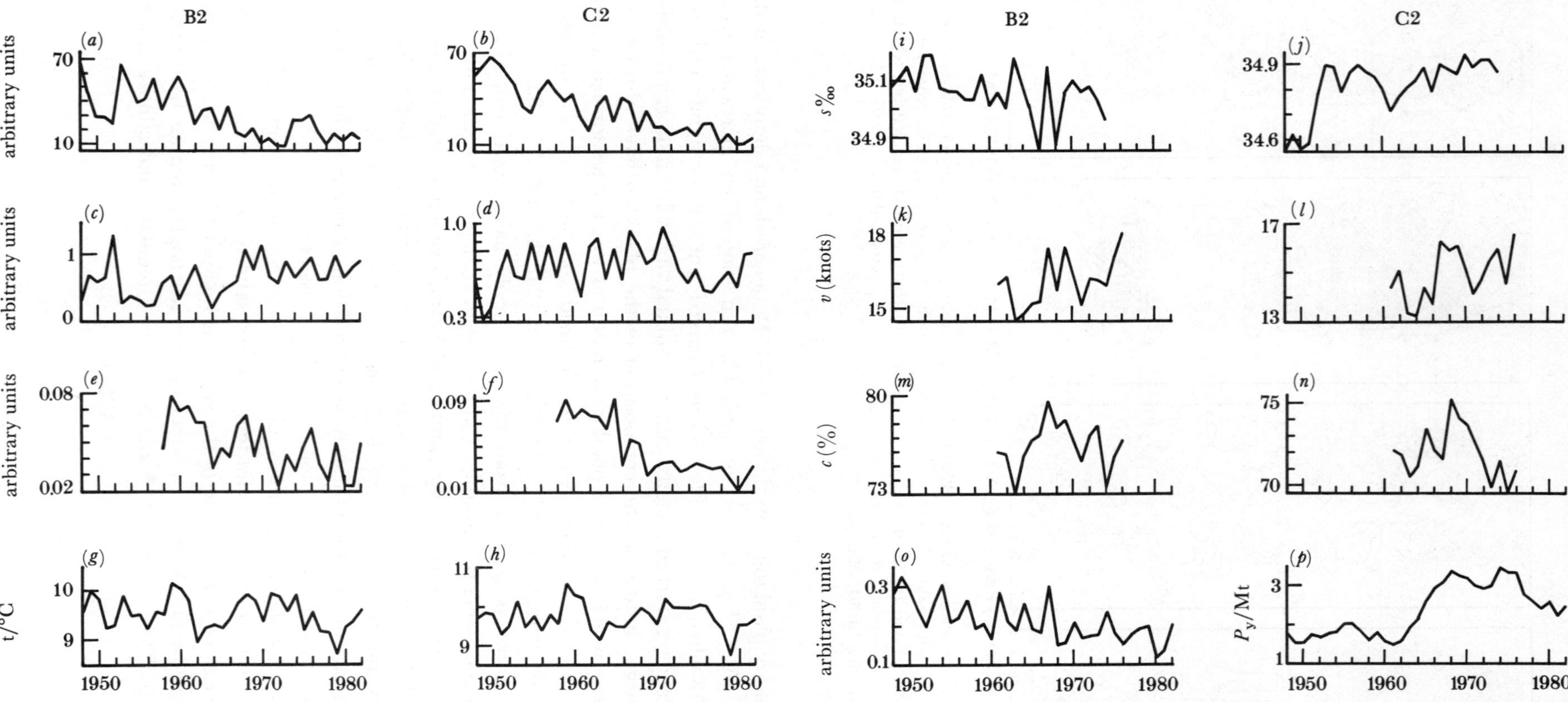

FIGURE 6. Graphs of annual means averaged for areas B2 and C2 for zooplankton biomass index (*a*, *b*), phytoplankton colour (*c*, *d*), diatom counts (*e*, *f*), sea surface temperature, *t* (*g*, *h*) and salinity, *s*, (*i*, *j*), scalar wind speed, *v* (*k*, *l*), and cloud cover, *c* (*m*, *n*) with plots of Lamb's westerly weather index (*o*) for the British Isles and total fish yields, P_y (*p*) from the North Sea.

−0.66***. Annual yields of North Sea fish are significantly correlated with the biomass index in both areas B2 and C2 (B2, −0.63***; C2, −0.66***). These inverse correlations may be fortuitous, but further suggest that grazing by fish may have had an increasing impact on the plankton in recent years. Alternatively, feedback loops from the plankton to the fish as suggested by Steele & Frost (1977) may exist, or both mechanisms may operate (see Reid 1984).

Plankton and hydrocarbon pollution

Information on the impact of oil on planktonic ecosystems was assessed by Davenport (1982) and further aspects have been reviewed by Corner (1978), Gray (1982), Johnston (1984) and Anon. (1985*b*). While laboratory and experimental mesocosm research document marked effects from hydrocarbon pollution at relatively high doses, there is little evidence for any permanent damage to open-sea pelagic environments from either prolonged low-level inputs of oil or large accidental spills. These conclusions must be qualified however, as there have been few field studies on plankton in the open-sea in relation to hydrocarbon concentrations outside major spill events. Moreover, Davies *et al.* (1980) have shown that zooplankton community structure is susceptible to oil at concentrations as low as 5–15 mg m^{-3}. This level is not much higher than the average concentration (4 mg m^{-3}) determined by Massie *et al.* (1985) in fluorescence surveys around the Brent, Beryl and Forties oilfields.

Hydrocarbons entering the marine ecosystem from offshore platforms come from three main sources (Read & Blackman 1980; Davies *et al.* 1984*b*): production water, displacement water and oil-based drilling muds. There is considerable variability and error in estimates of inputs from other anthropogenic sources (Whittle *et al.* 1982; Johnston 1984; Anon. 1985*b*), but there is a general consensus that inputs from rivers, the atmosphere, coastal refineries, sewage and shipping spills are far more important than sources from production platforms. The northwestern North Sea is a special case because the human population on the adjacent land has a low density and there is relatively little industrialization and minor river runoff with prevailing offshore winds. Here drilling muds were estimated to account for more than 50% of a total input of 18.4 kt hydrocarbon by Davies *et al.* (1984*b*). From 1981 to 1984 discharges of oil-based drilling muds increased from 5780 to 19600 t (U.K. Department of Energy, personal communication) and paralleled increasing production levels of oil. Any impact on the plankton from drilling muds is likely to be extremely localized because the hydrocarbons are bound to rock fragments and drop rapidly to the bottom close to platforms.

Conclusions

The plankton of the North Sea is a major renewable resource of which only a small fraction is harvested via fish for Man's use. As a source of food for the fish plankton is approximately five times as important as the benthos. Revised calculations of the energy flow from plankton to fish imply that the North Sea ecosystem is not as efficient as deduced by Steele (1974), although transfer efficiencies of at least 15% are still necessary to balance an energy budget of the ecosystem. Both empirical correlations and energy budget calculations suggest that changes in the composition and production of fish stocks may have partly caused an observed decline in the zooplankton of the North Sea by an increase in grazing pressure.

North Sea oil and gas are natural biological products, concentrated energy sources formed from the breakdown of plankton which was incorporated into sediments of a palaeo-North Sea basin. Like its antecedents, present-day plankton is capable of synthesizing a variety of hydrocarbons (Anon. 1985*b*). We know remarkably little about the rates and processes behind the formation of natural oils in the plankton, although natural oil-slicks associated with plankton blooms are a frequent phenomenon and certain planktonic organisms may have a large lipid content. Edyvean & Sneddon (1985) drew attention to the effects of lipids and natural oils from plankton on the blockage of pores in oil-bearing strata and filters used to clean the large volumes of seawater used on oil platforms. There is a need for further research on natural oils in plankton as cultured planktonic organisms may prove to be an alternative source of oil in the future.

The size and value of annual North Sea primary production is emphasized in a comparison with annual hydrocarbon yields which shows primary production to be approximately equivalent to both gas and oil production in 1984. Estimated reserves and oil and gas yields to date for the North Sea represent approximately 100 years of primary production at the rate determined for the present day. Yet only a minute fraction of the plankton is incorporated into sediment each year and is thus available for hydrocarbon formation, emphasizing the long temporal period required for diagenesis, maturation and concentration.

The suggestion that there is no visible effect of hydrocarbon pollution on plankton should not lead to complacency. Sampling strategies for hydrocarbons in plankton and water have ignored their physical and biological dynamics; surveys at the surface may bear little relation to the 'real world' of the plankton. A trend for higher concentrations of hydrocarbons in the thermocline was noted for example by Johnston (1984). There are a number of processes, including biodegradation by microorganisms, which flush the plankton clear of contaminants to concentrate them in bottom sediments. Faecal pelletization (Prahl & Carpenter 1979) and absorption to particulates leads to a rapid sedimentation of hydrocarbons. In addition, there is accumulating evidence (Peinert *et al.* 1982; Davies & Payne 1984) that much of the spring bloom of diatoms settles rapidly to the bottom to provide direct food to the benthos. Reworking of hydrocarbons concentrated in bottom sediments by tidal action (Jenness & Duineveld 1985) or winter storms (Dickson & Reid 1978) may lead to repeated, temporary high doses of pollutants especially in the shallow, southern North Sea.

If oil pollution in the North Sea is taken as increasing in proportion with hydrocarbon production (see Read & Blackman (1980)) any impact might be reflected in long-term changes in the plankton. Zooplankton biomass and diatoms are inversely and significantly correlated with both oil and gas production curves for the North Sea at the 0.5% level in area C2. An absence of any correlation between zooplankton and oil in area B2 and the many relations that have been determined between hydrobiological variables and long-term changes in the plankton suggest that the correlation only reflects an inverse similarity in trends and no real relation. Moreover, recent results from the Continuous Plankton Recorder Survey provide evidence for a possible reversal of the downward trend in the zooplankton (Colebrook *et al.* 1984).

Many planktonic organisms form resting eggs and cysts as overwintering mechanisms; they contain high levels of natural oils and lipids as energy stores which may concentrate carcinogenic and mutagenic substances. Evidence from mesocosm experiments (Davies *et al.* 1980) suggests that copepod eggs are particularly susceptible to oil pollution. Colebrook (1984) has shown a link between overwintering in the North Sea and plankton abundance the following year. Plankton may therefore be more susceptible to pollution during winter months,

I am indebted to H. G. Hunt, A. H. Taylor and J. M. Colebrook at I.M.E.R., D. K. Griffiths, Marine Biological Association and G. Pyne, British Petroleum Company for provision of data and assistance with calculations. This work forms part of the programme of the Institute for Marine Environmental Research, a component of the Natural Environmental Research Council, and is supported in part by the Ministry of Agriculture, Fisheries and Food.

References

Addy, J. M., Hartley, J. P. & Tibbetts, P. J. C. 1984 Ecological effects of low toxicity oil-based mud drilling in the Beatrice Oilfield. *Mar. Pollut. Bull.* **15**, 429–436.

Anon. 1980 *Solar radiation data for the United Kingdom 1951–75*. Bracknell: The Meteorological Office. (109 pages.)

Anon. 1985*a* *B.P. statistical review of world energy*. London: The British Petroleum Company plc.

Anon. 1985*b* *Oil in the sea, inputs, fates and effects*. Washington, D.C: National Academy Press.

Baars, M. A. & Fransz, H. G. 1984 Grazing pressure of copepods on the phytoplankton stock of the central North Sea. *Neth. J. Sea Res.* **18**, 120–142.

Burkill, P. H. 1982 Ciliates and other microplankton components of a nearshore food-web; standing stocks and production processes. *Annls Inst. océanogr., Paris* **58** (S), 335–350.

Colebrook, J. M. 1978 Continuous plankton records: zooplankton and environment North-East Atlantic and North Sea, 1948–1975. *Oceanol. Acta* **1**, 9–23.

Colebrook, J. M. 1984 Continuous plankton records: overwintering and annual fluctuations in the abundance of zooplankton. *Mar. Biol.* **84**, 261–265.

Colebrook, J. M. 1985 Sea surface temperature and zooplankton, North Sea, 1948 to 1983. *J. Cons. int. Explor. Mer* **42**, 179–185.

Colebrook, J. M, Robinson, G. A., Hunt, H. G., Roskell, J., John, A. W. G., Bottrell, H. H., Lindley, J. A., Collins, N. R. & Halliday, N. C. 1984 Continuous Plankton Records: a possible reversal in the downward trend in the abundance of the plankton of the North Sea and the Northeast Atlantic. *J. Cons. int. Explor. Mer* **41**, 304–306.

Corner, E. D. S. 1978 Pollution studies with marine plankton, Part 1. Petroleum hydrocarbons and related compounds. *Adv. mar. Biol.* **15**, 289–380.

Davenport, J. 1982 Oil and planktonic ecosystems. *Phil. Trans. R. Soc. Lond.* B **297**, 369–384.

Davies, J. M., Addy, J. M., Blackman, R. A., Blanchard, J. R., Ferbrache, J. E., Moore, D. C., Somerville, H. J., Whitehead, A. & Wilkinson, T. 1984*a* Environmental effects of the use of oil-based drilling muds in the North Sea. *Mar. Pollut. Bull.* **15**, 363–370.

Davies, J. M., Baird, I. E., Massie, L. C., Hay, S. J. & Ward, A. P. 1980 Some effects of oil-derived hydrocarbons in a pelagic food web from observations in an enclosed ecosystem and a consideration of their implications for monitoring. *Rapp. P.-v. Réun. Cons. int. Explor. Mer* **179**, 201–211.

Davies, J. M., Bell, J. S. & Houghton, C. 1984*b* A comparison of the levels of Hepatic Aryl Hydrocarbon Hydroxylase in fish caught close to and distant from North Sea Oil Fields. *Mar. Environ. Res.* **14**, 23–45.

Davies, J. M. & Payne, R. 1984 Supply of organic matter to the sediment in the northern North Sea during a spring phytoplankton bloom. *Mar. Biol.* **78**, 315–324.

Dickson, R. R. & Reid, P. C. 1978 Local effects of wind speed and direction on the phytoplankton of the Southern Bight. *J. Plankton Res.* **5**, 441–455.

Edinburgh Oceanographic Laboratory 1973 Continuous Plankton Records: a plankton atlas of the North Atlantic and North Sea. *Bull. mar. Ecol.* **7**, 1–174.

Edyvean, R. G. J. & Sneddon, A. D. 1985 A filtration of plankton from seawater. *Filtr. Sep.* **22**, 184–189.

Evans, F. 1977 Seasonal density and production estimates of the commoner planktonic copepods of Northumberland coastal waters. *Estuar coast. mar. Sci.* **5**, 223–241.

Faubel, A. & Hartwig, E. 1983 On the ecology of the benthos of sublittoral sediments, Fladen Ground, North Sea. I. Meiofauna standing stock and estimation of production. '*Meteor*' *Forsch. Ergebn.* D, no. 36, pp. 35–48.

Fransz, H. G. & Gieskes, W. W. C. 1984 The unbalance of phytoplankton and copepods in the North Sea. *Rapp. P.-v. Réun. Cons. int. Explor. Mer* **183**, 218–225.

Fransz, H. G., Miquel, J. C. & Gonzalez, S. R. 1984 Mesozooplankton composition, biomass and vertical distribution, and copepod production in the stratified central North Sea. *Neth. J. Sea Res.* **18**, 82–96.

Gieskes, W. W. C. & Kraay, G. W. 1984 Phytoplankton, its pigments, and primary production at a central North Sea station in May, July and September 1981. *Neth. J. Sea Res.* **18**, 51–70.
Goldsmith, R. A. & Bunker, A. F. 1979 *Woods Hole Oceanographic Institution collection of climatology and air/sea interaction (CASI) data.* Woods Hole Oceanographic Institution Technical Report WHOI-79-70. (75 pages.)
Gray, J. S. 1982 Effects of pollutants on marine ecosystems. *Neth. J. Sea Res.* **16**, 424–43.
Hartwig, E. & Thiel, H. 1983 On the ecology of the benthos of sublittoral sediments, Fladen Ground, North Sea. II. Quantitative studies on macrobenthic assemblages. '*Meteor*' *Forsch.Ergebn.* D, no. 36, pp. 49–64.
Heip, C., Herman, R. & Vincx, M. 1984 Variability and productivity of meiobenthos in the Southern Bight of the North Sea. *Rapp P.-v. Réun. Cons. int. Explor. Mer* **183**, 51–56.
Jenness, M. I. & Duineveld, G. C. A. 1985 Effects of tidal currents on chlorophyll *a* content of sandy sediments in the southern North Sea. *Mar. Ecol. Prog. Ser.* **21**, 283–287.
Johnston, R. 1984 Oil pollution and its management. In *Marine ecology*, vol. 5, part 3 (ed. O. Kinne), pp. 1433–1582. Chichester, New York, Brisbane, Toronto and Singapore: John Wiley & Sons.
Joint, I. R., Owens, N. J. P. & Pomroy, A. J. 1986 Seasonal production of photosynthetic picoplankton and nanoplankton in the Celtic Sea. *Mar. Ecol. Prog. Ser.* **28**, 251–258.
Joiris, C., Billen, G., Lancelot, C., Daro, M. H., Mommaerts, J. P., Bertels, A., Bossicart, M. & Nijs, J. 1982 A budget of carbon cycling in the Belgian coastal zone: relative roles of zooplankton, bacterioplankton and benthos in the utilization of primary production. *Neth. J. Sea Res.* **16**, 260–275.
Jones, R. 1984 Some observations on energy transfer through the North Sea and Georges Bank food webs. *Rapp. P.-v. Réun. Cons. int. Explor. Mer* **183**, 204–217.
Massie, L. C., Ward, A. P., Davies, J. M. & Mackie, P. R. 1985 The effects of oil exploration and production in the northern North Sea: Part 1 - The levels of hydrocarbons in water and sediments in selected areas, 1978–1981. *Mar. Environ. Res.* **15**, 165–213.
Monteith, 1973 *Principles of environmental physics.* (241 pages.) London: Edward Arnold.
Odum, E. P. 1953 *Fundamentals of ecology.* (384 pages.) Philadelphia: W. B. Saunders Company.
Peinert, R., Saure, P., Stegmann, P., Stienen, C., Haardt, H. and Smetacek, V. 1982 Dynamics of primary production and sedimentation in a coastal ecosystem. *Neth. J. Sea Res.* **16**, 276–289.
Pingree, R. D., Holligan, P. M. & Mardell, G. T. 1978 The effects of vertical stability on phytoplankton distributions in the summer on the northwest European Shelf. *Deep Sea Res.* **25**, 1011–1028.
Postma, H. & Rommets, J. W. 1984 Variations of particulate organic carbon in the central North Sea. *Neth. J. Sea Res.* **18**, 31–50.
Prahl, F. G. & Carpenter, M. R. 1979 The role of zooplankton fecal pellets in the sedimentation of polycyclic aromatic hydrocarbons in Dabob Bay, Washington. *Geochim. cosochim. Acta* **43**, 1959–72.
Rachor, E. 1982 Biomass distribution and production estimates of macro-endofauna in the North Sea. ICES CM 1982/L:2. (Unpublished manuscript.)
Read, A. D. & Blackman, R. A. A. 1980 Oily water discharges from offshore North Sea installations: A perspective. *Mar. Pollut. Bull.* **11**, 44–47.
Reid, P. C. 1984 Year-to-year changes in zooplankton biomass, fish yield and fish stock in the North Sea. ICES CM 1984/L:39. (Unpublished manuscript.)
Reid, P. C. & Budd, T. D. 1979 Plankton and environment in the North Sea within the period 1948–1977. ICES CM 1979/L:26. (Unpublished manuscript.)
Robinson, G. A. 1983 Continuous plankton records: phytoplankton in the North Sea 1958–1980, with special reference to 1980. *Br. phycol. J.* **18**, 131–139.
Steele, J. H. 1974 *The structure of marine ecosystems.* Oxford: Blackwell Scientific Publications.
Steele, J. H. & Frost, B. W. 1977 The structure of plankton communities. *Phil. Trans. R. Soc. Lond.* B **280**, 485–534.
Weichart, G. 1985 High pH values in the German Bight as an indication of intensive primary production. *Dt. hydrogr. Z.* **38**, 93–117.
Whittle, K. J., Hardy, R., Mackie, P. R. & McGill, A. S. 1982 A quantitative assessment of the sources and fate of petroleum compounds in the marine environment. *Phil. Trans. R. Soc. Lond.* B **297**, 193–218.
Wilde, P. A. W. J. De, Berghuis, E. & Kok, A. 1984 Structure and energy demand of the benthic community of the central North Sea. *Neth. J. Sea Res.* **18**, 143–159.
Yang, J. 1982 A tentative analysis of the trophic levels of North Sea Fish. *Mar. Ecol. Prog. Ser.* **7**, 247–252.

Discussion

G. M. DUNNET (*Zoology Department, University of Aberdeen, U.K.*). In Dr Reid's energy budgets for the North Sea, he did not specifically refer to seabirds. Given that, in crude approximation, say 10 million seabirds may be feeding in the North Sea throughout the year (equal to *ca.* 10 kt live mass) with higher metabolic rate, and therefore food requirements, can they be considered to be negligible in terms of global budgets for the North Sea?

Because of recent winter mortalities of large numbers of emaciated (not oiled) guillemots in northern Scotland, and recent poor breeding success of kittiwakes in Shetland, and declines in kittiwake numbers, does Dr Reid consider it likely that these can be attributed to shortages of small pelagic fish?

Secondly, Dr Reid introduced his talk by showing how varied the hydrology of the North Sea is, and then he gave global values for the whole system. What is known about the patchiness in the planktonic communities dependent food chains? Could local shortages of food now be beginning to affect seabird populations?

P. C. REID. In answer to the first part of Professor Dunnet's question, annual net production for seabirds in the North Sea is approximately 0.01 $kcal_{th}$ m^{-2} per year if it is taken as one third of Professor Dunnet's estimate of biomass and is thus an insignificant part of the total energy budget. Seabirds may, however, have an important impact on fish stocks as their food requirements are large. I have estimated this impact, using Professor Dunnet's estimate of biomass, in two ways, both of which give similar results.

First, I have applied the annual energy requirements for individual seabird species determined for Foula in the Shetlands by Furness (1978) to the total North Sea population, assuming the same porportions of biomass between species as in table 1 of Bourne (1983). Approximately 50% of North Sea seabirds as biomass (auks, gannets, terns, kittiwakes, shags, cormorants) feed almost totally on fish and most are selective for small fish. A small proportion of the diet of the remaining species, which are primarily scavengers, is also made up of fish. An estimated total of 60% (6 kt) of the seabird population feeding directly on fish is used here which equals a food requirement of 0.74 $kcal_{th}$ m^{-2} per year.

Second, by converting your biomass estimate to kilocalories (1.8 $kcal_{th}$ = 1 g live mass, a provisional conversion factor from Dunn & Brisbin (1980), and assuming the population eats fifty times its weight in a year (from Bourne (1983), considered an underestimate by Furness (1984)), and again considering 60% of the population as direct feeders on fish, a food requirement of 0.92 $kcal_{th}$ m^{-2} per year was determined.

The greater of these two estimates represents approximately 10% of the annual mean fish production in the period 1965 to 1969 and 13% in the period 1978 to 1982. Most of this predation would, however, have been on small pelagic species and would have represented 66% of the production of these species in 1965 to 1969 (a period when both stocks and fishing mortality were small) or a more realistic 19% in the period 1978 to 1982 when small 'industrial' fish species formed a larger and relatively more important component of total fish stocks. Seabirds may therefore have a major impact on North Sea fish stocks, an impact that is likely to increase if stocks of small pelagic species continue to decline.

To answer the second part of Professor Dunnet's question, guillemots and kittiwakes constitute approximately 40% of the total seabird biomass of the North Sea and have a diet that is primarily made up of small fish. Their population changes and breeding success are therefore likely to be closely linked to changes in fish stocks (Coulson & Thomas 1985; Birkhead 1986). Sprats, for example, which are a major winter food for kittiwakes (Coulson & Thomas 1985) have shown a major decline in the total North Sea stock due to poor recruitment (Anon. 1985); this decline has been especially evident off the east coast of Scotland since 1979 (McKay 1984).

Thirdly, little is known of the scales of variability and temporal change in patchiness of the plankton in the North Sea because of the logistics of sampling. Sampling by the Continuous Plankton Recorder smoothes out the effects of patchiness; trends determined from this data are consistent over the whole of the North Sea. Therefore any shortage of food linked to planktonic changes is likely to be experienced over a large area. Although there may be variability in sizes of local fish stocks between individual years, longer-term changes are also evident over the whole of the North Sea.

I am grateful to Mr A. W. G. John and Dr R. W. Furness for providing me with sources and discussion on North Sea birds.

References

Anon. 1985 Industrial Fisheries Working Group Report ICES CM 1985/Assess: 8. (Unpublished manuscript.)
Birkhead, T. 1986 Feeding ecology of common guillemots on Fair Isle, 1985; *Bull. Br. ecol. Soc.* **17**, 13–15.
Bourne, W. R. P. 1983 Birds, fish and offal in the North Sea. *Mar. Pollut. Bull.* **14**, 294–296.
Coulson, J. C. & Thomas, C. S. 1985 Changes in the biology of the Kittiwake *Rissa tridactyla*: a 31-year study of a breeding colony. *J. Anim. Ecol.* **54**, 9–26.
Dunn, E. H. & Brisbin, I. L. 1980 Age-specific changes in the major body components and caloric values of Herring Gull chicks. *Condor* **82**, 393–401.
Furness, R. W. 1978 Energy requirements of seabird communities: a bioenergetics model. *J. Anim. Ecol.* **47**, 39–53.
Furness, R. W. 1984 Seabird biomass and food consumption in the North Sea. *Mar. Pollut. Bull.* **15**, 244–248.
McKay, D. W. 1984 Sprat larvae off the east coast of Scotland. ICES CM 1984/:56. (Unpublished manuscript.)

Phil. Trans. R. Soc. Lond. B **316**, 603–623 (1987)
Printed in Great Britain

Molecular, cellular and physiological effects of oil-derived hydrocarbons on molluscs and their use in impact assessment

By M. N. Moore, D. R. Livingstone, J. Widdows, D. M. Lowe and R. K. Pipe

Natural Environment Research Council, Institute for Marine Environmental Research, Prospect Place, The Hoe, Plymouth, PL1 3DH, U.K.

The impact of pollutants on an organism is realized as perturbations at different levels of functional complexity. This presentation considers responses to oil-derived hydrocarbons at the molecular, subcellular, cellular and whole animal levels of organization, with particular emphasis on the use of marine molluscs as sentinel organisms for assessing pollutant effects.

A number of biological effects measurements are described which have been used in the development of early-warning systems based upon reactions to hydrocarbon-induced damage. These include those of the microsomal cytochrome P-450 dependent monooxygenase system involved in metabolism of organic xenobiotics, functional and structural responses of lysosomes to hydrocarbons, quantitative structural alterations in the cells of the digestive and reproductive systems and effects on physiological scope for growth and parallel correlations of this latter parameter with ecological parameters such as species diversity. Aspects of recovery processes are also considered.

Examples of both laboratory and field studies are cited to illustrate both the application of these approaches and the functional integration of the responses at the various levels of biological organization. This ability to link the various parameters in a functional manner is believed to strengthen the rationale for their use in impact assessment.

Introduction

The measurement of the impact of pollutants on marine organisms can be considered at different levels of functional complexity, namely, from the molecular level to the levels of the individual or population (Bayne *et al.* 1985). This presentation is concerned with the determination of the biological effects of petroleum-derived hydrocarbons in marine molluscs at the molecular, sub-cellular, cellular, tissue and individual physiological response levels, and their use in impact assessment.

The biological and toxicological bases of such responses have been described in considerable detail in several recent reviews and will not be reiterated in this presentation (Livingstone 1985; Moore 1985; Widdows 1985; Moore *et al.* 1987). Rather, we intend to focus on the use of such responses as indices of effect in relation to the environmental impact of petroleum hydrocarbons by using examples related to the North Sea where possible. The measurements of alterations in functional responses of organisms should form an important component of any environmental assessment programme; these range from relatively specific responses at the molecular and cellular levels, to more general (non-specific) responses to the sum of the environmental stimuli at the level of whole-animal physiological status (Livingstone 1985; Moore 1985; Widdows 1985). In this respect, measurements at the various levels of organization are considered to be complementary, and where possible they should have ecological significance, by implication, in terms of an adverse effect on growth, reproduction or survival of the individual and the population.

The detailed rationale for this approach has been outlined elsewhere on numerous occasions and does not require repetition (Bayne *et al.* 1979, 1982, 1985; Livingstone 1985; Moore 1985; Widdows 1985; Moore *et al.* 1987). Instead, we will outline a number of examples taken from laboratory experiments, experimental mesocosms and field situations, which will serve to exemplify the use of certain marine molluscs as 'indicator species' or 'sentinel organisms' in environmental monitoring. The fundamental concepts involved and many of the practical procedures are probably applicable to a wide range of species.

Molecular responses of the microsomal cytochrome P-450 monooxygenase system

Many contaminant organic compounds including petroleum hydrocarbons enter the marine environment and are dispersed by a variety of processes including uptake into the tissues of organisms (see review by Stegeman (1981)). Such compounds are often lipid soluble and toxic, and consequently mechanisms have evolved for their detoxication or elimination or both from the animal. This is facilitated by a number of apparently universally distributed enzyme systems that function to convert non-polar (lipid-soluble) compounds to more water-soluble and, hence, readily excretable metabolites. The metabolism of compounds such as polycyclic aromatic hydrocarbons, for example, has been classified into biotransformation (phase I) and conjugation (phase II) reactions. Phase I involves oxidation by various monooxygenase reactions including epoxidation, hydroxylation and dealkylation and is catalysed by the cytochrome P-450 monooxygenase or mixed function oxidase (MFO) system; the resulting products may then be converted to dihydrodiols or conjugated with glutathione and certain other small molecules or both (Stegeman 1981).

Paradoxically, during the course of these metabolic transformations, reactive electrophilic intermediates may be formed which are more toxic, mutagenic or carcinogenic than the parent compounds and the phase I and phase II systems must be viewed as part of a detoxication–toxication system, the protective value of which rests on the balance of the enzymes present and the reactive chemistry of the metabolites produced (Stegeman 1981; Bresnick 1982). An important feature of the system is that the activities of the enzymes and the concentration of cytochrome P-450, for example, may be increased by exposure of the animal to certain organic chemicals. This phenomenon is termed induction and it has been proposed to use this stimulation of components of the detoxication–toxication system as a possible method of monitoring the biological impact of organic chemical contaminants, particularly polycyclic aromatic hydrocarbons, in marine organisms (Payne 1977; Moore 1979; Stegeman 1981; Livingstone 1985).

There have been relatively few studies of the responses of the detoxication–toxication system of molluscs to organic chemical pollution but these encompass several species of mollusc and a number of contaminants, some of them oil-derived (see recent reviews by Livingstone (1985); Moore *et al.* (1987)). Increases in the content of cytochromes P-450 and b_5 in digestive gland microsomes (fragments of endoplasmic reticulum) of *Mytilus galloprovincialis* have been indicated in response to laboratory exposure to paraffins, anthracene, perylene and 3-methyl-cholanthrene, and in field exposures to hydrocarbons (Gilewicz *et al.* 1984). Enhanced aryl hydrocarbon hydroxylase activity has been reported in the digestive gland of oysters (*Crassostrea virginica*) in response to benzo(*a*)pyrene, 3-methylcholanthrene and PCBs (Anderson 1978), and in the digestive gland of mussels and other bivalves exposed to PCBs, polybrominated biphenyls and petroleum hydrocarbons (Payne *et al.* 1983).

The only investigation in which the responses both of the components (cytochrome P-450 and b_5 and cytochrome P-450 reductase) of the mixed function oxygenase (MFO) system and of the MFO activity (benzo(*a*)pyrene hydroxylase (BPH)) have been examined involved the exposure of mussels for four months to 30 p.p.b.† by volume diesel oil under conditions closely resembling the field situation (outdoor flow-through tanks, natural sea water and food, wave and tidal simulation) (Livingstone *et al.* 1985). The results are summarized in table 1 and show a doubling in the concentrations of cytochromes P-450 and b_5 and increased activities of NADPH cytochrome *c* reductase (NADPH-CYTCRED) and NADPH-neotetrazolium reductase (NTR). Cytochemically measured NTR activity and biochemically measured NADPH-CYTCRED activity are generally thought to measure some aspect of the *in situ* catalytic activity of NADPH-cytochrome P-450 reductase (Masters & Okita 1982). Although these findings are encouraging in the search for relatively specific effects indices, considerable work is still required before a generally acceptable diagnostic test for the effects of petroleum-derived chemical inducers is obtained from the responses of the molluscan detoxication–toxication system, as well as an understanding of the biological consequences of induction (Livingstone 1985).

The activity of NTR has been used extensively as a cytochemical indicator of induction by toxic organic chemicals in mammals (Smith & Wills 1981). Cytochemically determined activity of this enzyme in molluscan blood cells and digestive cells has been used as a relatively specific index of response to certain organic chemical contaminants, and can be measured in sections of supercooled tissues by microdensitometry (Chayen 1978; Moore 1979). It has been shown to be stimulated experimentally by a number of polycyclic aromatic hydrocarbons (PAHs), and by contamination in the immediate vicinity of a major oil terminal at Sullom Voe (Shetland Islands, U.K.) (Moore 1979; Moore *et al.* 1982, 1984, 1986; Widdows *et al.* 1984). These latter findings are illustrated in table 2.

A recent field investigation involving both biochemical and cytochemical approaches was done after the *Sivand* oil spill in the Humber estuary, U.K., in 1983. Cockles (*Cerastoderma edule*) were used in this study and the sample sites are illustrated in figure 1 and the methodology is described in Appendix 1.

The biochemical results are presented in table 3. Biochemical similarities were obvious between Horseshoe Point and Humberstone Fitties (the two sites on the southern coast farthest from the oil spill) and between Cleethorpes and Grimsby (the two sites nearest the oil spill and also located in industrialized areas). For statistical analysis the data were, therefore, pooled for each pair of sites and considered to be representative of a clean site (Horseshoe Point–Fitties) and a contaminated site (Cleethorpes–Grimsby). The data for Cleethorpes–Grimsby and for Spurn Bight (the previously pristine but oil-impacted site as a consequence of the spill) were compared with those of Horsehoe Point–Fitties by one-way analysis of variance. Microsomal protein yield was reduced at the contaminated sites ($p < 0.05$) and possibly also at Spurn Bight. Cytochrome P-450 was detected only at the contaminated sites but the '416' peak of

† In this paper one billion is used to represent 10^9.

TABLE 1. RESPONSES OF THE DIGESTIVE GLAND MICROSOMAL MFO SYSTEM OF (*MYTILUS EDULIS*) EXPOSED TO *CA*. 30 μg l^{-1} DIESEL-OIL HYDROCARBONS FOR FOUR MONTHS (MEAN ± STANDARD ERROR OF MEANS WITH NUMBER OF SAMPLES IN BRACKETS: EACH SAMPLE IS THE POOLED TISSUE OF EIGHT MUSSELS)

microsomal parameter	control	exposed
total protein†	5.21 ± 0.60 (6)	5.21 ± 0.29 (7)
NADH-FERRIRED‡	742 ± 33 (6)	772 ± 18 (7)
NADH-CYTCRED‡	95.4 ± 8.1 (4)	116.4 ± 4.3** (7)
NADPH-CYTCRED‡	10.3 ± 1.2 (4)	13.5 ± 0.9** (5)
P-450§	46.6 ± 14.3 (5)	92.0 ± 10.9** (7)
b_5§	25.8 ± 2.3 (5)	40.2 ± 5.8* (5)
BPH¶	54.1 ± 21.7 (6)	56.4 ± 3.8 (5)
NADPH-NTR‖	11.1 ± 1.0 (5)	15.3 ± 0.6** (5)

NADH-FERRIRED, NADH-ferricyanide reductase activity; NADPH-CYTCRED, NADPH-cytochrome *c* reductase activity; BPH, benzo(*a*)pyrene hydroxylase activity; NADPH-NTR-neotetrazolium reductase.
* $p \leqslant 0.1$.
** $p \leqslant 0.05$ (one-way analysis of variance).
† Milligrams per gram wet mass.
‡ Nanomoles per minute per milligram protein.
§ Picomoles per gram protein.
¶ Arbitrary fluorescence units per unit time.
‖ Relative units of absorbance.
Data from Livingstone *et al.* (1985). Mussels (4–5 cm length) were exposed in outdoor flow-through tanks at Solbergstrand Experimental Station, Norway.

the cytochrome P-450 spectra (there is evidence to suggest this is cytochrome P-420, the denatured form of P-450 (Livingstone *et al.* 1985)) was greatly elevated at Spurn Bight, i.e. sevenfold and fourfold greater than at Horseshoe Point–Fitties and Grimsby–Cleethorpes respectively. Cytochrome b_5 content and NADH-FERRIRED ($p < 0.1$) and NADH-CYTCRED ($p \leqslant 0.05$) activities were higher at Spurn Bight than at the two pooled groups, whereas NADPH-CYTCRED activity was elevated both at the contaminated sites ($p \leqslant 0.05$) and at Spurn Bight ($p \leqslant 0.1$). No differences were seen in BPH activities although only one measurement was available for Spurn Bight. The whole tissue PAH concentrations were markedly higher at both the contaminated sites and at Spurn Bight than those at Horseshoe Point (table 4, Oct. 1983). Elevated polycyclic aromatic hydrocarbon (PAH) concentrations, however, were also present at Humberside Fitties (table 4). The biochemical differences observed between the sites are consistent with results obtained so far for other gastropod and bivalve molluscs exposed to hydrocarbons, i.e. digestive gland microsomal cytochrome P-450 and b_5 content and NADPH-cytochrome *c* reductase activity increased in *Mytilus edulis* and *Littorina littorea* exposed to diesel-oil (Livingstone & Farrar 1985; Livingstone *et al.* 1985) and NADPH-cytochrome *c* reductase activity increased in the dog-whelk *Thais haemastoma* exposed to a water-soluble fraction of Louisiana crude-oil (Livingstone *et al.* 1986). The increased levels

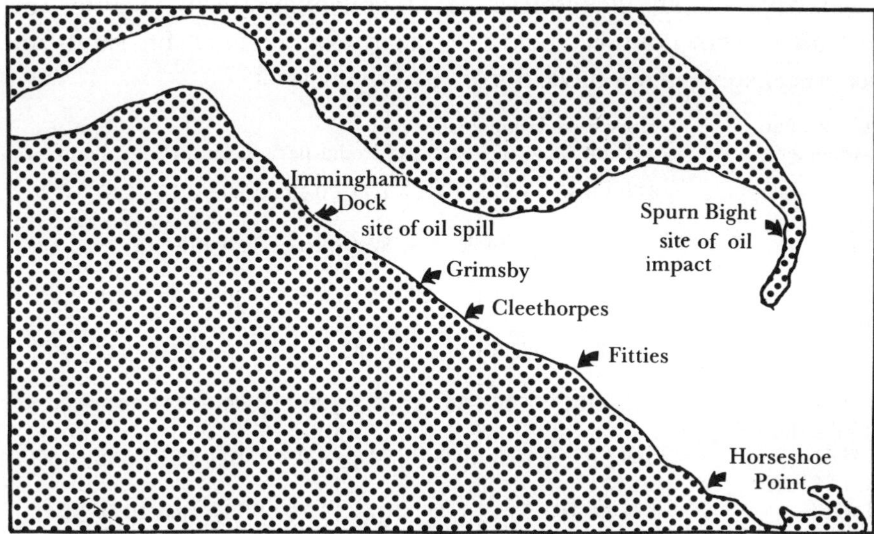

FIGURE 1. Humberside region showing location of populations of *Cerastoderma edule* at Grimsby, Cleethorpes, Humberstone Fitties, Spurn Bight (all sites of oil contamination), Horseshoe Point (uncontaminated reference site), and the site of the *Sivand* oil spill at Immingham Dock (October 1983).

TABLE 2. RESPONSES OF NADPH-NEOTETRAZOLIUM REDUCTASE (NTR) IN THE DIGESTIVE CELLS OF THE PERIWINKLE *LITTORINA LITTOREA* FROM THE VICINITY OF THE SULLOM VOE OIL TERMINAL

(Adapted from Widdows *et al.* 1984.)

sample site (July 1984)	NTR activity†
Gluss Voe (clean reference)	9.1 ± 5.9
Mavis Grind (contaminated reference)	18.3 ± 6.8†
Scatsta Voe	14.2 ± 3.6
Tanker Jetty 4 (contaminated reference)	30.7 ± 5.5‡
Ronas Voe	17.3 ± 7.5

† Relative absorbance ± standard deviation.
‡ $p \leqslant 0.05$, (Mann–Whitney U test, comparing with Gluss Voe reference site) ($n = 5$).
§ $p \leqslant 0.01$.

of the '416' peak in Spurn Bight cockles are interesting and may be due to elevated concentrations of cytochrome P-450 that has subsequently been denatured to cytochrome P-420, either *in vivo* or *in vitro* during homogenization, etc., possibly as a result of the action of hydrolases released from lysosomes destabilized by hydrocarbons. The correlation of elevated P-450 and P-450 reductase with tissue PAH body-burden was only partly due to the high PAH concentrations at Fitties. However, this site was nearer to the oil spillage than Horseshoe Point (figure 1) and it may be that some impact was made at the site but it was insufficient, or too early, to have affected the MFO system.

The activity of NTR was determined cytochemically in the digestive cells of cockle digestive glands. Sites were not pooled like the biochemical data. The results are presented in table 5. There were significant increases in activity at the Cleethorpes and Grimsby sites compared with

TABLE 3. DIGESTIVE GLAND MICROSOMAL MFO COMPONENTS AND ENZYME ACTIVITIES IN POPULATIONS OF THE COMMON COCKLE, *CERASTODERMA EDULE* L., FROM THE HUMBER ESTUARY, SAMPLED WITHIN 8 DAYS OF THE *SIVAND* OIL SPILL

microsomal parameter	Horseshoe Point	Humberstone Fitties	Cleethorpes	Grimsby	Spurn Bight
microsomal protein†	2.00±0.48 (2)	3.37±0.54 (3)	1.47±0.13 (3)	1.80±0.14 (3)	1.87±0.26 (3)
P-450‡	n.d. (1)	n.d. (2)	82.4±2.4 (2)	39.0±18.7 (2)	n.d. (2)
'416' peak§	25 (1)	24±5 (2)	61±12 (2)	36±3 (2)	178±76 (2)
b_5‡	79.7 (1)	92.8±17.1 (2)	92.5 (1)	93.7±2.1 (2)	147±19 (2)
NADH-FERRIRED¶	712±131 (2)	746±36 (3)	518±20 (3)	669±103 (3)	949±86 (3)
NADH-CYTCRED¶	109±8 (2)	98.3±9.9 (3)	90.2±7.0 (3)	110±5.6 (3)	142±16.3 (3)
NADPH-CYTCRED¶	6.74±2.16 (2)	5.26±1.29 (3)	10.54±1.19 (3)	9.13±1.66 (3)	10.56±2.14 (3)
BPH‖	9.1±6.9 (2)	5.8 (1)	8.2±2.3 (2)	10.2±0.3 (2)	6.1 (1)

Values are means ± standard error of means or ± range (numbers of samples in parentheses).
† Milligrams per gram wet mass.
‡ Picomoles per milligram protein.
§ Arbitrary units mg^{-1} protein.
¶ Nanomol per minute per milligram protein.
‖ Picomoles of total metabolites per minute per milligram protein.
n.d. Not detectable.

TABLE 4. CONCENTRATION OF TWO- AND THREE- RINGED POLYCYCLIC AROMATIC HYDROCARBONS IN WHOLE TISSUES OF *CERASTODERMA EDULE* FROM THE HUMBER

site	date	concentration of two- and three-ringed aromatic hydrocarbons†
Horseshoe Point	Oct. 83	2.35±0.01
	May 84	0.48±0.03
Humberstone Fitties	Oct. 83	20.1±3.0
Cleethorpes	Oct. 83	27.7±2.0
	May 84	2.36±0.11
Grimsby	Oct. 83	35.5±1.4
Spurn Bight	Oct. 83	14.5±1.2
	May 84	0.76

† Micrograms per gram wet mass (measured in terms of 2,3-dimethylnaphthalene and 1-methylphenanthrene).

the clean reference site at Horseshoe Point. Both of these sites had highly elevated levels of two- and three-ring polycyclic aromatic hydrocarbons (table 4). These findings for NTR are broadly similar to the effects reported for NADPH-cytochrome *c* (P-450) reductase (table 3).

Samples collected seven months after the *Sivand* oil spill showed that there was still significant elevation of NTR activity at Cleethorpes. However, the data should be interpreted with caution as there is apparently a seasonal change in activity based upon the results for the clean reference site at Horseshoe Point (table 5). Both two- and three-ring PAH values were still elevated at Cleethorpes compared with the clean reference site (table 4).

TABLE 5. NADPH-NEOTETRAZOLIUM REDUCTASE (NTR) IN HEPATOPANCREATIC DIGESTIVE CELLS IN *CERASTODERMA* FROM THE HUMBER

site	NTR activity† Oct. 1983	May 1984
Horseshoe Point	12.2±4.5	0±0
Humberstone Fitties	10.2±4.5	
Cleethorpes	26.6±5.4§	4.8±3.4‡
Grimsby	30.8±16.1‡	
Spurn Bight	20.5±14.3	16.5±15.6 ($n = 3$)

† Relative absorbance ± standard deviation.
‡ $p \leq 0.05$, Mann–Whitney *U*-test, comparing with Horseshoe Point reference ($n = 5$).
§ $p \leq 0.01$.

These findings are comparable with data from laboratory experiments, mesocosm studies (Solbergstrand) and Sullom Voe investigations on both mussels (*M. edulis*) and periwinkles (*L. littorea*) and are indicative of a definite effect of the oil spill on the cockles (Livingstone *et al.* 1985). Owing to the severe limitations in the data available, however, it would be imprudent to attempt to glean too much concerning the longer-term effects of the oil spill on cockle NTR activity.

FUNCTIONAL DISTURBANCES OF SUBCELLULAR ORGANIZATION: RESPONSES OF THE LYSOSOMAL SYSTEM

Subcellular and cellular pathology reflect perturbations of structure and function at the molecular level, hence, responses of lysosomes represent a logical hierarchical progression in terms of biological organization from the previous section. In many instances, the earliest detectable changes or 'primary events' are associated with a particular class of subcellular organelle (Slater 1978) such as the lysosome or endoplasmic reticulum (the microsomes of the preceding section). Investigations in mammals have revealed that much of the damaging action of environmental contaminant chemicals (xenobiotics) is produced by highly reactive metabolic derivatives. It is these activated chemical forms, generally produced by the cytochrome P-450 system, that are responsible for the initiation of what may be termed the primary intracellular disturbances. These may spread rapidly into a complex network of associated secondary and higher order disturbances which become progressively more difficult for the cell to reverse or modify. The degree to which metabolic derivatives of PAHs are responsible for toxicity, as opposed to the parent compounds, is difficult to ascertain given the current limitations of understanding of the rates and patterns of metabolic production in molluscs (see review by Moore *et al.* 1987). The numerous ways in which the structure and function of organelles and cells can be perturbed by contaminants such as PAHs have been reviewed in detail by Moore (1985) and Moore *et al.* (1987). In this presentation we are mainly concerned with injury induced by PAHs to the lysosomal system where membrane damage results in major changes in the structure and function of the lysosomal compartment (Moore 1985).

Mammalian lysosomes are noted for their responsiveness to many types of cell injury and molluscan lysosomes have also been shown to compartmentalize and accumulate a wide range of injurious agents including petroleum-derived PAH (Allison 1969; Baccino 1978; Hawkins 1980; Ericsson & Brunk 1975; Moore 1985; Moore *et al.* 1987). The cells of many molluscan tissues are especially rich in lysosomes (Sumner 1969; Owen 1972; Moore *et al.* 1987), particularly the digestive cells of the digestive gland.

There are a number of ways by which lysosomes can react to cellular injury; basically, these can be divided into three categories of decrease or increase in:

(1) lysosomal contents such as hydrolytic degradative enzymes and lipofuscin pigment;

(2) rate of membrane fusion events with either the cell membrane or other components of the vacuolar system, and

(3) lysosomal membrane permeability (Hawkins 1980).

For a number of reasons, however, lysosomal reactions to cell injury are not well understood owing to the many forms of cell injury and the wide variety of organisms studied. Moreover, one type of lysosomal alteration may assume the appearance of another type (Hawkins 1980; Ericsson & Brunk 1975). These three categories of lysosomal response have been discussed in detail by Moore *et al.* (1987) and consequently will only be summarized here.

Exposure of *L. littorea* to PAH has been demonstrated to result in increases in the activities of certain lysosomal enzymes, notably β-glucuronidase and acid phosphatase (Moore *et al.* 1982, 1986). A consequence of this pattern would be to prepare the cell for the degradation of particular macromolecules, hence making the products available for maintenance of the cell (Hawkins 1980; Moore *et al.* 1980). Regulation of the lysosomal system is dependent upon controlled fusion with other components of the vacuolar system such as phagosomes, primary lysosomes and the plasma membrane (Hawkins 1980). The digestive cells of molluscs are largely concerned with heterophagy and the digestion of food material (Owen 1972). Disturbances of the fusion processes involved could have marked consequences for the nutritional status of the organism by perturbing 'normal' intracellular digestion and the balance of autophagy to heterophagy (Moore 1980).

There are a number of indications that both experimental and field exposure to oil-derived PAHs induces profound alterations in the rate of fusion events in the lysosomal–vacuolar systems of molluscan digestive cells (see reviews by Moore (1985); Moore *et al.* (1987)). Ultrastructural studies show that the large secondary lysosomes (2–5 μm diameter approximately) in the digestive cells show marked increases in the presence of internalized membrane-bound vesicular components and that these secondary lysosomes become abnormally enlarged (up to 15 μm diameter) in mussels, oysters and periwinkles (Pipe & Moore 1986; Lowe *et al.* 1981; Moore *et al.* 1986; Couch 1984). Quantitative stereological analysis of these enlarged lysosomes (light microscopy) demonstrates that both lysosomal volume and surface area within the cells is significantly increased, while numerical density is decreased (Lowe *et al.* 1981). This is perhaps indicative of fusion of vacuolar components to produce these abnormally enlarged lysosomes. This type of response has also been observed to occur when mussels are exposed to an abrupt increase in salinity and has been linked to increased fusion of lysosomal vacuoles and autophagy (Pipe & Moore 1985*a*). These alterations are also associated with elevated intracellular protein catabolism and formation of amino acids as measured within the lysosomal cellular compartment (Bayne *et al.* 1981). The formation of these enlarged lysosomes has also been linked with atrophy of the digestive tubule epithelium which largely consists of digestive cells (Lowe *et al.* 1981) and this relationship will be discussed later in more detail.

The third category of lysosomal disturbance involves membrane permeability. This property can be investigated both biochemically and cytochemically. It is, however, more realistic to employ cytochemical procedures in molluscan digestive cells as the large secondary lysosomes do not readily lend themselves to the trauma of homogenization and subsequent fractionation (Bitensky *et al.* 1973). Cytochemical procedures are well established for a number of enzyme substrates in molluscs and these are all conceptually based on the Bitensky fragility test (see review by Bitensky *et al.* (1973)) for lysosomal stability as modified by Moore (1976, 1985). Lysosomal destabilization is measured as increased permeability to certain enzyme substrates whose products can be used to give a measurable final cytochemical reaction product (Moore 1976).

Destabilization of lysosomes has been demonstrated in molluscan digestive cells as a result of injury by 2-methyl naphthalene, 2,3-dimethyl naphthalene, anthracene, phenanthrene, diesel-oil emulsion, water accommodated fraction of crude oil (North Sea: Auk) and PAH-contaminated field samples from Shetland (table 6) (Moore 1985; Livingstone *et al.* 1985). Assessment of this type of injury has been confirmed as an extremely sensitive index of cellular condition and the destabilization of the lysosomal membrane appears to bear a quantitative relation to the magnitude of the stress response and this presumably contributes to the intensity of catabolic or degradative effects, as well as to the level of pathological change that results (Moore 1985).

TABLE 6. LYSOSOMAL STABILITY IN THE DIGESTIVE CELLS OF THE PERIWINKLE (*LITTORINA LITTOREA*) EXPOSED TO PAH IN AN EXPERIMENTAL MESOCOSM (SOLBERGSTRAND, OSLO FJORD) AND IN THE VICINITY OF THE SHETLAND OIL TERMINAL (SULLOM VOE, JULY 1982)

(Adapted from Moore *et al.* 1986*b* and Livingstone *et al.* 1985.)

experimental treatment or sample site		lysosomal membrane stability†
control		24 (20, 25)‡
ca. 30 μg l^{-1} total hydrocarbons	16 months exposure	2.6 (2, 5)§
ca. 130 μg l^{-1} total hydrocarbons	16 months exposure	2 (2, 2)§
Ronas Voe	(uncontaminated reference)	19.0 (15, 20)
Gluss Voe	(uncontaminated reference)	24.0 (20, 25)
Scatsta Voe		5.0 (5, 5)§
Tanker Jetty 4		3.8 (2, 5)§
Mavis Grind (contaminated reference)		8.0 (5, 10)§

† β-glucuronidase labilization period in minutes.
‡ Mean with data range in parentheses ($n = 5$).
§ $p \leqslant 0.01$, Mann–Whitney *U*-test comparing with either control data (experimental) or data for Ronas Voe (uncontaminated reference).

Further evidence of this type of effect has been obtained with cockles (*C. edule*) exposed to hydrocarbons in the Humber estuary following the *Sivand* oil spill. Contaminated cockles showed evidence of lysosomal destabilization (tables 4 and 7). Samples taken seven months later indicated that there was evidence of recovery at previously contaminated sites (tables 4 and 7); however, as all of these data represent small samples and two periods in time caution must be exercised in their interpretation.

TABLE 7. LYSOSOMAL MEMBRANE STABILITY IN HEPATOPANCREATIC DIGESTIVE CELLS OF *CERASTODERMA EDULE* FROM THE HUMBER

site	lysosomal membrane stability Oct. 1983	May 1984
Horseshoe Point	20 (20, 25)†	24 (20, 25)
Humberstone Fitties	23 (20, 25)	
Cleethorpes	15 (15, 15)‡	23 (20, 25)
Grimsby	10 (10, 10)‡	
Spurn Bight	17 (10, 20)	23.3 (20, 25)
		$n = 3$

† Labilization period of latent lysosomal β-glucuronidase in minutes (range in parentheses) $n = 5$.
‡ $p \leqslant 0.01$, Mann–Whitney U-test, comparing with Horseshoe Point (uncontaminated reference).

The consequences of destabilization of the secondary lysosomal compartment have been investigated in several experimental studies. Subcellular fractions rich in destabilized secondary lysosomes have been shown to contain significantly increased concentrations of amino acids compared with stable lysosomes (Bayne *et al.* 1981). This is indicative of enhanced intralysosomal protein catabolism. Ultrastructural investigations of cells with destabilized lysosomes indicate increased secondary lysosomal volume with evidence of increased autophagy and possible heterophagy of apoptotic cell fragments, thus providing further indications of elevated catabolic activity (Pipe & Moore 1985*a*, 1986).

A quantitative cytochemical approach such as the measurement of lysosomal permeability or stability, based on substrate penetrability (hydrolase latency), may also provide insight into the mechanisms of PAH-induced cell injury in molluscs. Caution is required, however, in the interpretation of lysosomal damage as a primary event, when it may in fact be a secondary- or higher-order alteration. Recent investigations of lysosomal responses to specific PAHs have demonstrated that the lysosomal disturbances are complex and differ markedly for PAHs which are structurally dissimilar, such as the isomeric three-ring forms anthracene and phenanthrene (Moore & Farrar 1985). The biochemical evidence of relatively low activity for cytochrome(s) P-450 monooxygenase in molluscan digestive-gland cells, when considered together with the ability of these cells to accumulate and retain very high concentrations of PAH, indicates that their loss by metabolic transformation is limited (Livingstone 1985). The fact that the secondary lysosomes are often lipid-rich, particularly after exposure to PAHs (M. N. Moore, unpublished data) would tend to argue for a direct effect on the lysosomes by these xenobiotics, rather than a secondary effect. Aromatic hydrocarbons and PAHs have been shown to penetrate synthetic phospholipid membranes and alter their physical and chemical properties including membrane fluidity and permeability (Roubal & Collier 1975; Nott *et al.* 1985).

Further evidence of lysosomal destabilization comes from several ultrastructural studies of the effects of phenanthrene on secondary lysosomes in digestive cells (Pipe & Moore 1986; Nott *et al.* 1985). These have demonstrated the presence of corrugation of the bounding membrane with possible associated blebbing activity. Increased frequency of membrane breaks has also been described, and although these breaks may be artefacts of fixation and tissue processing, their relative infrequency in control lysosomes is indicative of the greater fragility of lysosomes from cells exposed to phenanthrene. Apparent leakage of lysosomal β-glucuronidase has been demonstrated in the case of lysosomes from the digestive cells of phenanthrene-exposed *L. littorea*; extracellular release of lysosomal β-glucuronidase was also observed in these cells

(Pipe & Moore 1986). Szego (1975) described limited release of lysosomal enzymes into the cytosol and nucleoplasm after treatment of rat preputial-gland cells with 17β-oestradiol, which destabilizes the lysosomal membrane.

The consequences of such a release of lysosomal enzymes is uncertain but is believed to lead to enhanced cell damage and possibly cell death (Hawkins 1980; Ericsson & Brunk 1975). Evidence from rat preputial-gland cells indicates increased protein catabolism after oestrogen treatment, although enzyme release in these cells is non-injurious and precedes initiation of cell division (Szego 1975). This may represent a fundamental difference in the lysosomal response to physiological agonists as opposed to xenobiotics.

In summary, the evidence of both ultrastructural, quantitative cytochemical and morphometric approaches indicates that PAHs induce profound alterations in both structure and function. These involve all three categories of lysosomal response and there are some grounds for suggesting that membrane destabilization may represent the primary injury which could lead to the other events described above. Cytochemical demonstration of lysosomal membrane destabilization has proved to be a useful investigative tool both for PAH and other injurious agents. That this procedure does in fact measure membrane destabilization is further supported by evidence of reversibility and restabilization by treatment with hydrocortisone, an established membrane stabilizer (Moore 1976; Bayne *et al.* 1981).

Structural alterations in cells

Exposure of marine molluscs to single PAH and oil-derived hydrocarbon mixtures results in atrophy of the epithelium of the digestive tubules (Lowe *et al.* 1981; Couch 1984; Moore *et al.* 1987). This atrophy or epithelial 'thinning' involves structural changes in the digestive cells, the major component of the epithelium. These changes have been quantified in *M. edulis* (Lowe *et al.* 1981) by using image analysis of histological sections, and in the clam *Mercenaria mercenaria* by using morphometry (Tripp *et al.* 1984). There is evidence in both mussels and periwinkles that this may be a generalized response to toxic xenobiotics and other stressors such as starvation (Pipe & Moore 1985*a*; Moore *et al.* 1987).

Investigation of digestive cell atrophy has revealed that there is a significant increase in lysosomal volume as described in the previous section (Lowe *et al.* 1981; Moore & Clarke 1982). This increase in volume of the lysosomal compartment involved the formation of enlarged or 'giant' lysosomes and this alteration is associated with membrane destabilization and increased permeability (Moore & Clarke 1982). There is also evidence for increased lysosomal fusion events leading to the formation of the enlarged lysosomes (Pipe & Moore 1985*a*, 1986). As discussed in the previous section the consequences of these lysosomal disturbances would be increased autolytic and autophagic activity presumably leading to atrophy of the digestive cells.

Physiological investigation of mussels has shown that scope for growth is significantly reduced (and in some cases becomes negative) following exposure to PAH and oil-derived hydrocarbons (Widdows *et al.* 1982; Bayne *et al.* 1979). This situation is indicative of relatively enhanced tissue catabolism. Samples from these experiments demonstrate digestive cell atrophy and lysosomal disturbances as described above, arguing strongly for a mechanistic link from the lysosomal events through to the whole animal (Lowe *et al.* 1981; Moore & Clarke 1982; Widdows *et al.* 1982).

Turning to consideration of effects on reproduction at the cellular level, mussels exposed to both *ca.* 30 and *ca.* 130 μg l^{-1} hydrocarbons showed a reduction in the volume of storage cells in the mantle tissue, a reduction in volume of ripe gametes and increased degeneration or atresia of oocytes (table 8) (Lowe & Pipe 1985). These data indicate a direct impairment of the reproductive processes and the implication is that reproductive capability would be reduced, both by degeneration of oocytes and reduction in ripe gametes, as well as by a reduction

TABLE 8. REPRODUCTIVE TISSUE IN THE MUSSEL: EFFECTS OF HYDROCARBON EXPOSURE AND A RECOVERY PERIOD ON TISSUE VOLUMES OF THE COMPONENT CELLS

(Adapted from Lowe & Pipe 1985.)

condition	developing gametes	ripe gametes	adipogranular cells	vesicular cells	degenerating oocytes
control	0.10±0.03†	0.77±0.27	0.11±0.03	0.53±0.07	0.05±0.01
ca. 30 p.p.b. by volume§	0.13±0.04	0.44±0.16‡	0.05±0.04†	0.38±0.07 ‡	0.35±0.13‡
ca. 130 p.p.b. by volume§	0.13±0.03	0.30±0.11‡	0.01±0.01‡	0.14±0.02‡	0.31±0.09‡
recovery 53 days	0.28±0.08‡	0.66±0.18	0.11±0.05	0.54±0.04	0.01±0.01‡

† Mean (cubic millimetres) ±standard error, $n = 10$.
‡ $p < 0.05$; one-way analysis of variance, comparing with control.
§ One billion is used to represent 10^9.

in the energy reserves available for gametogenesis as supplied by the connective tissue storage cells (Lowe *et al.* 1982). Ultrastructural investigations designed to explore the mechanisms of oocyte degeneration have revealed that degradative lysosomal enzymes are associated with yolk granules and with pinocytotic phenomena that occur along the basal membrane of developing oocytes (Pipe & Moore 1985*b*). Lysosomal enzymes are also associated with the degradation (atresia) and resorption of oocytes, as well as the resorption of adipogranular storage cells (Lowe *et al.* 1982; Pipe & Moore 1985*b*). Future experiments will test whether polycyclic aromatic hydrocarbons have a detrimental effect on oocyte lysosomes leading to enhanced lysosomal autolytic processes.

PHYSIOLOGICAL RESPONSES: EFFECTS OF OIL EXPOSURE AND RECOVERY

Production of matter (growth and reproduction) is a fundamental property of all living organisms and one that is necessary if a population is to persist in a given environment. The amount of production represents the difference between an individual's or a population's intake and output of matter or energy, and this will vary under different environmental conditions. The measurement of individual physiological responses, such as rates of feeding, digestion, respiration, excretion and growth, and their integration by means of physiological energetics, can provide insight into the overall growth process and how it might be disrupted by environmental stress and pollution; these have been extensively reviewed elsewhere in the literature (Bayne *et al.* 1985; Widdows 1985; Moore *et al.* 1987).

The ultimate effect of petroleum hydrocarbons on rates of feeding and energy metabolism is to markedly reduce the energy available for growth and reproduction, often termed 'scope for growth' (see review by Widdows (1985)). The effects of petroleum hydrocarbons observed in laboratory experiments are also apparent in mesocosm experiments and in the field. For example, *Mya arenaria* subjected to oil spills showed a reduction in the energy available for growth (Gilfillan *et al.* 1976, 1977) and a decline in tissue and shell growth (Gilfillan & Vandermeulen 1978; MacDonald & Thomas 1982) which persisted for up to five years after the spill owing to the continued presence of oil in the sediments.

Although there have been many studies of the effects of hydrocarbons on bivalves, until recently there has been relatively little information concerning the rate of recovery from oil exposure and the extent to which physiological recovery is related to the depuration of hydrocarbons from the body tissues. A recent study (Widdows *et al.* 1985) has shown a marked reduction in the feeding rate and scope for growth of *M. edulis* exposed to low concentrations of diesel-oil (*ca.* 30 and *ca.* 130 $\mu g\ l^{-1}$) for eight months (figure 2). The high-oil exposed mussels had a negative scope for growth, indicating the need to utilize body reserves in order to satisfy the animal's energy requirements. This was confirmed by the observed degrowth in body tissues which resulted in a high mortality in this group (27% over eight months, (Lowe & Pipe 1987)). During recovery from chronic oil exposure the depuration of hydrocarbons from the mussel's body tissues was concomitant with the recovery of physiological performance (i.e. feeding and growth), thus demonstrating that the effect of oil on the mussel's performance is related to the concentration of hydrocarbons within the body tissues and is not directly related to the hydrocarbon concentration in the water.

High-oil exposed mussels were found to recover more rapidly than the low-oil exposed mussels, both in terms of hydrocarbon depuration and scope for growth (figure 2) (Widdows *et al.* 1985). This led to an 'overshoot' in the feeding rate and consequently the scope for growth by high-oil exposed mussels and thus accounted for the observed 'catch up' growth in body tissues during the two months of recovery from high-oil exposure (*ca.* 130 $\mu g\ l^{-1}$). The rate of recovery by low-oil exposed mussels was slower both in terms of tissue depuration and physiological performance, but both groups showed complete recovery after 55 days (i.e. their growth rates were not significantly different from the control).

Several field and laboratory studies have demonstrated a significant negative correlation between scope for growth and the concentration of specifically aromatic hydrocarbons in the tissues of molluscs (see review by Widdows (1985)). Figure 3 provides a synthesis of data derived from a mesocosm experiment and a field study of mussel populations in the vicinity of the Sullom Voe oil terminal and illustrates the significant negative correlation ($r^2 = -0.95$) between scope for growth and log of the concentration of two- and three-ringed aromatic hydrocarbons in the body tissues of *M. edulis* (this simply reflects the nature of the analytical procedure and the dominant component of the accumulated aromatic hydrocarbons rather than identifying these aromatic hydrocarbons as the sole toxic components). Such a relation demonstrates that hydrocarbons affect scope for growth over a wide range of tissue concentrations without an apparent threshold concentration of effect. Moreover, it illustrates the degree of contamination in the 'control' mussels in the mesocosm experiment conducted at Solbergstrand on the Oslo fjord (Norway) relative to the Shetland Islands (U.K.).

It is generally very difficult to make an inter-study comparison of the relation between biological effects measurements and tissue hydrocarbon concentrations because of the lack of compatibility, particularly in the chemical data. This is primarily because of the different extraction, analytical and quantification procedures adopted by different laboratories, which

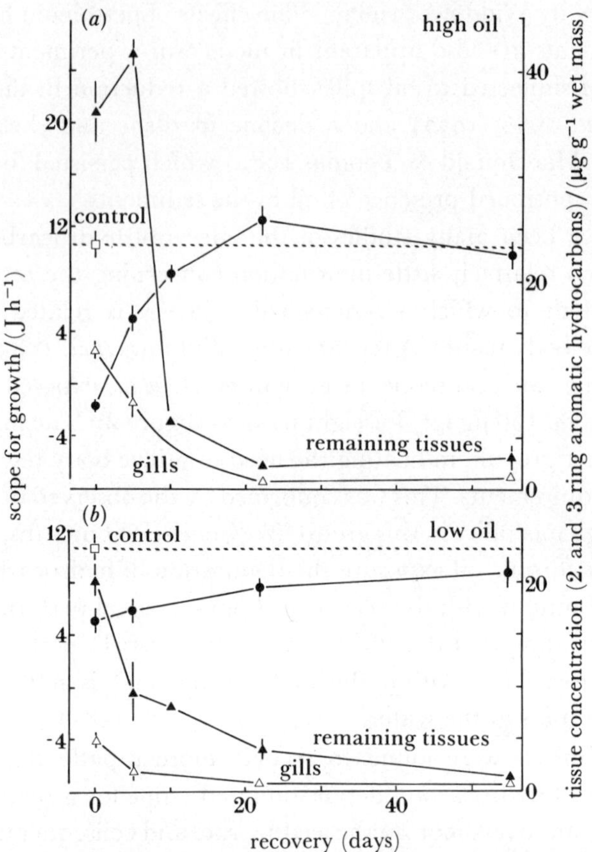

FIGURE 2. Scope for growth (● and □) (J h^{-1}; mean ± standard error) of *Mytilus edulis* (standard animal of 0.5 g dry mass) during recovery from eight months of exposure to *ca.* 130 µg l^{-1} (high-oil exposure) and *ca.* 30 µg l^{-1} (low-oil exposure). Short-term recovery (0, 5 and 10 days; May 1983) and long-term recovery (0, 22 and 55 days; May 1984) are expressed relative to control mussels (no significant differences between controls in 1983 and 1984). Depuration of two- and three-ringed aromatic hydrocarbons (µg g^{-1} wet mass) from gills (▲) and remaining tissues (△) (mean ± range of two pooled samples). Data from Widdows *et al.* 1985.

are usually semi-quantitative emphasizing either the lower or higher molecular mass fractions of aliphatic or aromatic hydrocarbons. A broad synthesis of data such as that achieved in figure 3, where there is consistency in both biological and chemical methodology, will not be possible until further standardization and intercalibration of procedures for the quantification of toxic hydrocarbons of low and high molecular mass has been agreed. However, to present the relation between scope for growth and the concentration of aromatic hydrocarbons in the tissues of mussels (figure 3) in the context of the degree of oil contamination in the North Sea we have inserted predicted tissue concentrations of two- and three-ring hydrocarbons based upon data for hydrocarbon concentrations in water (Massie *et al.* 1985). The conclusions drawn from these water and predicted tissue concentrations are that there is likely to be a very slight and probably measurable effect on the performance of mussels and other marine biota living in the water column in the vicinity of the North Sea oil platforms but that this degree of biological effect is not considered to result in a marked impact on the overall performance and survival of the mussel.

Reproductive processes provide the critical interface between the individual and the

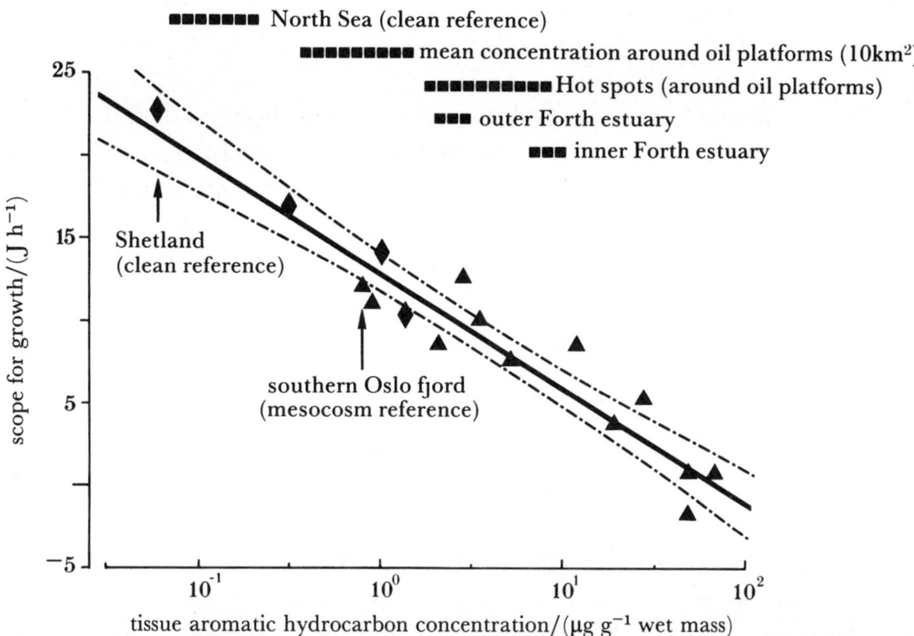

FIGURE 3. Relation between scope for growth (J h^{-1}; based upon a standard animal of 0.5 g dry mass, particulate organic matter of 0.5 mg l^{-1}, season of May–July) and the concentration of two- and three-ringed aromatic hydrocarbons (μg g^{-1} wet mass) in the body tissues of *Mytilus edulis*. (▲) Data from oil exposure and recovery experiments at Solbergstrand, Oslofjord, Norway (Widdows *et al.* 1985). (◆) Data from Sullom Voe, Shetland, U.K. (Widdows 1984). Horizontal bars represent an estimate of the tissue concentration based upon the observed concentration of hydrocarbons in the water at various sites in the North Sea (Massie *et al.* 1985).

population. Results of laboratory studies have indicated that stress, measured in terms such as scope for growth, has a significant effect on fecundity, egg size and subsequent growth rate of the larvae of *M. edulis* (Bayne *et al.* 1982). The reduction in energy available for growth and reproduction in response to hydrocarbon exposure (*ca.* 30 and *ca.* 130 μg l^{-1}) has been shown to result in a significant and concentration-dependent decline in both the mass of storage tissue in *M. edulis* and the mass of gametes produced (fecundity) (Lowe & Pipe 1985; Lowe & Pipe 1987). *Macoma balthica* exposed to 30 μg petroleum hydrocarbons per litre also showed gamete resorption and abnormal gamete development at 300 μg l^{-1} (Stekoll *et al.* 1980).

Therefore a decline in the overall energy available for production as a response to petroleum hydrocarbons results in a reduction in the energy allocated to reproduction (reproductive effort), which, when combined with an increased mortality, ultimately leads to a reduction in the residual reproductive value (Bayne *et al.* 1982) or the long-term prospects of producing successful offspring in future years.

POPULATION AND COMMUNITY EFFECTS

The damaging effects of petroleum hydrocarbons on cellular and physiological responses, which are ultimately reflected in the growth, reproduction and survival of the individual, suggests a major reduction in population fitness and the ability to maintain the population or colonize new habitats, as well as increased susceptibility to predation, parasitism and disease. Such population effects have been observed in both field (*Mya arenaria*: Gilfillan *et al.* (1977);

Gilfillan & Vandermeulen (1978)) and mesocosm studies (*Mytilus edulis*: Widdows *et al.* (1985); Bakke & Sorensen (1985)) in the form of reduced recruitment, increased mortality and reduced production.

The effects of oil on populations and communities have been studied in two major benthic mesocosm experiments (Grassle *et al.* 1981; Bakke & Sorensen 1987) with the aim of bridging the gap between laboratory based short-term toxicity experiments and field observations after oil pollution, and examining the effects of oil on trophic interactions in the ecosystem. The findings from these experiments have been reviewed by Moore *et al.* (1987). The results of mesocosm studies have largely confirmed the conclusions of earlier studies that bivalves, including mussels, are sensitive to petroleum hydrocarbons when chronically exposed.

The ecological consequences of changes in populations will be a shift in species composition and diversity within the community. The effects of petroleum hydrocarbons on community structure have been demonstrated in many field studies, e.g. effects of an oil spill (Sanders *et al.* 1980) industrial petroleum waste (McLusky 1982) and offshore oil platforms (Davies *et al.* 1984).

The community response to oil pollution is a classical successional response to a point-source of organic pollution (Pearson & Rosenberg 1978). There is a gradient of increasing species diversity, beginning with an abiotic zone in the immediate vicinity of a point-source of heavy contamination, followed by a zone characterized by a low number of species, low abundance and low biomass, and then a zone showing a progressive increase in the number of species and a dramatic increase in abundance and biomass of a few dominant species. Finally, with increasing distance from a discharge the community becomes more diverse but abundance and biomass declines.

In relation to North Sea oil activities there is significant offshore contamination of sediments close to platforms using oil-based drilling muds (Davies *et al.* 1981). A review of the environmental effects of these oil-contaminated sediments around oil platforms (Davies *et al.* 1984) has concluded that, beyond the immediate area of physical smothering by oil-based mud cuttings, the major deleterious effects on the benthic community (e.g. low species diversity and high numbers of opportunistic species such as *Capitella capitata*) occurred within 500 m of the platforms and beyond which there was a transition zone, generally within 400–1000 m, where 'subtle' biological effects (e.g. slight changes in species diversity and community structure) could be detected. Species diversity declined progressively with a logarithmic increase in oil concentration in the sediments and there was no evidence of a theshold of effect. Where there were quantitative gas chromatography–mass spectrometry data for two oil-fields (Beryl and Thistle), a linear relation between species diversity and log of naphthalene concentration in the sediment was also recorded. It is interesting to note that where hydrocarbon concentrations (regardless of units) span several orders of magnitude above background levels there is a parallel relation and good agreement between biological effects such as scope for growth, measured at the individual level of organization using an indicator species, and changes in species diversity at the community level (see review by Moore *et al.* 1987). This supports the notion that adverse effects measured at the cellular and individual levels ultimately manifest themselves at the population and community levels of organization. Cellular and physiological responses therefore serve to complement ecological surveys of community structure, not only by providing an early detection of effects, often before statistically significant changes occur at the community level, but also by providing some insight into the toxicological causes of changes in population and community structure. A particularly important attribute of sublethal cellular and physiological responses is that they are amenable to both laboratory and field measurement, unlike traditional toxicity testing based upon LC50 and ecological surveys based upon community structure. On the one hand they can be used to derive 'concentration–response' relations to help identify and predict the effects of potential pollutant levels; on the other hand they can be used to monitor the effects in the environment.

Conclusions

The application of an integrated approach in the use of indices of effect for the assessment of sublethal effects of petroleum-derived hydrocarbons in molluscs has been described. This rationale has been productive in enabling us to develop conceptual links between responses at various levels of biological organization, certainly from the molecular level to that of the individual. Study of the responses to oil-derived hydrocarbons has extended our understanding of certain aspects of physiological processes and has also highlighted the capacity for PAHs to disturb biochemical, cellular and physiological processes.

Although there is promising evidence of oil-induced effects on the cytochrome P-450 monooxygenase system, our understanding of its role in molluscs is still somewhat limited and considerable research will be required to characterize the enzymic components and patterns of hydrocarbon metabolism before we can assess the significance of this system in toxicity and detoxication. We have, perhaps, a better understanding of lysosomal function in molluscs and the consequences of disturbance by PAHs. Here, we see a logical progression from lysosomal dysfunction leading to enhanced catabolism resulting in epithelial atrophy manifesting in turn as a reduction in 'scope for growth' and perhaps a reduction in viable oocytes.

There is also a considerable requirement for a better understanding of the physiological and molecular processes involved in the uptake, retention, compartmentation and loss of oil-derived PAHs in molluscan tissues. Such an understanding should provide a toxicological interpretation of the chemical data for tissue residues.

Prediction of ecological consequences has been discussed, with some attempt at synthesis, and this is probably one of the most difficult problems facing those involved in biological effects measurements. This does not, however, invalidate the use of indices of biological effect as an 'early warning system' for detection of environmental deterioration. A further complicating factor arises from the fact that contaminant PAHs seldom occur in environmental isolation and this therefore poses the question of interactive biological effects resulting from multiple xenobiotic challenge (Malins & Collier 1981; Moore *et al.* 1984). Indices of biological effects must also be capable of resolving such a situation, particularly those indices that are regarded as being relatively specific for particular types of xenobiotic. Although subtle interactive effects have been observed in experimental studies there is no evidence as yet that they are obscuring measurement of environmental impact (Moore *et al.* 1984).

Appendix 1

Materials and methods relating to Sivand oil spill

Cockles, *Cerastoderma edule* L. (3–4 cm length) were collected within 6–8 days of the *Sivand* oil-spill in Immingham Dock (October 1983) from the sites shown in figure 1. Digestive glands (hepatopancreas) were dissected out, damp-dried, frozen in liquid nitrogen and stored at

-70 °C before biochemical analysis. The pooled tissues of about six cockles per sample were used and, because of the limited number of cockles available, three or less samples only were taken per site. Microsomes were prepared and assayed for cytochrome P-450 and b_5 content, NADPH- and NADH-cytochrome *c* (NAD(P)H-CYTCRED) and NADH-ferricyanide (NADH-FERRIRED) reductase activities and benzo(*a*)pyrene hydroxylase (BPH) activity as described by Livingstone & Farrar (1984); microsomal protein was measured by the method of Lowry *et al.* (1951). Whole tissues of cockles were also taken for hydrocarbon analysis and two pools of five animals each were used per site. The shells of the cockles were cleaned with acetone, the tissues removed, damp-dried and similarly stored at -70 °C before chemical analysis. PAHs were extracted by steam distillation, analysed by high performance liquid chromatography and quantified in terms of two-ring (2,3-dimethyl-naphthalene equivalents) and three-ring (1-methylphenanthrene equivalents) aromatic hydrocarbon compounds as described by Donkin & Evans (1984).

For cytochemical determinations of NADPH-neotetrazolium reductase (NTR) and lysosomal stability (based on the latency of β-glucuronidase), digestive glands ($n = 5$) were excised and processed as described by Moore *et al.* (1982) and Lowe & Moore (1985).

A further sample of cockles was taken seven months after the spill occurred (May 1984) from Horseshoe Point (reference site), Cleethorpes and Spurn Bight. These were also processed for NTR, lysosomal stability and analysis of PAHs (as above).

References

Allison, A. C. 1969 Lysosomes and cancer. In *Lysosomes in biology and pathology* (ed. J. T. Dingle & H. B. Fell), vol. 2, pp. 178–204. Amsterdam, London and New York: Elsevier.

Anderson, R. S. 1978 Benzo[*a*]pyrene metabolism in the American oyster *Crassostrea virginica*. *EPA Ecol. Res. Ser. Monogr.* (EPA-600/3-78-009.)

Baccino, F. M. 1978 Selected patterns of lysosomal response in hepatocytic injury. In *Biochemical mechanisms of liver injury* (ed. T. F. Slater), pp. 518–557. New York and London: Academic Press.

Bakke, T. & Sorensen, K. 1987 Oil in rocky shore mesocosms: a six years experiment. In *Proc. 1985 Oil Spill Conference sponsored by Am. Pet. Inst.* (In the press.)

Bayne, B. L., Dixon, D. R., Ivanovici, A., Livingstone, D. R., Lowe, D. M., Moore, M. N., Stebbing, A. R. D. & Widdows, J. 1985 *The effects of stress and pollution on marine animals.* (384 pages.) New York: Praeger Scientific.

Bayne, B. L., Moore, M. N., Widdows, J., Livingstone, D. R. & Salkeld, P 1979 Measurement of the responses of individuals to environmental stress and pollution: studies with bivalve molluscs. *Phil. Trans. R. Soc. Lond.* B**286**, 562–581.

Bayne, B. L., Moore, M. N. & Koehn, R. K. 1981 Lysosomes and the response by *Mytilus edulis* to an increase in salinity. *Mar. Biol. Lett.* **2**, 193–204.

Bayne, B. L., Widdows, J., Moore, M. N., Salkeld, P., Worrall, C. M. & Donkin, P. 1982 Some ecological consequences of the physiological and biochemical effects of petroleum compounds on marine molluscs, *Phil. Trans. R. Soc. Lond.* B**297**, 219–239.

Bitensky, L., Butcher, R. S. & Chayen, J. 1973 Quantitative cytochemistry in the study of lysosomal function. In *Lysosomes in biology and pathology* (ed. J. T. Dingle), vol. 3, pp. 465–510. Amsterdam, New York and Oxford: Elsevier.

Bresnick, E. 1982 The molecular biology of the induction of the hepatic mixed function oxidases. In *International encyclopedia of pharmacology and therapeutics*, section 108 (*Hepatic cytochrome P-450 monooxygenase system*) (ed. J. B. Schenckman & D. Kupfer), pp. 191–225. Oxford: Pergamon Press.

Chayen, J. 1978 Microdensitometry. In *Biochemical mechanisms of liver injury* (ed. T. F. Slater), pp. 254–292. New York and London: Academic Press.

Couch, J. A. 1984 Atrophy of diverticular epithelium, as an indicator of environmental irritants in the oyster *Crassostrea virginica*. *Mar. environ. Res.* **14**, 525–526.

Davies, J. M., Hardy, R. & McIntyre, A. D. 1981 Environmental effects of North Sea oil operations. *Mar. Pollut. Bull.* **12**, 412–416.

Davies, J. M., Addy, J. M., Blackman, R. A., Blanchard, J. R., Ferbrache, J. E., Moore, D. C., Somerville, H. J., Whitehead, A. & Wilkinson, T. 1984 Environmental effects of the use of oil-based drilling muds in the North Sea. *Mar. Pollut. Bull.* **15**, 363–370.

Donkin, P. & Evans, S. V. 1984 Applications of steam distillation to the analysis of petroleum hydrocarbons in water and mussels (*Mytilus edulis*) from dosing experiments with crude oil. *Analytica chim. Acta* **156**, 207–219.

Ericsson, J. L. E. & Brunk, U. T. 1975 Alterations in lysosomal membranes as related to disease processes. In *Pathobiology of cell membranes* (ed. B. F. Trump and A. V. Arstila), vol. 1, pp. 217–253. New York, San Francisco and London: Academic Press.

Gilewicz, M., Guillaume, J. R., Carles, D., Leveau, M. & Bertrand, J. C. 1984 Effects of petroleum hydrocarbons on the cytochrome P-450 content of the mollusc bivalve *Mytilus galloprovincialis*. *Mar. Biol.* **80**, 155–159.

Gilfillan, E. S., Jiang, L. C., Donovan, D., Hanson, S. & Mayo, D. W. 1976 Reduction in carbon flux in *Mya arenaria* caused by a spill of No. 2 fuel oil. *Mar. Biol.* **37**, 115–123.

Gilfillan, E. S., Mayo, D. W., Page, D. S., Donovan, D. & Hanson, S. 1977 Effects of varying concentrations of petroleum hydrocarbons in sediments on carbon flux in *Mya arenaria*. In *Physiological responses of marine biota to pollutants* (ed. F. J. Vernberg, A. Calabreese, F. P. Thurberg & W. B. Vernberg), pp. 299–314. New York: Academic Press.

Gilfillan, E. S. & Vandermeulen, J. H. 1978 Alterations in growth and physiology in chronically oiled soft-shelled clams, *Mya arenaria*, chronically oiled with Bunker C from Chedabucto Bay, Nova Scotia, 1970–1976. *J. Fish. Res. Bd Can.* **35**, 630–636.

Grassle, J. F., Elmgren, R. & Grassle, J. P. 1981 Response of benthic communities in MERL experimental ecosystems to low level, chronic additions of No. 2 Fuel Oil. *Mar. environ. Res.* **4**, 279–297.

Hawkins, H. K. 1980 Reactions of lysosomes to cell injury. In *Pathobiology of cell membranes* (ed. B. F. Trump and A. V. Arstila), vol. 2, pp. 252–285. New York, San Francisco and London: Academic Press.

Livingstone, D. R. 1985 Responses of the detoxication/toxication enzyme system of molluscs to organic pollutants and xenobiotics. *Mar. Pollut. Bull.* **16**, 158–164.

Livingstone, D. R. & Farrar, S. V. 1984 Tissue and subcellular distribution of enzyme activities of mixed-function oxygenase and benzo[a]pyrene metabolism in the common mussel *Mytilus edulis* L. *Sci. tot. Envir.* **39**, 209–235.

Livingstone, D. R. & Farrar, S. V. 1985 Responses of the mixed function oxidase system of some bivalves and gastropod molluscs to exposure to polynuclear aromatic and other hydrocarbons. *Mar. environ. Res.* **17**, 101–105.

Livingstone, D. R., Moore, M. N., Lowe, D. M., Nasci, C. & Farrar, S. V. 1985 Responses of the cytochrome P-450 monooxygenase system to diesel oil in the common mussel. *Mytilus edulis* L., and the periwinkle *Littorina littorea* L. *Aquat. Toxicol.* **7**, 79–91.

Livingstone, D. R., Stickle, W. B., Kapper, M. & Wang, S. 1986 Microsomal detoxication enzymes of the marine snail, *Thais haemastoma*, to laboratory oil exposure. *Bull. environ. Contam. Toxicol.* **36**, 843–850.

Lowe, D. M. & Moore, M. N. 1985 Cytological and cytochemical procedures. In *The effects of stress and pollution on marine animals* (ed. B. L. Bayne *et al.*), pp. 179–204. New York: Praeger Scientific.

Lowe, D. M. & Pipe, R. K. 1985 Cellular responses in the mussel *Mytilus edulis* following exposure to diesel oil emulsions: reproductive and nutrient storage cells. *Mar. environ. Res.* **17**, 234–237.

Lowe, D. M. & Pipe, K. R. 1987 Mortality and quantitative aspects of storage cell utilization in mussels, *Mytilus edulis*, following exposure to diesel oil hydrocarbons. *Mar. environ. Res.* (In the press.)

Lowe, D. M., Moore, M. N. & Clarke, K. R. 1981 Effects of oil on digestive cells in mussels: quantitative alterations in cellular and lysosomal structure. *Aquat. Toxicol.* **1**, 213–226.

Lowe, D. M., Moore, M. N. & Bayne, B. L. 1982 Aspects of gametogenesis in the marine mussel *Mytilus edulis*. *J. mar. biol. Ass. U.K.* **62**, 133–145.

Lowry, O. H., Rosebrough, N. J., Farr, A. L. & Randall, R. J. 1951 Protein measurement with the Folin phenol reagent. *J. biol. Chem.* **193**, 265–275.

MacDonald, B. A. & Thomas, M. L. H. 1982 Growth reduction in the soft-shell clam *Mya arenaria* from a heavily oiled lagoon in Chedabucto Bay, Nova Scotia. *Mar. environ. Res.* **6**, 145–156.

McLusky, D. S. 1982 The impact of petrochemical effluent on the fauna of an intertidal estuarine mudflat. *Estuar. coast. Shelf Sci.* **14**, 489–499.

Malins, D. C. & Collier, T. K. 1981 Xenobiotic interactions in aquatic organisms: effects on biological systems. *Aquat. Toxicol.* **1**, 257–268.

Massie, L. C., Ward, A. P. & Davies, J. M. 1985 The effects of oil exploration and production in the North Sea: part 1 – the levels of hydrocarbons in water and sediment in selected areas, 1978–1981. *Mar. environ. Res.* **15**, 165–213.

Masters, B. S. S. & Okita, R. T. 1982 The history, properties and function of NADPH-cytochrome P-450 reductase. In *Hepatic cytochrome P-450 monooxygenase system* (ed. J. B. Schenkman & D. Kupfer), pp. 343–360. Oxford: Pergamon Press.

Moore, M. N. 1976 Cytochemical demonstration of latency of lysosomal hydrolases in digestive cells of the common mussel, *Mytilus edulis*, and changes induced by thermal stress. *Cell Tissue Res.* **175**, 279–287.

Moore, M. N. 1979 Cellular responses of polycyclic aromatic hydrocarbons and phenobarbital in *Mytilus edulis*. *Mar. environ. Res.* **2**, 255–263.

Moore, M. N. 1980 Cytochemical determination of cellular responses to environmental stressors in marine organisms. In *Biological effects of marine pollution and the problems of monitoring organisms* (ed. A. D. McIntyre & J. B. Pearce). *Rapp. P.-v. Réun. Cons. int. Explor. Mer.* **179**, 7–15.

Moore, M. N. 1985 Cellular responses to pollutants. *Mar. Pollut. Bull.* **16**, 134–139.
Moore, M. N. & Clarke, K. R. 1982 Use of microstereology and quantitative cytochemistry to determine the effects of crude oil-derived aromatic hydrocarbons on lysosomal structure and function in a marine bivalve mollusc *Mytilus edulis. Histochem J.* **14**, 713–718.
Moore, M. N. & Farrar, S. V. 1985 Effects of polynuclear aromatic hydrocarbons in lysosomal membranes in molluscs. *Mar. environ. Res.* **17**, 222–225.
Moore, M. N., Koehn, R. K. & Bayne, B. L. 1980 Leucine aminopeptidase (aminopepidase-1), *N*-acetyl-β-hexosaminidase and lysosomes in the mussel, *Mytilus edulis* L., in response to salinity changes. *J. exp. Zool.* **214**, 239–249.
Moore, M. N., Livingstone, D. R. & Widdows, J. 1987 Hydrocarbons in marine molluscs: biological effects and ecological consequences. In *Metabolism of Polynuclear Aromatic Hydrocarbons (PAHs) by Aquatic Organisms* (ed. U. Varanasi). Boca Raton, Florida: CRC Press Inc. (In the press.)
Moore, M. N., Pipe, R. K. & Farrar, S. V. 1982 Lysosomal and microsomal responses to environmental factors in *Littorina littorea* from Sullom Voe. *Mar. Pollut. Bull.* **13**, 340–345.
Moore, M. N., Pipe, R. K., Farrar, S. V., Thomson, S. & Donkin, P. 1986 Lysosomal and microsomal responses in *Littorina littorea*: further investigations of environmental effects in the vicinity of the Sullom Voe Oil Terminal and the effects of experimental exposure to phenanthrene. In *Oceanic processes in marine pollution – biological processes and waste in the ocean* (ed. J. M. Capuzzo & D. R. Kester), vol. 1, pp. 89–96. Melbourne, Florida: Kriege Publishing.
Moore, M. N., Widdows, J., Cleary, J. J., Pipe, R. K., Salkeld, P. N., Donkin, P., Farrar, S. V., Evans, S. V. & Thomson, P. E. 1984 Responses of the mussel *Mytilus edulis* to copper and phenanthrene: interactive effects. *Mar. environ. Res.* **14**, 167–183.
Nott, J. A., Moore, M. N., Mavin, L. J. & Ryan, K. P. 1985 The fine structure of lysosomal membranes and endoplasmic reticulum in the digestive gland of *Mytilus edulis* exposed to anthracene and phenanthrene. *Mar. environ. Res.* **17**, 226–229.
Owen, G. 1972 Lysosomes, peroxisomes and bivalves. *Sci. Prog., Oxf.* **60**, 299–318.
Payne, J. F. 1977 Mixed function oxidase in marine organisms in relation to petroleum hydrocarbon metabolism and detection. *Mar. Pollut. Bull.* **8**, 112–116.
Payne, J. F., Rahimtula, A. & Lim, S. 1983 Metabolism of petroleum hydrocarbons by marine organisms: the potential of bivalve molluscs. In *26th Annual Meeting of the Canadian Federation of Biological Societies*, vol. 26, no. PA-235.
Pearson, T. H. & Rosenberg, R. 1978 Macrobenthic succession in relation to organic enrichment and pollution of the marine environment, *Oceanogr. Mar. Biol.* **16**, 229–311.
Pipe, R. K. & Moore, M. N. 1985*a* Ultrastructural changes in the lysosomal–vacuolar system in digestive cells of *Mytilus edulis* as a response to increased salinity. *Mar. Biol.* **87**, 157–163.
Pipe, R. K. & Moore, M. N. 1985*b* The ultrastructural localization of acid hydrolases in developing oocytes of *Mytilus edulis. Histochem. J.* **17**, 939–949.
Pipe, R. K. & Moore, M. N. 1986 An ultrastructural study on the effects of phenanthrene on lysosomal membranes and distribution of the lysosomal enzyme β-glucuronidase in digestive cells of the periwinkle *Littorina littorea. Aquat. Toxicol.* **8**, 65–76.
Roubal, W. T. & Collier, T. K. 1975 Spin-labelling techniques for studying mode of action of petroleum hydrocarbons on marine organisms. *Fishery Bull.* **73**, 299–305.
Sanders, H. L., Grassle, J. F., Hampson, G. R., Morse, L. S., Garner-Price, S. & Jones, C. C. 1980 Anatomy of an oil spill: long term effects from the grounding of the barge Florida off W. Falmouth, Massachusetts, *J. mar. Res.* **38**, 265–380.
Slater, T. E. 1978 Biochemical studies on liver injury. In *Biochemical mechanisms of liver injury* (ed. T. F. Slater), pp. 2–44. New York: Academic Press.
Smith, M. T. & Wills, E. D. 1981 The effects of dietary lipid and phenobarbitone on the production and utilization of NADPH in the liver: a combined biochemical and quantitative cytochemical study. *Biochem. J.* **200**, 691–699.
Stegeman, J. J. 1981 Polynuclear aromatic hydrocarbons and their metabolism in the marine environment. In *Polycyclic hydrocarbons and cancer* (ed. H. V. Gelboin & P. O. P. T'so), vol. 3, pp. 1–60. New York: Academic Press.
Stekoll, M. S., Clement, L. E. & Shaw, D. G. 1980 Sublethal effects of chronic oil exposure on the intertidal clam *Macoma balthica. Mar. Biol.* **57**, 51–60.
Sumner, A. T. 1969 The distribution of some hydrolytic enzymes in the cells of the digestive gland of certain lamellibranchs and gastropods. *J. Zool.* **158**, 277–291.
Szego, C. M. 1975 Lysosomal function in nucleocytoplasmic communication. In *Lysosomes in biology and pathology*, (ed. J. T. Dingle & R. Dean), vol. 4, pp. 385–477. Amsterdam, London and New York: Elsevier.
Tripp, M. R., Fries, C. R., Craven, M. A. & Grier, C. E. 1984 Histopathology of *Mercenaria mercenaria* as an indicator of pollutant stress. *Mar. environ. Res.* **14**, 521–524.
Widdows, J. 1985 Physiological responses to pollution. *Mar. Pollut. Bull.* **16**, 129–134.

Widdows, J., Bakke, T., Bayne, B. L., Donkin, P., Livingstone, D. R., Lowe, D. M., Moore, M. N., Evans, S. V. & Moore, S. 1982 Responses of *Mytilus edulis* on exposure to the water accommodated fraction of North Sea oil. *Mar. Biol.* **67**, 15–31.
Widdows, J., Cleary, J. J., Dixon, Dr R., Donkin, P., Livingstone, D. R., Lowe, D. M., Moore, M. N., Pipe, R. K., Salkeld, P. N. & Worrall, C. M. 1984 Sublethal biological effects monitoring in the region of Sullom Voe, Shetland, July, 1983. (34 pages.) Aberdeen: Shetland Oil Terminal Environmental Advisory Group.
Widdows, J., Donkin, P. & Evans, S. V. 1985 Recovery of *Mytilus edulis* L. from chronic oil exposure. *Mar. environ. Res.* **17**, 250–253.

Discussion

R. E. Jones (*School of Ocean Sciences, University College of North Wales, Menai Bridge, U.K.*). The sites deemed to be affected by oil in the Humber estuary are situated near a very large discharge of acid waste (rich in metals) from a titanium dioxide processing installation; could such a discharge account at least in part for the observed changes? In addition, metals such as copper are associated with oil installations and may exert toxic effects at concentrations as low as 10 $\mu g\ l^{-1}$ on *Mytilus edulis*. Can such effects be separated from those induced by oil?

M. N. Moore. The biological effects observed in cockles after the *Sivand* oil spill are essentially what we would have predicted based upon evidence from laboratory and field investigations in mussels and periwinkles. Although the lysosomal destabilization is a generalized stress response, and can be induced by metals such as copper, the stimulation of the components of the cytochrome P-450 monooxygenase system would be expected from exposure to oil-derived polycyclic aromatic hydrocarbons and not from metals. Moreover, there is evidence of recovery of lysosomal stability after depuration of hydrocarbons at the Cleethorpes site.

Widdows, J., Bakke, T., Bayne, B. L., Donkin, P., Livingstone, D. R., Lowe, D. M., Moore, M. N., Evans, S. V. & [illegible] *Mytilus edulis* [illegible] North Sea oil. *Mar. Biol.* [illegible]
Widdows, J., [illegible], Dixon, D. R., Donkin, P., Livingstone, D. R., Lowe, D. M., Moore, M. N., Pipe, R. K., Salkeld, P. N. & Worrall, C. M. [illegible] Sullom Voe [illegible]
Widdows, J., [illegible] *Mar. Ecol.* [illegible]

Discussion

R. L. [illegible] (*School of Ocean Sciences, University College of North Wales, Menai Bridge, U.K.*). [illegible]

M. N. Moore. The biological effects observed [illegible] [illegible]

Phil. Trans. R. Soc. Lond. B **316**, 625–640 (1987)
Printed in Great Britain

Experimental studies of the effects of drilling discharges

By M. J. Leaver[1], D. J. Murison[1], J. M. Davies[1] and D. Rafaelli[2]

[1] *Department of Agriculture and Fisheries for Scotland, Marine Laboratory, P.O. Box* 101, *Victoria Road, Torry, Aberdeen AB*9 8*DB, U.K.*

[2] *Department of Zoology, University of Aberdeen, Tillydrone Avenue, Aberdeen AB*9 2*TN, U.K.*

The long-term effects of diesel and four different low-toxicity oil-based drilling-mud cuttings on the chemistry and benthic fauna of a marine sediment were compared.

Eighteen tanks (mesocosms) containing beach sediment were deployed in a control and five treatment, three replicate experimental set-up. The drill cuttings used had total oil concentrations ranging from 6 to 15% and were added to the tanks in quantities calculated to be representative of hydrocarbon levels measured in sediments 400–500 m from platforms in the North Sea.

Selected parameters monitored at varied intervals throughout the experiment were: redox profiles, sulphide concentrations, hydrocarbon concentration and meiofaunal abundance. Numbers of macrofaunal organisms evacuating the sediments in the seven days after treatment application were also recorded.

Redox measurements showed all sediments other than controls to be significantly reduced, but no differences were observed between treatments. Sediments became most reduced after 3 months and showed recovery thereafter, approaching control values after 15 months.

Sulphide levels peaked at about 3 months and declined thereafter, but were still elevated above control levels after 15 months. Highest concentrations were recorded in treatments with the highest oil content.

Mesobenthic meiofaunal abundance was significantly reduced in all cuttings treatments, but effects from individual treatments were indistinguishable. Epi/endobenthic copepod abundance increased markedly in all low-toxicity treatments, and particularly in those with lower oil content, but remained similar to control levels in diesel treatments throughout the experiment.

In the seven days after treatments, sediment evacuation rates by *Tellina* were highest in diesel and tended to reflect oil concentrations in low-toxicity treatments.

The results show that in equal oil concentrations diesel and low-toxicity oil-based drilling-mud cuttings have indistinguishable effects on sediment chemistry, but that, even after 15 months of weathering, diesel-based cuttings are demonstrably more toxic to benthic fauna.

Introduction

A number of North Sea surveys have demonstrated that sediment hydrocarbon levels around drilling platforms are elevated for distances of up to 3000 m (Grahl-Nielson *et al.* 1980; McIntosh *et al.* 1983; Addy *et al.* 1984; Massie *et al.* 1985). Effects on benthic fauna have been observed up to 1000 m from platforms, although the most deleterious effects occur within 500 m, and these have been correlated with sediment oil concentrations (Davies *et al.* 1984). Abnormally low redox potentials (*Eh*) in the surface layers of sediments close to platforms also appear to be linked with large oil-concentrations (Addy *et al.* 1984).

Elevated hydrocarbon levels in sediments close to platforms are primarily due to the discharge of oiled drill-cuttings which are separated from circulating oil-based muds during drilling operations. In an attempt to minimize environmental damage there has, in recent years, been a progressive switch from the use of diesel in oil-based muds to oils with lower acute toxicities. By 1984, only 5% of wells in the North Sea were drilled with diesel-based muds whereas 70% were drilled with low-toxicity oil-based muds. These low-toxicity muds may be over 200 times less toxic, in 96 h LC_{50} tests to the shrimp *Crangon* than diesel muds (Blackman *et al.* 1982). However, Davies *et al.* (1984) concluded that, up to 1983, there was insufficient evidence to distinguish between the ecological effects of diesel and low-toxicity oiled cuttings in the field. Addy *et al.* (1984) have pointed out that the high levels of hydrocarbons detected close to platforms are not necessarily toxic to the incumbent fauna, because the effects observed are consistent with changes due to organic enrichment. Similarly Dow (1984), using an experimental tank/sediment system, found no differences between diesel and first generation low-toxicity oiled cuttings in effects on benthic meiofauna, sediment physico-chemistry and microbial activity.

Here we describe the treatment of a model marine sediment, contained in onshore tanks, with diesel and four different low-toxicity oiled cuttings. The physico-chemistry and meiofauna of the sediments were monitored for a period of 15 months. The major objectives of the experiment were to distinguish between organic enrichment and toxicity effects and to determine whether relatively long-term exposures to different (i.e. diesel and low-toxicity) oiled cuttings produce different biological effects in the sediments.

Materials and methods

1. *Experimental tank system*

In April 1984 18 cylindrical glass fibre tanks 1.3 m in diameter and 0.8 m deep (figure 1) were filled to a depth of 20 cm with a natural marine sediment collected from the low-water springs zone at Aultbea, Loch Ewe, Scotland. The sediment was overlain with 40 cm of seawater, half of which was changed daily. Analysis of sediment particle size-distribution was done on five 2 cm diameter cores taken to 16 cm depth from the field site, as described by McIntyre & Murison (1973). Meiofaunal analyses were performed as described below.

In July 1984 equal masses (530 g) of oiled cuttings were spread evenly on the sediment surfaces to give a coating of about 1 mm. This gives an oil loading of 400 p.p.m. by mass in the top 2 cm, assuming an average cuttings oil content of 10%, and is representative of hydrocarbon concentrations detected at 500 m from several North Sea platforms (Grahl-Nielson *et al.* 1980; McIntosh *et al.* 1983; Addy *et al.* 1984; Massie *et al.* 1985). Diesel (D) and

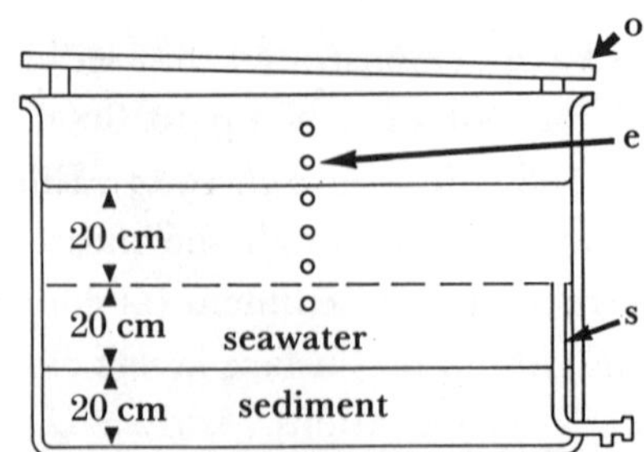

FIGURE 1. Section through a Loch Ewe sediment tank. o, Opaque lid; e, entry ports for seawater supply; s, standpipe.

four low-toxicity oiled cuttings (1, 2, 3, 4), obtained from North Sea drilling operations, were distributed randomly among five triplicate sets of tanks. A sixth undisturbed set acted as controls (C).

2. *Sampling programme*

Sediment redox profiles (*Eh*), sediment soluble sulphide concentrations, total hydrocarbons analysed by u.v.-fluorescence, *n*-alkanes by gas chromatography–flame ionization detector (GC–FID) and aromatics by gas chromatography–mass spectrometry (GC–MS) were determined before cuttings application and at 1 week and 1, 3, 5, 7, 10 and 15 months after application. Meiofauna was sampled on six occasions from two of the three tanks in each treatment over the 15 month experimental period. Care was taken to ensure that redox profiles and cores for sulphide, hydrocarbon and meiofaunal determinations were obtained from predesignated, undisturbed areas of sediment.

3. *Methods*

Redox profiles (up to six replicates) were generated by the method of Pearson & Stanley (1979) by using combination glass *Eh* electrodes (Russel pH Ltd., Type CMPT/RoD/SA/ 1.5 mm), by noting *Eh* at 1 cm intervals to a depth of 15 cm.

Soluble sulphide concentrations were determined by a method described in the Orion sulphide ion electrode, model 94-16, instruction manual (Orion Research Inc., Cambridge, Massachusetts, Form 94-16 IM/9730) and by Stanley *et al.* (1981) as modified by Dow (1984). Determinations were made on the top 5 cm of sediment (up to three replicate cores per tank) obtained with a truncated 5 ml syringe.

Sediment cores for hydrocarbon analyses, 2.2 cm in diameter and 2 cm deep, were obtained in duplicate for each tank by using a large truncated syringe and were immediately weighed into 500 ml screw top glass jars. Cores were also taken from each tank for dry mass analysis of the sediment.

Sediment cores and 15–20 g subsamples of fresh mixed cuttings were extracted into dichloromethane by the method of Massie *et al.* (1985). The volumes of the cuttings extracts were noted and 20 μl samples taken for further analysis before filtering and evaporating until all volatile solvent was removed. The remaining residues were weighed and then used as standard oils for u.v.-fluorescence analysis of the corresponding sediment samples.

Separation of sediment extracts and 20 μl cuttings extracts into the alkane and aromatic fractions was performed by high-performance liquid chromatography (P. R. Mackie, unpublished observations).

C_{15} to C_{33} *n*-alkanes, pristane and phytane were determined by capilliary GC–FID (Hewlett Packard 5880 with 2250 data system) as described by Mackie *et al.* (1978). Concentrations in original samples were determined from recovery of squalane standard with which the original extraction mixture was spiked.

Aromatic hydrocarbons were quantified by GC–MS (VG micromass 16F with VG 2250 data system) as described by Massie *et al.* (1985). Concentrations of aromatics in the original samples were determined from recovery of deuterated naphthalene, biphenyl, dibenzanthracene, anthracene and pyrene with which the original extracts were spiked.

u.v.-Fluorescence measurement of sediment oil concentrations was performed by the method of Massie *et al.* (1985) calibrating against oil extracted from the corresponding cuttings.

Meiofaunal samples were collected from 12 of the 18 tanks (i.e. 2 tanks per treatment) as duplicate 16 cm deep by 2.2 cm diameter cores by using transparent acrylic tubes. One core was divided into four horizontal sections of 4 cm each and the other retained intact. Samples were preserved in 5% neutral formalin with the addition of a small quantity of rose bengal stain until extraction of meiofauna by tapwater elutriation in a modified Boisseau apparatus. Enumeration of taxa was done on a static grid under a horizontal-scan stereo microscope.

Statistical analysis of meiofaunal abundance data was confined to the numerically dominant taxa, Nematoda and Harpacticoida (epi- and endobenthic). Transformed data were tested for homogeneity of cell mean-variances by Bartlett's test and overall differences between means were tested for significance by two-way analysis of variance.

Results

Particle size-analysis of field site sediments indicated a moderately well-sorted fine sand (skewed to the right and leptocurtic) with median diameter 2.63 ϕ (0.162 mm).

Redox profiles taken immediately before cuttings addition, three months after setting up the sediment system, showed that all tanks had similar profiles to those later observed in control tanks throughout the experiment. Pretreatment sulphide levels were also similar to controls. Gravimetric analyses of fresh cuttings revealed oil contents as follows: diesel, 9%; low toxicity 1, 6%; low toxicity 2, 11%; low toxicity 3, 12%; low toxicity 4, 15%. 96 h LC_{50} tests of the muds to *Crangon* by R. A. A. Blackman (personal communication) were as follows: diesel, 55 p.p.m. by mass dry drilling mud; low toxicities 1, 2, 3 and 4, all greater than 10000 p.p.m. by mass dry drilling mud.

Redox profiles (which are plotted as means of up to six replicates) taken after cuttings additions, showed marked changes in the physico-chemical status of the sediment in all tanks to which cuttings were added. A rapid reduction in redox potential, reaching the least point in all treatments after three months (October), was followed by a slow recovery period. Fifteen months after the addition of cuttings, redox values at 1 cm below sediment surfaces were approaching those of control tanks (figure 2). Values at 9 cm showed a similar though less marked trend (figure 3) whereas at 15 cm there was little or no recovery. No differences could be discerned in redox profiles, at any time, between different treatments but all treatments were clearly different from controls which remained remarkably stable throughout the experiment (figure 4*a*–*c*).

Soluble sulphide in the top 5 cm of sediment of all cuttings treatments increased dramatically between one month and three months after cuttings addition. Thereafter sulphide levels fell in all treated tanks, but remained high when compared with those in controls. A slight increase occurred after 10–15 months in some cuttings treatments. The greatest sulphide concentrations appeared in the tanks with the highest oil loading (low toxicity 4), and the lowest in the diesel-cuttings treatment (excepting control tanks). The remaining treatments showed sulphide levels between those of diesel and low toxicity 4 and could not be separated (figure 5).

Total oil levels, determined from u.v.-fluorescence of sediment extracts, showed a steady decline in all cuttings treatments throughout the experiment (figure 6).

GC–MS data from all fresh cuttings and sediment cores 7 d after addition reveal very similar profiles, naphthalenes accounting for 65–75% of all polycyclic aromatic hydrocarbons (PAHs) analysed. By contrast, after 400 d naphthalenes were much reduced in all treatments and consequently anthracenes and phenanthrenes accounted for a larger percentage of total PAHs

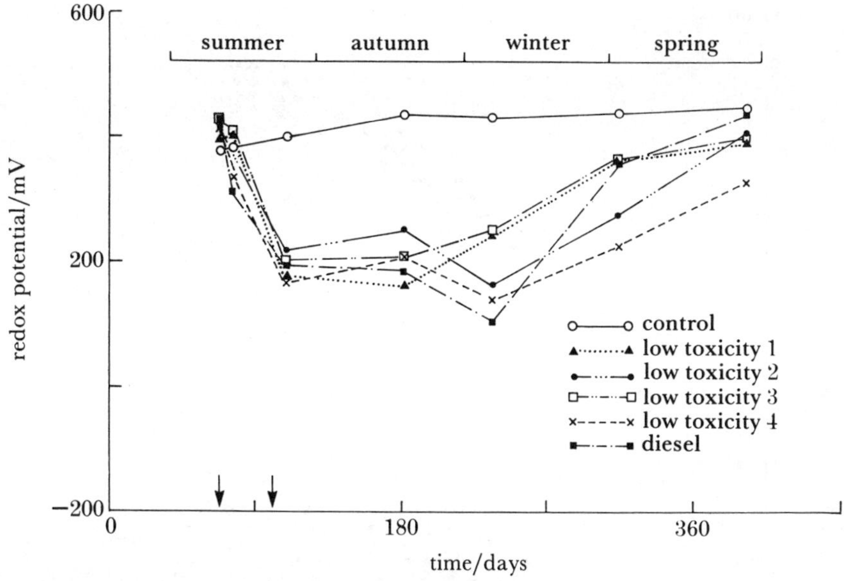

FIGURE 2. Redox potential of sediments at 1 cm depth over the period of the experiment. The arrows indicate times of treatments.

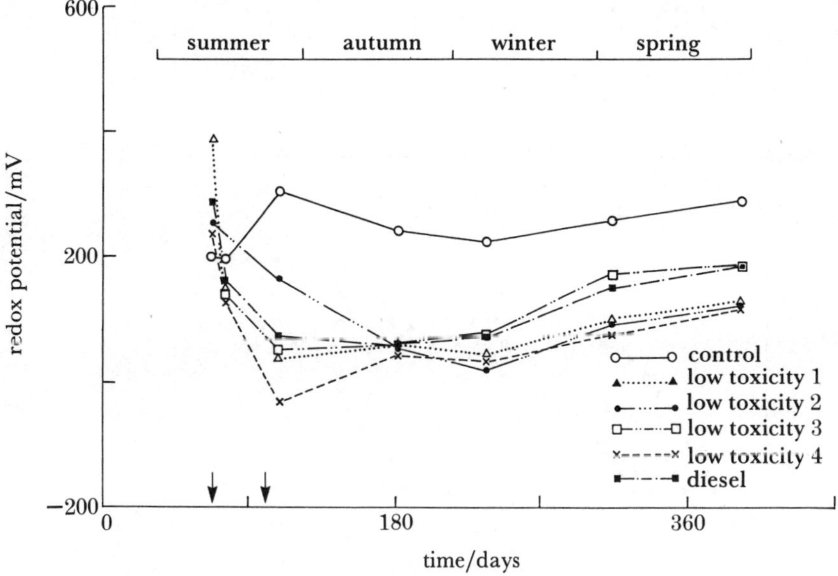

FIGURE 3. Redox potential of sediments at 9 cm depth over the period of the experiment. The arrows indicate times of treatments.

than previously. Diesel treatments showed the greatest total amounts of PAHs, in fresh cuttings, 7 d after addition and 400 d after addition, followed by low toxicities 4, 3, 2 and 1. Total PAHs were less after 400 d than after 7 d in all cuttings treatments (figure 7).

Alkane profiles generated from GC–FID data showed similar patterns in all cuttings treatments (figure 8 *a–c*). Sediment cores taken from tanks 400 d after cuttings additions showed a general reduction in all *n*-alkanes when compared with fresh cuttings. This was obvious from examination of the relative amounts of essentially non-degradable pristane and phytane

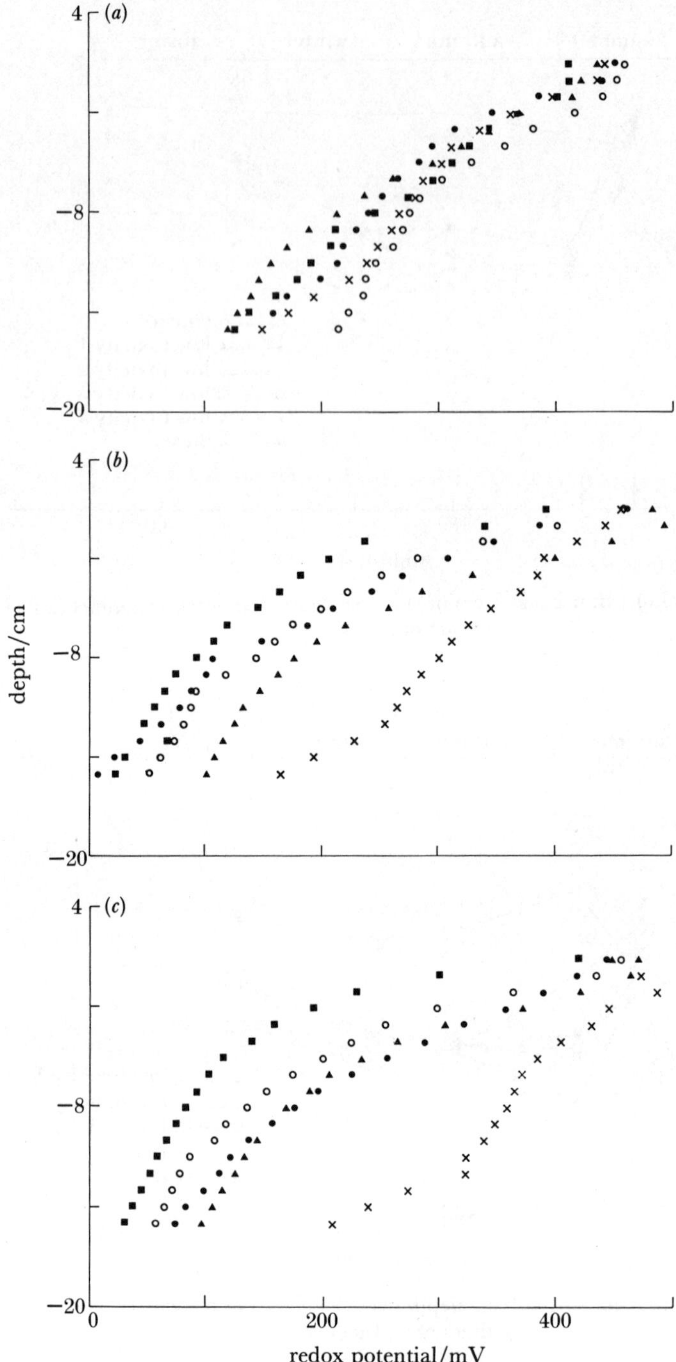

FIGURE 4. (*a*) Redox profiles of control tanks over period of experiment. (*b*) Redox profiles of low toxicity 4 tanks over the period of the experiment. (*c*) Redox profiles of diesel tanks over the period of the experiment. Symbols used: ×, predose; ●, 7 d; ■, 114 d; ○, 248 d; ▲, 470 d.

compared with C_{17} and C_{18} *n*-alkanes. The proportion of unresolved complex mixture (UCM) and more complex resolved aliphatic components also increased, particularly in low toxicity 4 treatments, which contained the greatest total level of aliphatics.

Laboratory examination of five field-site meiofaunal samples (day 0) indicated an abundant and fairly diverse community, numerically dominated by nematodes (84 %) with harpacticoid

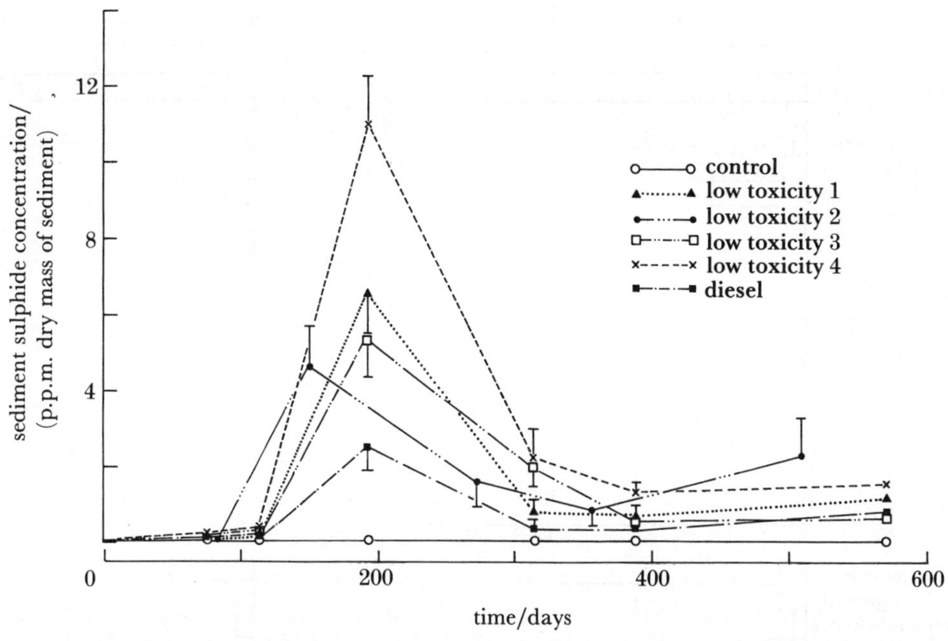

FIGURE 5. Sulphide concentrations of the tank sediments over the period of the experiment. The bars represent + and − standard error of the mean.

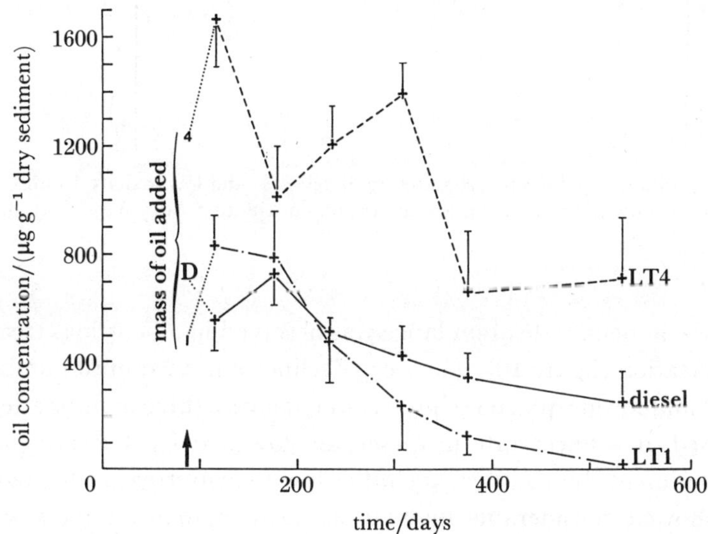

FIGURE 6. Total oil concentrations in the tank sediments over the period of the experiment. The bars represent + and − standard error of the mean. The arrow indicates the time of cuttings addition. LT1 and LT4 stand for low toxicity 1 and low toxicity 4.

copepods (epi-, endo- and mesobenthic), turbellarians, gastrotrichs and ciliates well represented (figure 9). Smaller numbers of other taxa including tardigrades, oligochaetes, archiannelids, ostracods, halacarids and temporary meiofaunal stages of polychaetes and bivalves were also recorded. Pretreatment meiofaunal abundance was recorded in the 12 tanks after a 9 week acclimation period (day 63) and showed a similarly diverse community with mean abundance reduced by about 25% (figure 9). Nematodes remained dominant (83%) and for this reason were used as a convenient index of tank meiofaunal abundance after addition of cuttings.

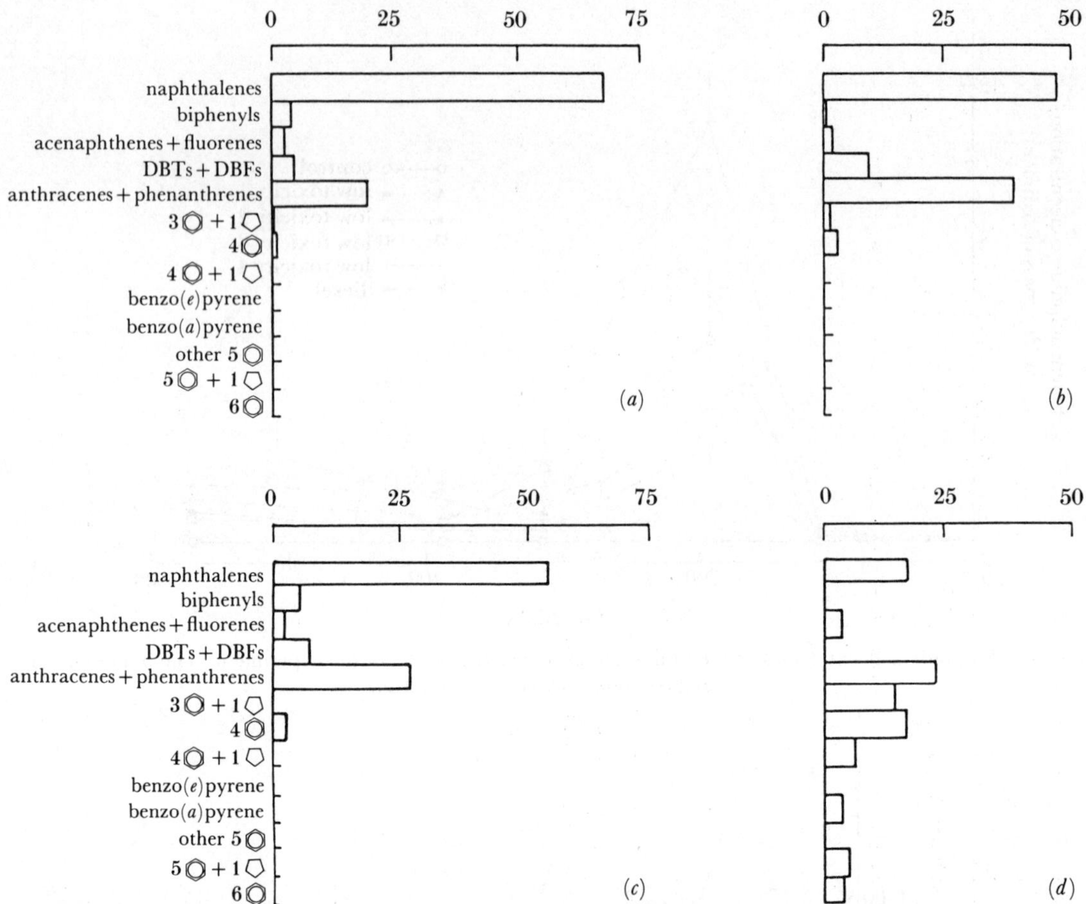

FIGURE 7. Percentage composition of polycyclic aromatics in diesel (*a*, *b*) and low toxicity 1 tanks (*c*, *d*) at 7 d (*a*, *c*) and 400 d (*b*, *d*) after cuttings addition. Total aromatics content (in $\mu g\ g^{-1}$ dry mass): (*a*), 6.75; (*b*) 2.14; (*c*) 0.90; (*d*) 0.53.

A substantial decline in nematode abundance was observed in all cuttings treatments in the first month after application (figure 10). The steep decline continued until numbers had fallen in all tanks to approximately one quarter of pretreatment levels, three months after application. This decline continued at a lower rate in diesel and low toxicity 4 treatments during the remainder of the experiment. By contrast, the other low toxicity treatments, particularly low toxicities 1 and 3, showed considerable fluctuations in nematode numbers with the mean abundance in the former exceeding that of controls after 11 months. Compared with treatments, controls maintained a significantly greater, though declining, mean abundance throughout the experiment. Although this decline may be associated with ecological factors inherent in tank isolation experiments, it is probable that the effect reflects the natural abundance cycle of meiofauna in this region (figure 10, adapted from McIntyre & Murison (1973)).

Mesobenthic (interstitial) copepods were eliminated in all treated tanks within 100 d of additions, and controls exhibited a general decline in abundance over much of the experimental period (figure 11). By contrast epi- and endobenthic forms showed a more varied response (figure 12). All low-toxicity treatments exhibited erratic fluctuations reflecting substantially

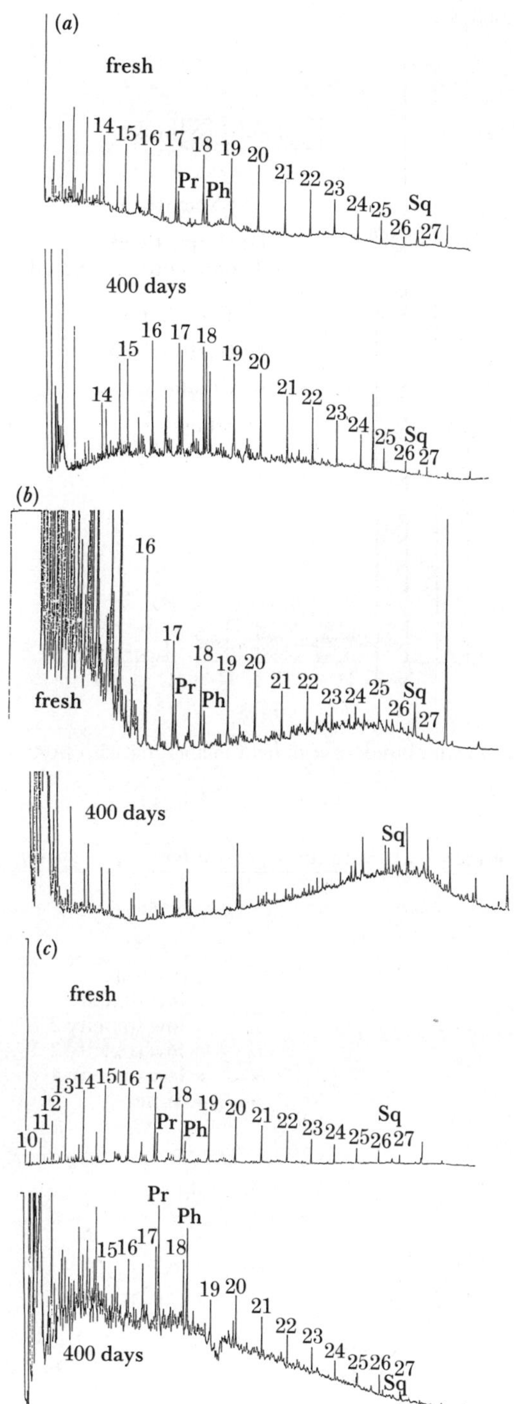

FIGURE 8. (*a*). GC–FID trace of aliphatics from fresh diesel-cuttings and sediment 400 d after cuttings addition. Squalane (Sq) in fresh-cuttings = 0.22 mg g^{-1}; squalane in sediment = 55.2 ng g^{-1} dry mass. (*b*) GC–FID trace of aliphatics from fresh low toxicity 1 cuttings and sediment 400 d after cuttings addition. Squalane (Sq) in fresh cuttings = 0.17 mg g^{-1}; squalane in sediment = 61.6 ng g^{-1} dry mass. (*c*) GC–FID trace of aliphatics from fresh low toxicity 4 cuttings and sediment 400 d after cuttings addition. Squalane (Sq) in fresh cuttings = 17.9 mg g^{-1}; squalane in sediment = 57.1 ng g^{-1} dry mass.

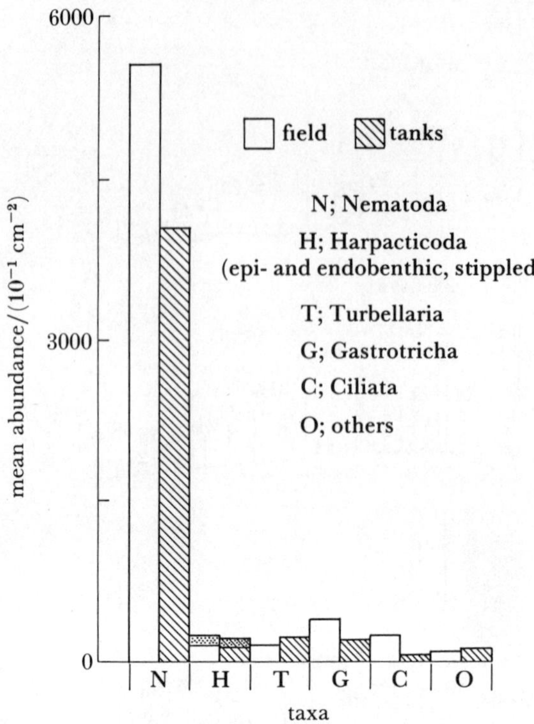

FIGURE 9. Comparison of mean abundances of field meiofauna with pretreatment tank meiofauna.

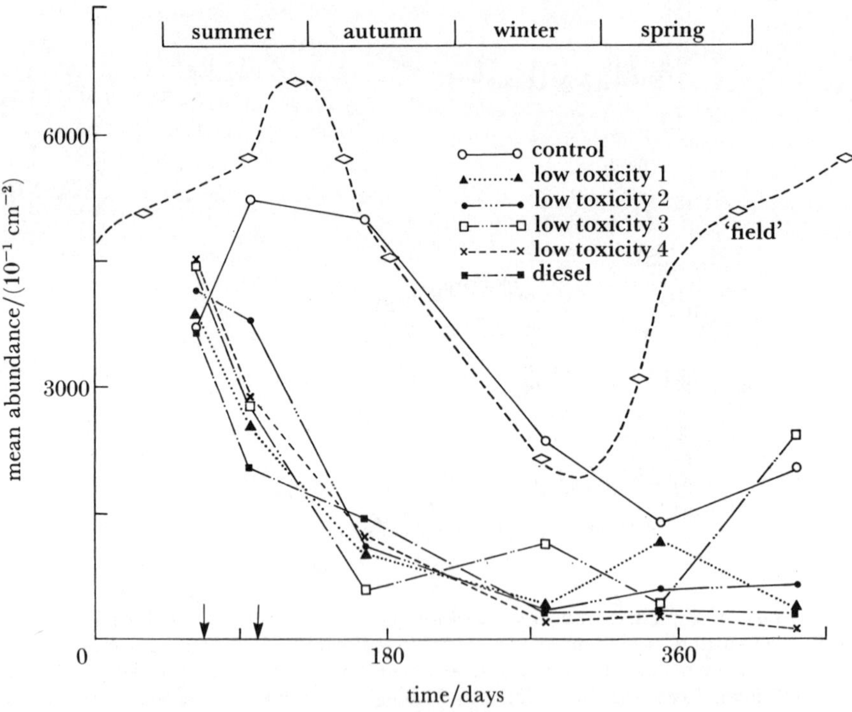

FIGURE 10. Nematode abundance in control and treated tanks over the period of the experiment. Also shown is the local field abundance cycle. Arrows indicate times of treatments.

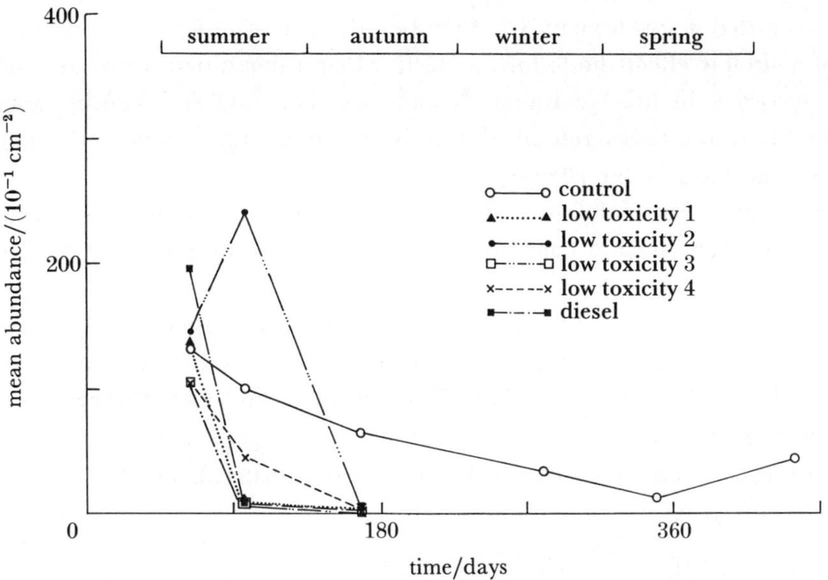

FIGURE 11. Interstitial (mesobenthic) copepod abundances in control and treated tanks over the duration of the experiment.

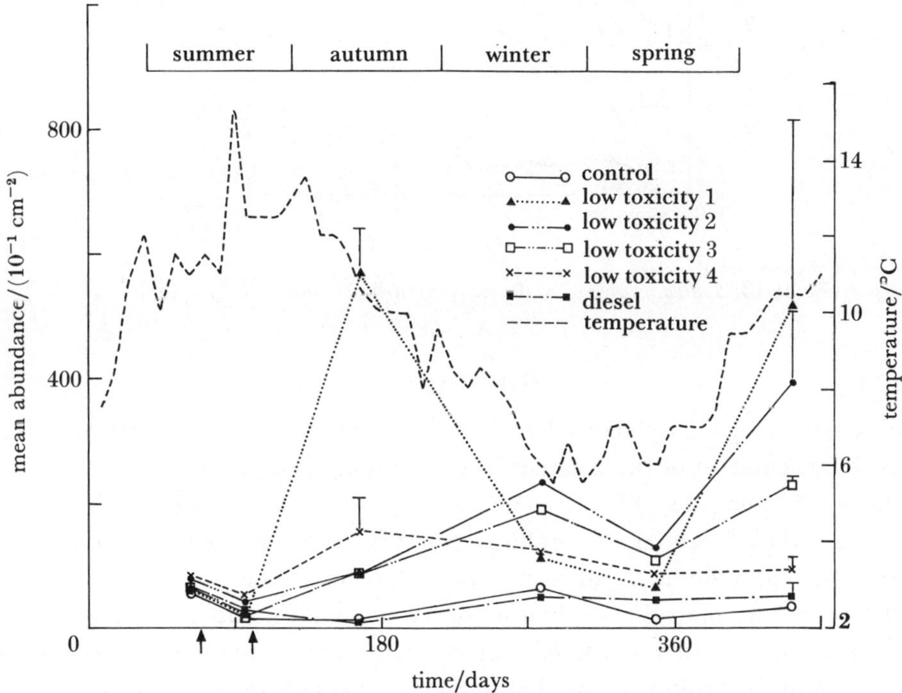

FIGURE 12. Epi- and endobenthic copepod abundances in control and treated tanks over the period of the experiment. Arrows indicate times of treatments.

enhanced populations at certain times during the experiment. Peak abundance (576 individuals per 10 cm^2) was recorded in low toxicity 1 in October, three months after cuttings application, with evidence of a similar effect the following June when a mean density of 517 individuals per 10 cm^2 was recorded in this treatment. Numbers of epi- and endobenthic copepods in control and diesel-cuttings tanks remained relatively small and constant throughout the experiment (13–70 individuals per 10 cm^2).

Although highly significant meiofaunal effects were indicated from statistical analyses no clear comparisons between cuttings could be shown because of complex time and treatment interactions.

The macrofauna of the sediments has not yet been analysed because the tanks are still being monitored, but counts of *Tellina* coming to the surface in the seven days immediately after cuttings addition (figure 13) indicated that diesel-treated tanks had the greatest rates of evacuation, followed by low toxicity 4 and then the other cuttings treatments. No *Tellina* were observed exposed on the sediment surfaces of control tanks during this period.

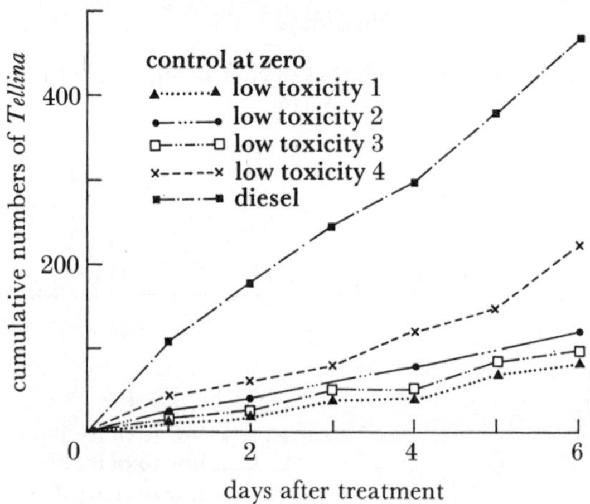

FIGURE 13. *Tellina* mortality in the days immediately after cuttings addition.

DISCUSSION

The tank system appears, from consideration of sediment physico-chemistry and meiofaunal populations, representative of the original field situation. Redox profiles taken immediately before cuttings addition (day 65, i.e. 65 d after collection of sediment) showed oxidized sediments in all tanks, similar to profiles recorded from the sediment collection area. Pretreatment tank sulphide levels, as in natural sediments of this type, were universally low (less than 10 p.p.b.† by mass dry drilling mud). Meiofaunal numbers, although 25 % lower than field populations, showed unaltered high diversity, with nematodes accounting for 83 % of the population in the tanks compared with 84 % in field samples after 63 d.

The rapidity of the change in redox profiles immediately after cuttings addition argues against organic enrichment as the main agent. Dow (1984) noted an equally swift response in sediments experimentally exposed to oiled cuttings in winter when microbial activity was at a minimum. The effects were more likely to have been caused by smothering of the sediment and severe limitation of water and gas exchange because, for equal masses of cuttings and

† In this paper one billion is used to represent 10^9.

despite differing oil levels, the magnitude and rate of change of redox in all cuttings treated tanks was essentially the same.

However, the changes in sediment soluble sulphide levels did appear to be associated with organic enrichment. Soluble sulphide in marine sediments is overwhelmingly the product of bacterial sulphate reduction, the activity of which reflects the amount and biological lability of organic matter available for decomposition (Goldhaber & Kaplan 1975). Sulphide levels in cuttings-treated tanks reached a maximum three months after application, but rose only very slightly in the first two months, before increasing rapidly between two and three months. Levels then dropped rapidly to a constant level of about ten times that of controls. This pattern is consistent with the observations of Westrich & Berner (1984) that organic material can be divided into two classes: easily degradable material capable of supporting high levels of sulphate reduction, and more refractory material able to support only small levels of sulphate reduction. The tank with the greatest sulphide levels, low toxicity 4, also contained the greatest total oil and *n*-alkane concentrations. Diesel-oiled cuttings-treated tanks generated the smallest sulphide concentrations and contained the smallest alkane levels but not the smallest total oil levels. *n*-Alkanes are known to be more readily degraded than aromatics (Delaune *et al.* 1980) and thus appeared to act as substrates for rapid growth of the sulphate reducers once conditions were appropriate.

The nematode abundance fell rapidly in all cuttings-treated tanks immediately after application. This was most likely to have been a response to the change in the physico-chemical status of the sediments because the decrease in these organisms paralleled the change in redox profiles more closely than the other parameters measured. Numbers continued to fall at a reduced rate in diesel and low toxicity 4 tanks but fluctuated in other treated tanks, sometimes to levels greater than those in controls. In isolated experimental systems, such as the tanks described here, physico-chemical recovery may not be associated with the corresponding degree of biological recovery because of the constraints on recruitment. Thus the increasing fluctuations in nematode numbers observed in the low toxicity tanks may be indicative of potential for recovery, true recovery being inhibited by lack of available recolonizers, and may represent single species at peak reproductive times.

Numbers of epi- and endobenthic copepods fluctuated erratically in all treatments except controls and diesel, which remained relatively constant throughout the experiment. The greatest fluctuations occurred in the low toxicity 1 treatment (that with the smallest oil concentration). Similarly elevated densities of benthic harpacticoid copepods have been observed in organically enriched field and experimental environments such as in an accumulating sewage sludge dumping ground (Moore & Pearson 1986); in oil sediments near a North Sea platform (Moore *et al.*, this symposium); in fuel-oil treated MERL mesocosms (Grassle *et al.* 1981); and in organically enriched (powdered *Ascophyllum nodosum*) mesocosms (Gee *et al.* 1985). Physically disturbed sediments have also been shown to induce an opportunistic response from an epibenthic copepod species (Alongi 1985). By contrast, the opportunistic response of these copepods does not seem to have occurred at all in the diesel treatment or to such an extent in the low toxicity 4 treatment. The lack of a response in the diesel treatment may be due to the relatively high concentrations of PAH (as shown by the GC–MS data) which are primarily responsible for the toxic effects of diesel (Anderson *et al.* 1974). Similarly, the reduced response in the low toxicity 4 treatment may also be due to the high PAH levels incurred by the large oil concentration in this treatment. Investigations into the effects of natural crude-oil seepage in the Santa Barbara Channel have indicated that the meiofaunal community may act as an important intermediate in the transfer of energy from microbial hydrocarbon degrading complexes to higher trophic levels (Montagna & Spies 1985). Dow (1984) noted increased numbers of aerobic heterotrophic bacteria in tanks dosed with comparable levels of diesel and low-toxicity oiled cuttings, but did not see an increase in glucose or amino acid utilization. Naphthalene mineralization rates were greater, showing that the larger numbers of bacteria were using the oil as a food source. The greatest copepod abundances occurred in the low toxicity 1 treatment, the treatment with the least PAH levels and the least total oil levels. The low abundance and suppressed seasonal response of epi- and endobenthic copepods in the control treatment compared with the normal field situation may be due to a potential food limitation imposed by the artificial water-exchange system. These results suggest a 'playoff' between organic enrichment and toxicity in the effects of oiled cuttings on epi- and endobenthic copepods.

The change in the ratio of nematodes to copepods under different pollution loadings has received much attention recently (see, for example, Boucher 1985; Moore & Pearson 1986; Sandulli 1987). The technique was originally suggested for monitoring pollution on coarse-particle (bathing) beaches, but has frequently been extended to silty sublittoral sediments. The copepod fauna of sublittoral muds is quite unlike that of sandy beaches and the effects of organic loading on sediment chemistry and the nematode fauna are also likely to be substantially different. Not too surprisingly, the responses of nematodes and copepods to pollution differ markedly in the two habitats. In common with other sublittoral pollution studies, the nematode/epibenthic copepod ratio declined in cuttings treatments; nematodes decreased and epibenthic copepods were either unaffected or increased (table 1). As expected, the ratio of nematodes to mesopsammic (interstitial) copepods increased under stress, these copepods being more vulnerable to smothering effects of the cuttings. There were no discernable between-treatment differences in the behaviour of these ratios.

TABLE 1. RATIOS OF NEMATODES TO MESOPSAMMIC COPEPODS IN CUTTINGS TREATED AND CONTROL TANKS OVER THE DURATION OF THE EXPERIMENT

treatment time		control	low toxicity 1	low toxicity 2	low toxicity 3	low toxicity 4	diesel
day 63	E	67.6	72.0	53.6	77.5	54.2	52.6
	M	28.6	28.6	29.0	43.2	43.8	18.8
day 96	E	377.1	141.2	82.9	139.5	52.9	79.4
	M	53.9	363.1	15.8	348.4	67.7	229.3
day 167	E	388.2	1.8	12.3	6.6	7.6	143.0
	M	81.4	—	—	—	—	—
day 279	E	36.8	3.0	1.5	5.8	1.8	6.6
	M	73.6	—	—	—	—	—
day 349	E	87.8	16.9	4.5	3.3	3.3	7.2
	M	140.4	—	—	—	—	—
day 433	E	56.9	0.7	1.6	10.3	1.3	5.7
	M	48.8	—	—	—	—	—

field ratio: epi- and endobenthic (E) 70.5, mesopsammic (M) 40.1

The diesel-oiled cuttings appeared more toxic to the macrofauna in the tanks during the seven days after addition. However, no data are available on macrofaunal populations immediately before addition, although the original field-site has been sampled. It will be interesting to examine the macrofaunal populations when the experiment has run its course.

Conclusions

The results presented here would seem to support the contention that diesel-oiled cuttings differ from low toxicity cuttings in their long-term impact. The higher short-term acute toxicity of diesel was apparent from counts of macrofauna evacuating sediments. Greater long-term toxicity may be a factor in the inability of diesel treatments to support large populations of epi- and endobenthic copepods despite having similar total oil levels to low toxicity treatments which supported elevated copepod abundances.

Effects on the physico-chemistry of the sediments appeared to be twofold: an initial smothering effect dependent more on the total quantity of cuttings than their oil contents, and an organic enrichment effect dependent on the biological lability and quantity of oil. Organic enrichment, although contributing to high toxic-sulphide levels, seems to be a relatively short-term effect compared with the long-term toxicity of certain hydrocarbons. On balance, therefore, it is likely that the discharge of low-toxicity oiled cuttings during drilling operations will have a less toxic long- and short-term impact than the discharge of equivalent amounts of diesel-oiled cuttings.

References

Addy, J. M., Hartley, J. P. & Tibbetts, P. J. C. 1984 Ecological effects of low toxicity oil-based mud drilling in the Beatrice oilfield. *Mar. Pollut. Bull.* **15**, 429–436.

Alongi, D. M. 1985 Effect of physical disturbance on population dynamics and trophic interactions among microbes and meiofauna. *J. mar. Res.* **43**, 351–364.

Anderson, J. W., Neff, J. M., Cox, B. A., Tatem, H. E. & Hightower, G. M. 1974 Characteristics of dispersions and water-soluble extracts of crude and refined oils and their toxicity to estuarine crustaceans and fish. *Mar. Biol.* **27**, 75–85.

Blackman, R. A. A., Fileman, T. W. & Law, R. J. 1982 Oil-based drill-muds in the North Sea – the use of alternative base-oils. ICES CM: 1982/E: 13. (Unpublished manuscript.)

Boucher, G. 1985 Long term monitoring of meiofaunal densities after the *Amoco Cadiz* oil spill. *Mar. Pollut. Bull.* **16**, 328–333.

Davies, J. M., Addy, J. M., Blackman, R. A., Blanchard, J. R., Ferbrache, J. E., Moore, D. C., Somerville, H. J., Whitehead, A. & Wilkinson, T. 1984 Environmental effects of the use of oil-based drilling muds in the North Sea. *Mar. Pollut. Bull.* **15**, 363–370.

Delaune, R. D., Hambrick, G. A. & Patrick, W. H. 1980 Degradation of hydrocarbons in oxidised and reduced sediments. *Mar. Pollut. Bull.* **11**, 103–106.

Dow, F. K. 1984 Studies on the environmental effects of production water and drill cuttings from North Sea oil installations. Ph.D. thesis, University of Aberdeen.

Gee, J. M., Warwick, R. M., Schaanning, M., Berge, J. A. & Ambrose, W. G. 1985 Effects of organic enrichment on meiofaunal abundance and community structure in sublittoral soft sediments. *J. exp. mar. Biol. Ecol.* **91**, 247–262.

Goldhaber, M. B. & Kaplan, I. R. 1975 Controls and consequences of sulphate reduction in recent marine sediments. *Soil Sci.* **119**, 42–55.

Grahl-Nielson, O., Sunby, S., Westrheim, K. & Wilhelmsen, S. 1980 Petroleum hydrocarbons in sediment resulting from drilling discharges from a production platform in the North Sea. In *Proceedings of the Symposium on Research on Environmental Fate and Effects of Drilling Fluids and Cuttings, Lake Buena Vista, Florida, 21–24 January* 1980, vol. 1, pp. 541–561. Washington, D.C.: U.S. Government Printing Office.

Grassle, J. F., Elmgren, R. & Grassle, J. P. 1981 Response of benthic communities in MERL experimental ecosystems to low-level chronic additions of No. 2 fuel oil. *Mar. environ. Res.* **4**, 279–297.

Mackie, P. R., Hardy, R. & Whittle, K. J. 1978 Preliminary assessment of the presence of oil in the ecosystem at Ekofisk after the blowout, 22–30 April 1977. *J. Fish. Res. Bd Can.* **35**, 544–551.

Massie, L. C., Ward, A. P., Davies, J. M. & Mackie, P. R. 1985 The effects of oil exploration and production in the northern North Sea. Part 1. The levels of hydrocarbons in water and sediments in selected areas, 1978–1981. *Mar. environ. Res.* **15**, 165–213.

McIntosh, A. D., Massie, L. C. & Mackie, P. R. 1983 A survey of hydrocarbon levels and some biodegradation rates in water and sediments around North Sea oil platforms, 1981, 1982. ICES CM: 1983/E: 42. (Unpublished manuscript.)

McIntyre, A. D. & Murison, D. J. 1973 The meiofauna of a flatfish nursery ground. *J. mar. biol. Ass. U.K.* **53**, 93–118.
Montagna, P. A. & Spies, R. B. 1985 Meiofauna and chlorophyll associated with *Beggiatoa* mats of a natural submarine oil seep. *Mar. environ. Res.* **16**, 231–242.
Moore, C. G. & Pearson, T. H. 1986 Response of a marine benthic copepod assemblage to organic enrichment. In *Proceedings of the Second International Conference on Copepoda, Ottawa, Canada, 13–17 August 1984.* (ed. G. Schriever, H. K. Schminke & C.-t. Shih), National Museum of Natural Sciences, Ottawa. *Syllogeus* no. 58, 369–373.
Pearson, T. H. & Stanley, S. O. 1979 Comparative measurement of the redox potential of marine sediments as a rapid means of assessing the effect of organic pollution. *Mar. Biol.* **53**, 371–379.
Sandulli, R. 1987 Preliminary report of a laboratory experiment on the effects of organic pollution on meiofauna. *Acta XVII Congresso S.I.B.M. Ferrara, 1985* in *Nova Thalassia.* (In the press.)
Stanley, S. O., Leftley, J. W., Lightfoot, A., Robertson, N., Stanley, I. M. & Vance, I. 1981 The Loch Eil project: sediment chemistry, sedimentation and the chemistry of the overlying water in Loch Eil. *J. exp. mar. Biol. Ecol.* **55**, 299–313.
Westrich, J. T. & Berner, R. A. 1984 The role of sedimentary organic matter in bacterial sulphate reduction: the G model tested. *Limnol. Oceanogr.* **29**, 236–249.

Discussion

W. A. Hamilton (*Department of Microbiology, University of Aberdeen, U.K.*). Dr Davies puts considerable stress on redox as a component of his test system. Would he care to comment on the fact that whereas he was recording values of around −200 mV in the Beryl field, in his tanks all values were above zero?

J. M. Davies. I suggest that whereas the gross environment might be of positive redox, there would none the less be microenvironments that would be of lower redox and that these would be the site of sulphide production.

W. A. Hamilton. I agree with Dr Davies's suggestion regarding microenvironments and sulphide production, but this limitation of probe measurements would apply equally to all systems studied and therefore the point still holds that the tank model system did not truly mirror the conditions under cutting piles on the seabed.

J. M. Davies. The experiment was not designed to simulate the conditions under the cuttings pile in the immediate vicinity of the platform but rather the conditions 400–500 m from the platform where oil concentrations in the top 2 cm of the sediment are of the order of 500 p.p.m. by mass dry sediment rather than 150000 p.p.m. by mass dry sediment in the cuttings pile.

R. A. A. Blackman (*MAFF Fisheries Laboratory, Burnham-on-Crouch, U.K.*). Could Dr Davies say what role *Tellina* played in the economy of the tanks? Was it a competitor for food, making resources available through faeces and pseudo-faeces, reworking surface sediments, etc.? (It was assumed that the graph displayed showed only the mortality rates for *Tellina* surfacing in the tanks, but that some remained alive in the sediments.)

J. M. Davies. *Tellina* are filter feeders, predominantly filtering plankton from the water column, although they have been shown to remove sediment detritus in the immediate vicinity of their siphons.

The total numbers of *Tellina* remaining in the tanks will not be determined until the end of the experiment when the tanks are emptied.

Phil. Trans. R. Soc. Lond. B **316**, 641–654 (1987)
Printed in Great Britain

Oil pollution studies of the Solbergstrand mesocosms

BY J. S. GRAY

University of Oslo, Department of Biology, Section of Marine Zoology and Chemistry, 0316 Blindern, Oslo 3, Norway

Two medium-scale ecosystems (mesocosms) were built on the Oslofjord: one a hard-bottom intertidal system and the other a subtidal soft-sediment system. The hard-bottom mesocosm consists of four basins, two controls and two which were dosed with diesel-oil (129 $\mu g\ l^{-1}$ a high oil (HO) dose and 29 $\mu g\ l^{-1}$ a low oil (LO) dose). Both oil doses caused high mortality of *Mytilus edulis* and growth was reduced in the macroalgae *Ascophyllum nodosum* and *Laminaria digitata*. Recruitment of *Littorina littorea* was also affected by oil so that populations declined over time.

Subtidal benthic communities have been established in the mesocosm and show variations in sediment chemistry within the range found in the field. Although recruitment of benthic macrofauna is reduced, dominant species and species structure remain closely similar to that in the field over six months. Bioturbation effects studied in the mesocosm have shown the important influence of large, rare species in structuring benthic communities, a finding which would not be possible in nature by diving or by the use of submersibles. Preliminary results from a community taken from 200 m depth and established in the mesocosm suggest that it is now possible to do detailed manipulation experiments on communities simulating the whole continental shelf.

INTRODUCTION

Mesocosms are facilities that allow experimentation at scales intermediate between the laboratory and the field. In marine research, mesocosms have been used to enclose large bodies of seawater with their constituent plankton communities (the Loch Ewe experiments in Scotland and Saanich Inlet in Canada, reviewed in Grice & Reeve (1982)), as enclosures linking water column and benthos (the Marine Ecosystem Research Laboratory at Rhode Island U.S.A. (Pilson *et al.* 1977) and the Kiel Bay experiments (von Bodungen *et al.* 1976)), and as enclosures simulating tidal flat communities (the Texel facility in Holland (De Wilde & Kuipers 1977) and the German caissons (Farke *et al.* 1984)). No mesocosms have been used to study intertidal hard bottom communities or subtidal benthic communities. The Solbergstrand facilities cover the last two aspects and are therefore unique.

Mesocosms have been primarily used for studies of effects of pollutants on marine organisms. The advantages of mesocosms over the natural field-habitats are that the environment is usually enclosed and thus easier to experiment in a controlled manner rather than being subject to the vagaries of climate. Experiments with pollutant materials in the natural environment are often uncontrolled (e.g. dosing).

Although effects of oil on marine systems have been much studied, one relatively little-studied aspect is the potential long-term effect of small amounts of oil discharged continuously. Such a scenario would apply to offshore oil platforms, in refineries or indeed at most estuaries.

With financial assistance from British Petroleum (B.P. Norge A/S) it was possible to convert an old trout-farm owned by the Norwegian Institute for Water Research (NIVA) at Solbergstrand on the Oslofjord, Norway into a research facility. A joint research programme was initiated between NIVA, B.P. and the University of Oslo, and has run from 1981 to 1986.

The primary focus of the research has been on long-term effects of small amounts of diesel-oil on individuals and populations of intertidal rock communities (algae, bivalves, gastropods and Crustacea) and on the feasibility of simulating subtidal soft-sediment communities within a mesocosm with the aim of studying effects of, for example, drill-cuttings on natural communities.

The hard bottom mesocosm

Material and methods

Four basins (measuring 8.4 m × 5.0 m × 1.25 m) contain seawater pumped from 1 m depth in the fjord outside the mesocosm (figure 1). During the coldest periods water is pumped from 13 m depth to prevent formation of ice within the basins. Turnover time for the water in each basin is four hours. Over the period July 1982 to October 1984 the minimum monthly mean

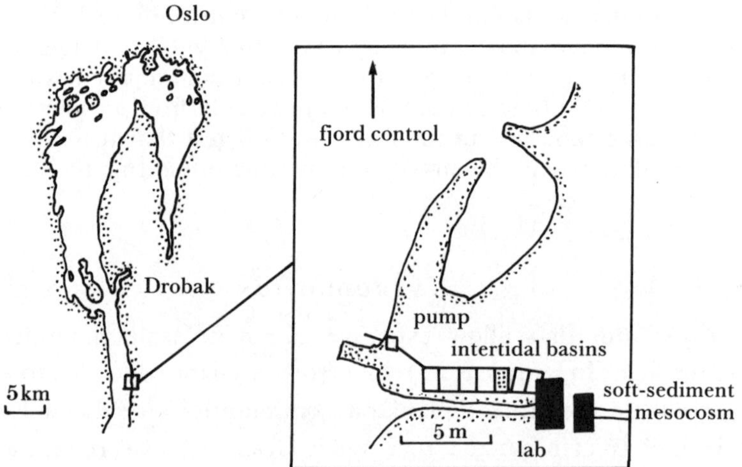

FIGURE 1. Location of Solbergstrand mesocosms.

temperature was 2 °C and the maximum 23 °C. Salinity varied from 14‰ in June 1983 to 31‰ in December 1983 and January 1984. Water flows continuously into the basins at a rate of 10 m^3 h^{-1} and the outflow is adjusted in height by a mechanical pulley system to give a simulated 12 h 25 min tidal cycle. On one side of each basin is a wave generator, running along the full length of the basin, which works mechanically and produces waves of 25 cm in height with a period of *ca.* 2 m. On the opposite side of each basin to the wave generator is a series of steps 42.5 cm in width and 19 cm in height with a slope of 19 cm across the basin. Thus the left-hand side of one step is level with the right-hand side of the step above. The communities were established in 1979 by transporting into each of the four basins stones containing an attached algal flora and constituent fauna. The basins were allowed to develop by natural recruitment, which was large, for three years before oil dosing began. A typical zonation pattern for the sheltered Oslofjord ranging from *Laminaria* through *Ascophyllum* and the genus *Fucus* resulted.

Choice of the type of oil to be used in the experiments was the subject of considerable debate. The water-accommodated fraction (WAF) of diesel oil was chosen to ensure consistency of the dose. Our initial concentrations projected were 200 (high oil, HO) and 50 (low oil, LO) $\mu g\ l^{-1}$ total hydrocarbons dosed to two basins. The two remaining basins acted as controls. The dosing system is shown in figure 2. Inflowing seawater was heated to 12.5 °C before being mixed

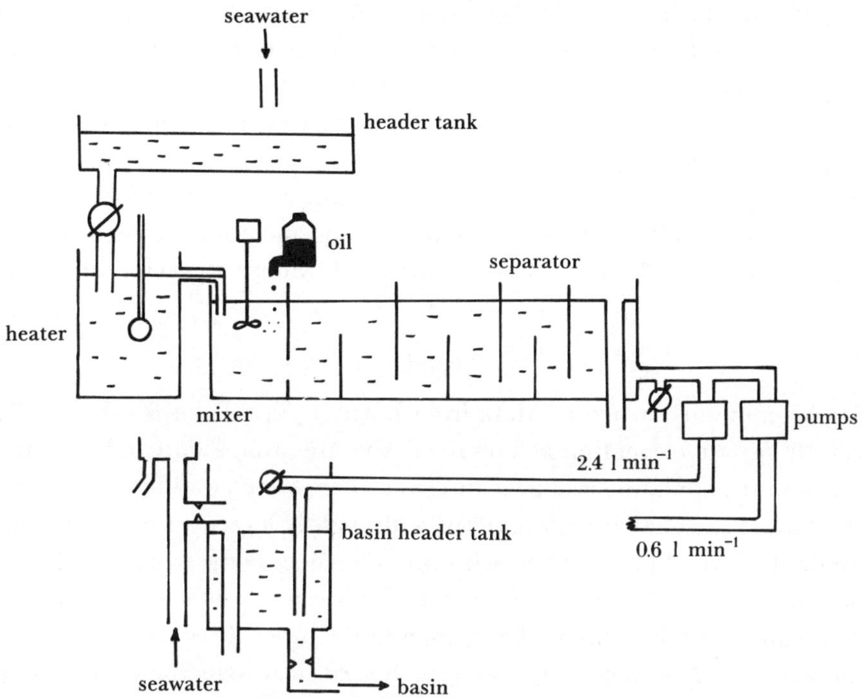

FIGURE 2. Oil dosing system giving 127 $\mu g\ l^{-1}$ (HO) and 27.9 $\mu g\ l^{-1}$ (LO) of water-accommodated fraction (WAF) of oil in dosed basins.

vigorously in the mixing chamber. The separator skimmed off oil droplets and the baffles ensured that only WAF was pumped to the basins†. The concentration of hydrocarbons in the basins was measured routinely by u.v.-fluorescence with periodic checks being made by gas chromatography–mass spectrometry (GC–MS) analyses. After subtracting values for the control basins the added hydrocarbon concentration for the period September 1982 to September 1984 had a mean of 127.3 $\mu g\ l^{-1}$ in the HO basin and 27.9 $\mu g\ l^{-1}$ in the LO basin, giving fourfold dose differences. The background hydrocarbon level was 5.6 $\mu g\ l^{-1}$ (Bakke & Sørensen 1985).

Most studies were done at the individual and population level rather than the community and ecosystem level. Algal communities were studied by placing transects over the steps within each basin and measuring percentage cover for the algae and numbers for the fauna (Bokn & Moy 1987). Growth of individual plants of *Laminaria digitata* and *Ascophyllum nodosum* was studied by marking tips of individual plants. Bokn (1984, 1985) should be consulted for further details.)

The gastropod *Littorina littorea* was a major component of the basins and was studied in detail. Over 3500 individuals were marked with identifying numerical codes and colour codes for catch records, both marks were covered by a cyanoacrylate film before release. Individuals were

† Oil was dosed to the basins from May 1982 to September 1982.

measured with a digital caliper (accuracy ± 0.03 mm) when caught. In addition to the two basin controls a control station was established in Oslofjord near Solbergstrand (figure 1). Sampling was done monthly from July 1982 to September 1983. Data were obtained on individual growth rates, population mortality, recruitment and fecundity. (Details of methods are not presented and will appear elsewhere.)

Mytilus edulis was also a common component of the basins and population dynamics were studied by separating the initial basin population into size classes, placing them in nylon-mesh tubes (as used in mussel culture) and attaching the tubes to ropes strung across the basins. Regular sampling was done to assess growth, mortality and fecundity.

Population genetics studies on enzyme polymorphisms were done on *Littorina littorea*, *Mytilus edulis* and *Semibalanus balanoides*.

Finally, physiological studies of scope for growth, feeding rates, respiration rates and excretion rates and biochemical studies of lysozymal enzymes and membrane structures were done on *L. littorea* and *M. edulis* by IMER & NIVA scientists (see Moore, this symposium; Bakke 1985).

Results

Macroalgae community structure (data from Bokn (1984); Bokn & Moy (1987)) showed surprisingly little variation over time as a result of exposure to oil. Figure 3 shows that *A. nodosum* had almost constant population size over time and there were no differences between basins. *Fucus serratus* showed seasonal growth patterns with slightly less cover in the HO basin whereas *L. digitata* showed seasonal patterns but with no differences between basins. The understorey algae *Phymatolithon lenormandii*, *Chondrus crispus*, *Cladophora rupestris* and *Ulva lactuca* showed significantly greater growth in the LO basin than in the other three basins, (figure 3).

Individually marked *L. digitata* and *A. nodosum* showed clear reductions in growth rate in both oiled basins when compared with the controls (figure 4), but only in the second and third years of growth (Bokn 1985).

L. littorea populations showed over 12 months of exposure to oil a statistically significant decline in population density in all basins (figure 4). Calculations of mortality (table 1) showed that the absolute mortality rate (M) was significantly greater in HO than LO but the LO population showed no differences from the controls.

The basins were not closed populations and some recruitment could have occurred from outside. To investigate this, measurements were taken of the numbers of *L. littorea* eggs in the inlet and outlet of each basin. Between 16 and 30% of recruits came from outside the basins and the total numbers discharged clearly related to the population size within the respective basins. Basin C4 had the greatest number of larvae discharged and the highest population density as estimated by mark and recapture data. Throughout the study period the high-oil basin (HO) had the least number of individuals less than 6 mm in length, followed by the low oil (LO) and the two controls had greatest numbers. Survival of the smallest individuals was poorest in the oiled basins.

The growth rate pattern of an individually marked *L. littorea* is shown in figure 5. Maximal growth occurs from July to September. Figure 5 shows that growth rate is lower for individuals of similar starting size in HO and LO than in C2. The growth rate in C4 and in the fjord control with higher population densities than in the other basins was, however, even less. The K value from fitted von Bertalanffy curves (table 2) shows that for all individuals measured in July 1983 no clear differences were found between oiled and control basins. The growth rate studies show

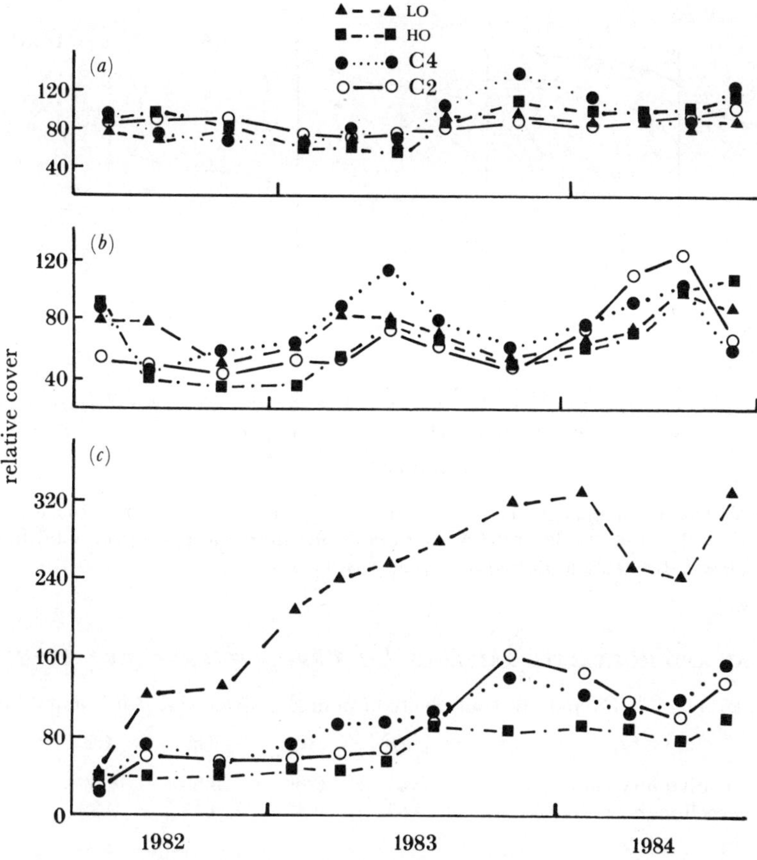

FIGURE 3. Macroalgae cover after addition of oil (WAF) at 127 $\mu g\ l^{-1}$ (MO) and 27.9 $\mu g\ l^{-1}$ (LO). C2 and C4 are two control basins. (Data from Bokn (1985).) (*a*) *Ascophyllum nodosum*. (*b*) *Laminaria digitata*. (*c*) *Phymatolithon lenormandii*.

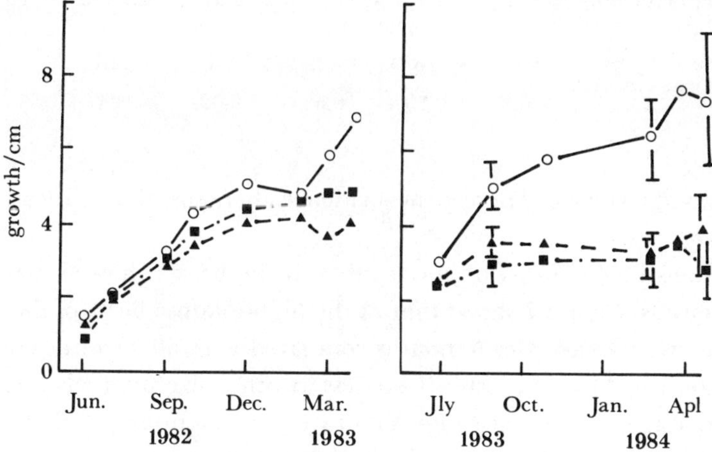

FIGURE 4. Growth of *Ascophyllum nodosum* dosed with oil (WAF). HO = 127 $\mu g\ l^{-1}$, LO, 27.9 $\mu g\ l^{-1}$, C2, control. Symbols as in figure 3.

that oil did not have an appreciable effect and that density differences among the studied populations were more important than oil in affecting growth rate.

Analyses of size–frequency distributions of *L. littorea* suggested that stabilizing selection might have occurred where the modal size is selected at the expense of the smallest and largest individuals (figure 6). A comparison with the controls, however, shows that the populations

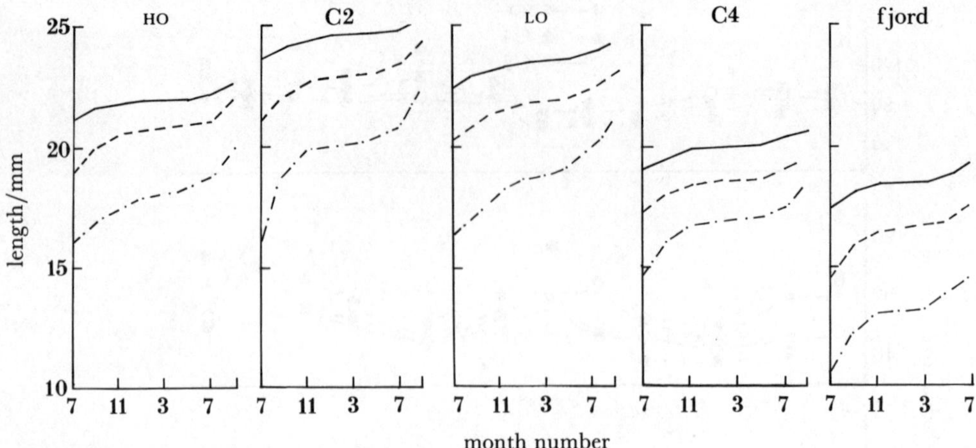

FIGURE 5. Growth rates of *Littorina littorea* exposed to oil (WAF), data from E. Lystad & K. Moe, unpublished results. HO = 127 μg l^{-1}, LO = 27.9 μg l^{-1}. C2 and C4, basin controls; fjord, fjord control. Solid line, year class 3; broken line, year class 2; broken and dotted line, year class 1.

TABLE 1. SURVIVAL AND MORTALITY RANGE OF *LITTORINA LITTOREA* (SEP. 1982, SEP. 1983)

(HO, 127 μg l^{-1}, LO, 27.9 μg l^{-1}; C2 and C4, controls (data from E. Lystad & K. Moe, unpublished results))

	HO	LO	C2	C4
relative survival (S)	0.51	0.66	0.65	0.68
absolute mortality (M)	0.67	0.42	0.43	0.39

TABLE 2. GROWTH RATE OF *LITTORINA LITTOREA* (JULY 1982 TO JULY 1983)

(K values from the Von Bertalanffy equation. HO, 127 μg l^{-1}, LO, 27.9 μg l^{-1}; C2 and C4, controls; fjord, fjord control)

	HO	LO	C2	C4	fjord
K	0.62	0.75	0.53	0.62	0.34

here also showed similar trends. There is no evidence therefore that oil leads to stabilizing selection on *L. littorea.*

M. edulis populations placed in nylon mesh tubes in the basins showed markedly different mortalities among basins. Figure 7 shows that at the high-oil dose 95% of the 10–15 mm size fraction transplanted were dead after 5 months compared with 50% in the control basin and 30% in the fjord control. Mortality rates were less at other size intervals but in HO reached over 95% in a year. With such marked mortality rates one might expect that survivors would show differences in genetic composition to the initial population.

Electrophoretic analyses on enzyme polymorphisms showed that *M. edulis* had an average heterozygosity of 11.5±3.7% for 30 loci studied (Fevolden & Garner 1986). Fevolden and Garner's analyses showed no significant deviation from Hardy–Weinberg expectation indicating that Oslofjord is inhabited by one single randomly mating stock. Six polymorphic loci were compared among basin populations and with the fjord control. A classification analysis of the polymorphic gene loci showed (figure 8) that between the basin and fjord populations

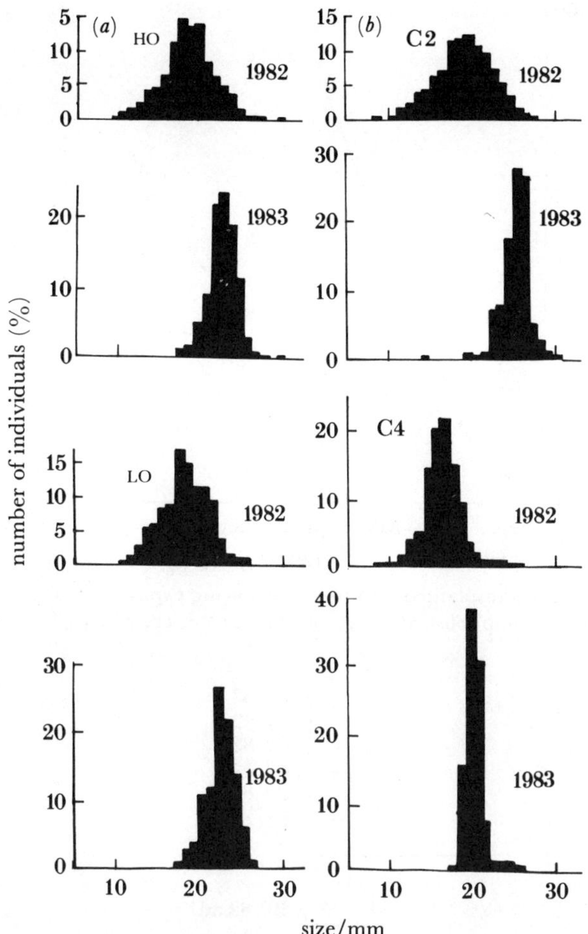

FIGURE 6. Size-frequency distribution of *L. littorea* exposed to oil (WAF) over 12 months, data from E. Lystad & K. Moe (unpublished results). (*a*) HO, 127 μg l^{-1}, LO, 27.9 μg l^{-1}. (*b*) C2, C4, control basins.

adults differed at the *LAP*-3 and *PGI* loci, whereas the new recruits differed at the *IDH*, *LAP*-2 and *PGM/C*-3 loci in one year, but not in the next year. One locus, *IDH*, showed differences between oiled and non-oiled recruits in 1983 but not in recruits in 1984. Thus year to year variations in allele frequencies are probably more important than effects of oil.

Genetic studies done on *L. littorea* (S. E. Fevolden, unpublished data) and *Semibalanus balanoides* (Fevolden & Sigurdsson, unpublished data) show similar results with little evidence that oil leads to dramatic short-term selection on the enzymes studied.

Extensive studies have also been done on the physiology, biochemistry and cell biology of *M. edulis* and *L. littorea* by scientists from IMER (see M. N. Moore *et al.*, this symposium, for a detailed discussion of results). The suspension-feeding rate of *M. edulis* was 67 % of the control in LO and only 34 % of the control in HO. Food-absorption efficiency was also less in exposed animals. Oxygen consumption did not, however, vary among exposed and control animals whereas ammonium excretion was greater in HO but the same as the controls in LO (Widdows & Donkin 1985). Lysosomal membrane stability was reduced in digestive cells of *M. edulis* and *L. littorea* at both HO and LO (Moore *et al.* 1984). NAPDH–neotetrazolium reductase, a cytochrome marker for NAPDH–cytochrome P-450 reductase, a component of the microsomal

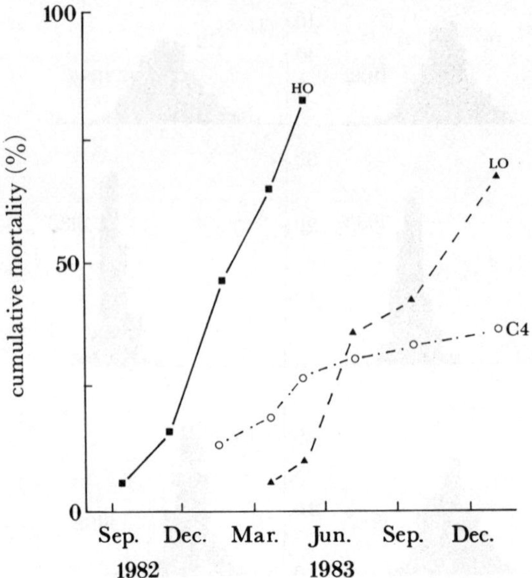

FIGURE 7. Mortality of *Mytilus edulis* transplanted into basins, following exposure to oil over 12 months, data from Walday & Thome (unpublished data). HO, 127 μg l^{-1}; LO, 27.9 μg l^{-1}; C4, control.

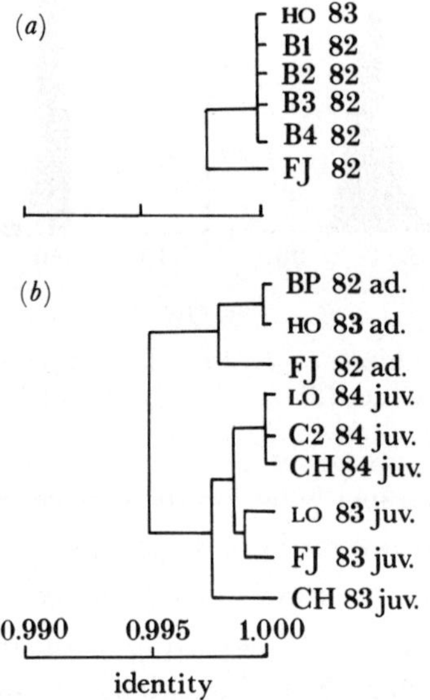

FIGURE 8. Classification analyses of population genetics data on *Mytilus edulis* using Nei's unbiased identity coefficients (Nei 1978). Data from Fevolden & Garner (1986). (*a*) Comparison of 20 gene loci in adult *M. edulis* from basins (B1–B4) and fjord (FJ) before oil exposure (82, 83, 1982, 1983). (*b*) Comparison of polymorphic gene loci after exposure to oil for 12 months. BP, pooled basin samples; HO, 127 μg l^{-1}; LO, 27.9 μg l^{-1}; C2 and C4, controls; FJ, fjord control; ad., adult; juv., juveniles.

detoxification–toxification system was elevated in *M. edulis* and *L. littorea* at both oil doses (Moore *et al.* 1984). Yet although the negative effects occurred rapidly, in one day, recovery was possible with lysosomal stability being restored in 3–4 d.

Effects on hard-bottom system

The results of the exposure of intertidal systems to chronic doses of oil over a period of 25 months showed that except for the high mortalities in populations of *M. edulis*, effects were not dramatic. The macroalgal community structure remained fairly constant with no change in *A. nodosum* or in *L. digitata*. *F. serratus* showed a slightly decreased population density in the higher-dosed basin. These findings are in keeping with known effects on algal communities. After the *Torrey Canyon* oil-spill there was no evidence of a severe and direct impact of oil on intertidal or subtidal algal communities (Chasse 1980). Similarly in Milford Haven, an area subject to continual small oil-spills, no changes that could be attributed to oil were observed in macroalgal communities over a 20 year period (Dicks & Hartley 1982).

Some algae, notably the *r*-selected opportunist species (*P. lenormandii*, *C. crispus*, *C. rupestris* and *U. lactuca*) in fact increased in abundance in the low-oil dose. This increase could not be related to a smaller population density of the herbivorous gastropod *L. littorea* as no comparable increase occurred in the control basins (Bokn & Moy 1985). Small amounts of oil probably directly enhance growth in these species although the mechanism of enhancement is unknown.

At the individual level there were marked effects of oil on growth rates of *A. nodosum* and *L. digitata*. Thus effects of chronic oil-pollution are likely to be found after many years of exposure and could lead to severe problems where commercial exploitation of these species occurs, as in the Norwegian alginate industry, where there is an annual harvest of 150 kt from a standing stock of 1.8 Mt (Bokn 1985).

At the high-oil dose there was increased mortality in *L. littorea* populations compared with the controls and the low-oil basin. However, in all basins the populations declined indicating that there is a direct effect of enclosing the system within the mesocosm. Lower recruitment was the primary cause of the decline but in addition to this effect the high-oil dose had a significantly lower level of recruitment. *L. littorea* showed large mortality after the *Torrey Canyon* oil-spill where an estimated 54% of the population was killed (Chasse 1980). However, no measurable changes in populations occurred in Milford Haven with irregular oil-spills over a 20-year period, suggesting that recruitment was not reduced. In the oiled basins *M. edulis* did not secrete byssus threads and fell to the floor where they were eaten by *Carcinus maenas*. The lack of byssus secretion was oil-induced and mortality was not just an effect of predation as caged animals not exposed to crab predation also suffered high mortalities (IMER, personal communication). This high *M. edulis* mortality was induced at a dose of only 27.9 $\mu g\ l^{-1}$. This response has been noted previously by Linden (1977), who found reduced secretion of byssus at oil concentrations of 130 μg hydrocarbons l^{-1} equivalent to the highest dose used in our experiments.

M. edulis is a major component of the fouling community of North Sea oil rigs and platforms and greatly increases stress on metal structures by giving greater surface area for wave impingement. Perhaps a solution to this problem is to discharge small amounts of oil continuously at concentrations up to 130 $\mu g\ l^{-1}$ (that used here) so that byssus secretion is inhibited and high mortality occurs.

With this high mortality it was surprising to us that there was no evidence of short-term genetic selection. Correlations between balancing selection and allozyme or genotype frequencies in *M. edulis* have been demonstrated in response to salinity (Koehn *et al.* 1976, 1980*a*, *b*; Thiesen 1978; Gartner-Kepkay *et al.* 1983), to temperature (Koehn *et al.* 1976, 1980*a*; Levinton & Suchanek 1978) and degree of exposure (Gosling & Wilkins 1981). Battaglia *et al.* (1980) found that the Mediterranean mussel, *Mytilus galloprovincialis*, had higher frequencies of the alleles for 6-phosphoglucosedehydrogenase and *PGI* along a pollution gradient in the lagoon of Venice. This was not the case with the Oslofjord data for *M. edulis*, *L. littorea* or *S. balanoides*. It is possible, as Nei (1983) suggested, that there was not sufficient genetic variability on which selection could operate, or alternatively it may be that although a wide range of enzymes (30) were tested, others not tested were selected.

The physiological and biochemical data show that sublethal effects do occur in both *M. edulis* and *L. littorea* but that these effects are reversible on return to clean seawater.

The soft sediment mesocosm

The advantages of a subtidal soft-sediment mesocosm are probably greater than those of an intertidal, in that experiments cannot be done in a controlled manner in the subtidal even by diving and by the use of submersibles. Only the MERL system (Pilson *et al.* 1977) includes subtidal sediment but at MERL the design is for an integrated water-column sediment system and sampling of the benthos is done blindly through a 5.5 m high water-column. The Solbergstrand system is designed to establish benthic communities at medium scales (i.e. 0.25 m^2 boxes) where the surface of the sediment can be sampled at fine scales without disturbing adjacent areas.

The mesocosm consists of two indoor concrete basins each coated with epoxy resin (4.9×21.5 m and 4.6–5.0×21.5 m). Each basin is further divided into three sections by removable vertical walls. Across each basin runs a moveable bridge to facilitate loading of sediment boxes in the basins. Seawater from 42 m depth in the fjord flows into each section by a horizontal pipe with 2.5 mm holes placed 5 cm apart, 60 cm above the tank bottom. Water flows in a laminar fashion across the basin to a horizontal outflow pipe with 4.5 mm diameter holes, 5 cm apart. The outflow pipe has adjustable height and water depth can be varied from 0 to 1.7 m. Water flow across the basins is approximately 1 cm s^{-1}. Over the first two years of operation the salinity was never below 30‰ and the temperature varied from 5 to 11 °C. Light can be varied from 0 to 0.2 μE m^{-2} s^{-1}† and follows a natural light–dark cycle.

Initially experiments were designed to clarify the amount of food material that needed to be added to keep the benthic community alive and functioning approximately normally. Powdered *Ascophyllum nodosum* was added at two levels equivalent to 50 and 200 g C m^{-2} with untreated boxes left as controls. Sediment was obtained by sampling with a 0.25 m^2 box-corer (a modified USNEL spade-corer built by Adolf Wuttke, Hamburg). The sediment was placed in plastic boxes (0.54 m^2 surface area, depth 20 cm) with the result that much disturbance of the surface occurred.

In the second experiment a removable liner was placed within the box-corer which enclosed the sample. On retrieving the sampler the liner was placed on a bottom plate for transport and placement within the basins. Samples obtained with the liner usually contained a 30 cm depth of undisturbed sediment and a supernatant water layer of 10 cm. In this experiment natural plankton was added in suspended form to the sediment four times over a six-week period giving total doses of 5 g C m^{-2} and 20 g C m^{-2}. Three boxes were used for each treatment and three remained as controls.

Samples were taken for analysis of sediment and water chemistry, bacteria, meio- and macrofauna.

† 1 Einstein = 1 mol photons; 1 μE = 6×10^{17} photons.

Results

With the addition of *A. nodosum* powder there was little difference between the experiment and the field in relation to profiles of water content, carbon and nitrogen (figure 9). The change at 4–6 cm depth can undoubtedly be related to disturbance of the sediment on filling the boxes. The high dose of *A. nodosum* led to a marked decrease in *Eh* and pH (down to 2.0) compared with the field. In the natural phytoplankton addition, pH remained between 7.2 and 7.4. Oxygen consumption measured by bell-jar experiments was 214, 262 and 438 μmol m^{-2} h^{-1} in the control, 5 g C m^{-2}, and 20 g C m^{-2} respectively (M. Schaaning, unpublished data).

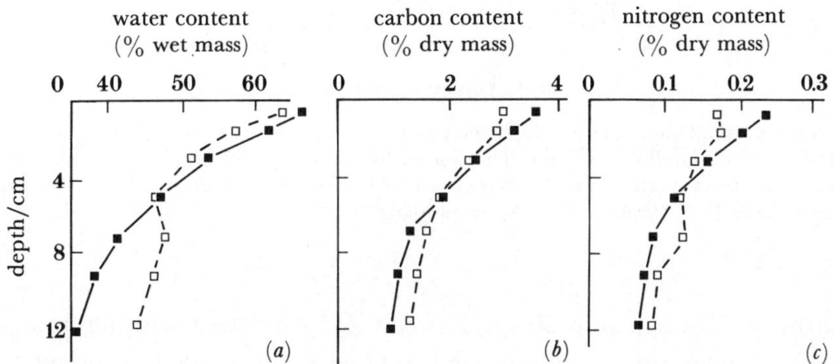

FIGURE 9. Profiles of (*a*) water content, (*b*) carbon and nitrogen content of sediment in mesocosm (broken line) one year after transplantation without using a core liner, compared with the field (solid line), data from M. Schaanning (unpublished results).

Bacterial production as measured by [^{3}H]thymidine incorporation into bacterial DNA showed no significant difference between field and mesocosm over a six-month period (G. M. Skeie, personal communication).

Four box-core samples were taken at 30 m depth in Bjornhodebukta, Oslofjord in December 1984. Two were immediately sieved through a 1 mm sieve and the remaining two sieved after six months in the mesocosm. Table 3 shows that 50% of the total number of species found occurred in both field and mesocosm. The 31% found only in the field represented only 2% of the number of individuals. The 10% found only in the mesocosm represented only 1% of the number of individuals. Figure 10 shows the relation between densities in field and mesocosm. The dominant species show similar abundances in both, whereas motile species such as the amphipods, *Ophiura affinis* and two sedentary spionid polychaetes were less abundant in the mesocosm. Surface and subsurface deposit feeders have almost identical abundances in field and mesocosm whereas filter feeders and carnivores had only 50% of the field abundance in mesocosm (Berge *et al.* 1986).

TABLE 3. COMPARISON OF MESOCOSM, SEVEN MONTHS AFTER ESTABLISHMENT, WITH FIELD DATA

(Data from two replicate 0.25 m^2 box-core samples.)

	species	individuals
in mesocosm and field	41 (50%)	2752 (97%)
in mesocosm only	10 (12%)	29 (1%)
in field only	31 (38%)	59 (2%)
total	82	2840

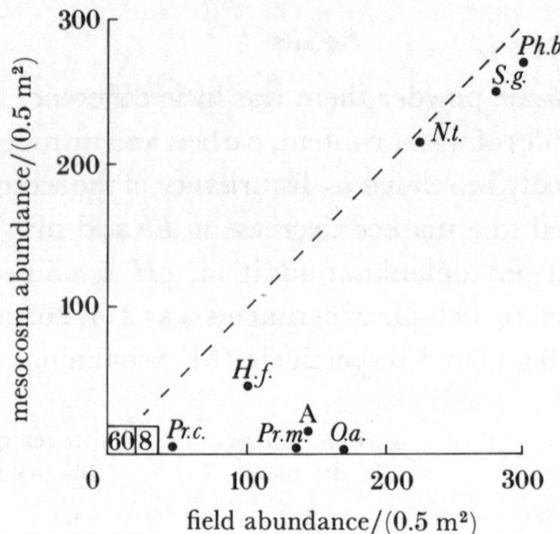

FIGURE 10. Comparison of faunal abundances in field and mesocosm. The numbers within boxes indicate the number of species found at corresponding densities. The dotted line shows perfect similarity. Abbreviations: *Ph.b.*, *Philomedes brenda*; *S.g.*, *Sosane gracilis*; *N.t.*, *Nuculoma tenuis*; *H.f.*, *Heteromastus filiformis*; *Pr.m.*, *Prionospio malmgreni*; *Pr.c.*, *Prionospio cirrifera*; *O.a.*, *Ophiura affinis*; A, Amphipoda.

Over 19 months in the mesocosm 30 species recruited compared with 66 in a similar box placed in the field, whereas only 77 individuals recruited to the mesocosm compared with 891 individuals in the field.

In addition to the above, studies of local-scale bioturbation on meiofauna by the large-sized but rare polychaete *Streblosoma bairdii* have been done (Warwick *et al.* 1986). These studies showed that the tentacle feeding area of the deposit feeder had a reduced population density and a higher diversity of meiofauna compared with the unaffected sediment. The nematodes in the unaffected sediment were primarily diatom feeders whereas bacterial feeding nematodes were confined to feeding mounds. Time-lapse studies of *S. bairdii* feeding patterns have been done and a paper is in preparation (Rumohr *et al.* 1987).

We conclude from these results that the observations on sediment chemistry in the sediment transported to the mesocosm are within the range of variation found in the natural environment. Although recruitment of macrofauna is reduced, the benthic community in terms of the dominant species and community structure closely follows that of the natural environment. The huge advantage of the mesocosm is the ease of sampling and manipulation of the communities in an almost undisturbed state. The important role that bioturbation plays in the flux of nutrients and pollutant materials to and from the sediment can be studied in detail. The system is now being used to simulate and experiment on communities taken from 200 m depth, giving us the possibility of assessing effects of pollution over analogue communities to the whole continental shelf. For the first time effects can be studied directly and not be subject to the inefficiency of remote and blind sampling with subsequent *a posteriori* data interpretation.

CONCLUSIONS

Use of enclosed meso-scale systems for the study of effects of pollutants allows examination of effects of controlled doses of pollutants over long-time periods in a manner not possible in the natural environment. Care must be taken, however, to ensure that the effects of enclosing the system are carefully documented. Although recruitment of macroalgae was not affected in the Solbergstrand hard-bottom mesocosm, molluscs were affected. Moreover, the basins were not identical in faunal and floral composition at the outset of the experiments with the result that controls were different and results hard to interpret. Nevertheless, clear long-term effects occurred as a result of oil pollution, notably on algal growth, recruitment of *L. littorea* and byssus secretion of *M. edulis*, all of which factors in the natural environment would lead to severe consequences for the communities concerned.

In the soft-sediment mesocosm, after reliable methods for establishing the community were established, it was possible to hold alive for periods of six months communities simulating the continental shelf. Thus unique opportunities are now available for assessing effects of pollutants on important processes such as fluxes of material into and out of the sediment, the effects of bioturbation on sediment processes and the role of key organisms in controlling community structure and processes.

I thank B.P. Petroleum Development (Norway) for financing this project, which is a joint project between the Norwegian Institute for Water Research, the University of Oslo and B.P. I am indebted to my colleagues associated with the project for so freely allowing me to use their data. In particular I thank Torgeir Bakke and Tor Bokn (NIVA), John-Arthur Berge, Morten Schaaning, Svein Fevolden, Geir Scheie, Einar Lystad, Kjell Moe and Fridtjof Moy.

References

Bakke, T. 1985 Energy balance in *Littorina littorea*. *A. Rep. Solbergstrand exp. Stn* **1984**, 13.1–13.11.

Bakke, T. & Sørensen, K. 1985 Chemical analyses of hydrocarbon content in water. *A. Rep. Solbergstrand exp. Stn* **1984**, 2.1–2.14.

Battaglia, B., Bisol, P. M. & Rodino, E. 1980 Experimental studies on some genetic effects of marine pollution. *Helgoländer wiss. Meeresunters.* **33**, 587–595.

Berge, J. A., Schaaning, M., Bakke, T., Sandøy, K. A., Skeie, G. M. & Ambrose, W. G., Jr 1986 A soft bottom sublittoral mesocosm by the Oslofjord: descriptions, performance and examples of application. *Ophelia* **26**, 37–54.

Bodungen, B. v., von Brockel, K., Smetacek, V. & Zeitschel, B. 1976 The plankton tower: structure to study water/sediment interactions in enclosed water columns. *Mar. Biol.* **34**, 369–372.

Bokn, T. 1984 Effects of diesel oil on recolonization of benthic algae. *Hydrobiologia* **116/117**, 383–388.

Bokn, T. 1985 Effects of diesel oil on commercial benthic algae in Norway. In *Proceedings of the 1985 Oil Spill Conference (Prevention, Behaviour, Control, Cleanup)*, Los Angeles, pp. 491–496. Washington, D.C.: American Petroleum Institute.

Bokn, T. & Moy, F. 1987 Effects of diesel oil on marine littoral rock communities. (In preparation.)

Chasse, C. 1980 Les leçons d'un bilan écologique modéré, quoique coûteux et menaçant. In *Impact socioéconomique de la marée noire provenant de l'Amoco Cadiz* (ed. F. Bonnieux, P. Dauce & P. Rainelli), pp. 74–92. Rennes Rapport UVLOE-INRA.

De Wilde, P. A. W. J. & Kuipers, B. R. 1977 A large indoor tidal mud-flat ecosystem. *Helgoländer wiss. Meeresunters.* **30**, 334–342.

Dicks, B. & Hartley, J. P. 1982 The effects of repeated small oil spillages and chronic discharges. *Phil. Trans. R. Soc. Lond.* B **297**, 285–307.

Farke, H., Schulz-Bades, K., Ohm, K. & Gerlach, S. A. 1984 Bremerhaven Caisson for intertidal field studies. *Mar. Ecol. Progr. Ser.* **16**, 193–197.

Fevolden, S. E. & Garner, S. P. 1986 Population genetics of *Mytilus edulis* from Oslofjord in oil polluted and non-polluted water. *Sarsia* **71**, 247–258.

Gartner-Kepkay, K. E., Zouros, E., Dickie, L. M. & Freeman, K. R. 1983 Genetic differentiation in the face of gene flow: a study of mussel populations from a single Nova Scotia embayment. *Can. J. Fish. aquat. Sci.* **40**, 443–451.

Gosling, E. M. & Wilkins, N. P. 1981 Ecological genetics of the mussels *Mytilus edulis* and *M. galloprovincialis* on Irish coasts. *Mar. Ecol.* **4**, 221–227.

Grice, G. D. & Reeve, M. R. 1982 Introduction and description of experimental ecosystems. In *Marine Mesocosms* (ed. G. D. Grice & M. R. Reeve), pp. 2–9. New York: Springer Verlag.

Koehn, R. K., Milkman, R. & Mitton, J. B. 1976 Population genetics of marine pelecypods. IV. Selection, migration and genetic differentiation in the blue mussel *Mytilus edulis*. *Evolution* **30**, 2–32.

Koehn, R. K., Newell, R. I. E. & Immerman, F. 1980*a* Maintenance of an aminopeptidase allele frequency cline by natural selection. *Proc. natn. Acad. Sci. U.S.A.* **77**, 5385–5389.

Koehn, R. K., Bayne, B. L., Moore, M. N. & Siebenhaller, J. F. 1980*b* Salinity-related physiological and genetic differences between populations of *Mytilus edulis*. *Biol. J. Linn. Soc.* **14**, 319–334.

Levinton, J. S. & Suchanek, T. H. 1978 Geographic variation, niche breadth and genetic differentiation at different geographic scales in the mussels *Mytilus californianus* and *M. edulis*. *Mar. Biol.* **49**, 363–375.

Linden, D. 1977 Sublethal effects of oil on mollusc species from the Baltic Sea. *Wat. Air Soil Pollut.* **8**, 305–313.

Moore, M. N., Livingstone, D. R. & Lowe, D. 1984 Sublethal cellular and molecular effects and short-term recovery of mussels (*Mytilus edulis*) and periwinkles (*Littorina littorea*) following chronic exposure to petroleum hydrocarbons. *A. Rep. Solbergstrand exp. Stn* **1984**, 14-1–14-37.

Nei, M. 1978 Estimation of average heterozygosity and genetic distance from a small number of individuals. *Genetics* **89**, 583–590.

Nei, M. 1983 Genetic polymorphism and the role of mutation in evolution. In *Evolution of genes and proteins* (ed. M. Nei & R. K. Koehn), pp. 165–190. Sunderland, Massachusetts: Sinauer Associates.

Pilson, M. E. D., Vargo, G. A., Gearing, P. & Gearing, J. N. 1977 The Marine Ecosystem Research Laboratory: a facility for the investigation of effects and fates of pollutants. In *Proceedings of the Second National Conference on the Interagency Energy/Environmental R and D Program*, pp. 513–516. Washington, D.C.: U.S. Environmental Protection Agency.

Rumohr, H., Ambrose, W. G. & Berge, J. A. 1987 Feeding in *Streblosoma bairdii* studied by time-lapse photography. (In preparation.)

Thiesen, B. F. 1978 Allozyme clines and evidence of strong selection at three loci in *Mytilus edulis* L. (Bivalvia) from Danish waters. *Ophelia* **17**, 135–142.

Warwick, R. M., Gee, J. M., Berge, J. A. & Ambrose, W. G. 1986 Effects of the feeding activity of the polychaete *Streblosoma bairdii* (Malmgren) on meiofaunal abundance and community structure. *Sarsia* **71**, 11–16.

Widdows, J. & Donkin, P. 1985 Sublethal biological effects and recovery of mussels (*Mytilus edulis*) following chronic exposure to petroleum hydrocarbons: physiological responses. *A. Rep. Solbergstrand exp. Stn* **1984**, 12.1–12.12.

Phil. Trans. R. Soc. Lond. B **316**, 655–668 (1987)
Printed in Great Britain

Environmental monitoring of the Beatrice oilfield development

BY J. M. ADDY

Britoil plc, P.O. Box 120, *Britoil House, Hill of Rubislaw, Anderson Drive, Aberdeen AB*9 8*XB, U.K.*

The Beatrice oilfield is the first nearshore development in United Kingdom waters. Situated in an area of major ecological and nature-conservation importance, the development of Beatrice has demanded thorough attention to environmental planning and monitoring through all stages.

Assessment of the environmental implications of the development was a major concern from the earliest stages of project planning. A range of baseline studies was performed to provide a basis for future monitoring. During the planning phase, as later in the field's development, consultation and communication with the statutory bodies and local community was a priority. The environmental assessment and its component studies were discussed with all concerned. As well as providing the detailed scientific data, the work was also presented in summary form to the local communities. The local community as well as the authorities needed to be reassured that the environment would be protected.

Monitoring of the marine environment of the Moray Firth is a continuing activity involving a wide range of studies by specialist teams in four main areas: (1) ornithological studies; (2) intertidal monitoring; (3) sublittoral studies; (4) others, including structural fouling by marine growth. The long-term monitoring programme is built on extensive baseline data gathered before oil was produced at Beatrice. The scope of this work will be outlined in this paper and examples given of some of the results. In addition to the monitoring aspects, the data is often of considerable academic interest, and Britoil encourages publication of scientific papers arising from the programme. The overall conclusions are that only very localized environmental impact has been caused beneath and in the immediate vicinity of the drilling platforms and that development and operation of the Beatrice field has not damaged the marine environment and resources of the Moray Firth.

INTRODUCTION

The Beatrice field was discovered in 1976 and at a distance of only 12 miles off the Scottish mainland is currently Great Britain's nearest inshore oilfield (figure 1). The development consists of four offshore platforms, a 49 mile long pipeline and a storage and tanker loading terminal at Nigg Bay in the Cromarty Firth. These various facilities are operated by Britoil on behalf of a consortium consisting of: Britoil plc, Deminex U.K. Oil & Gas Ltd, Hunt Overseas Oil Inc., Kerr–McGee Oil (U.K.) Ltd and Lasmo North Sea plc.

Developed within sight of land, and surrounded by an environmentally significant coastline, the Beatrice Field and associated facilities demanded that special attention be given to protect and monitor the environment. From the early days of the development, indeed before the development plan was approved, biological data has been obtained from a wide range of studies throughout the area.

The aims of this work were twofold: to establish a series of environmental baselines

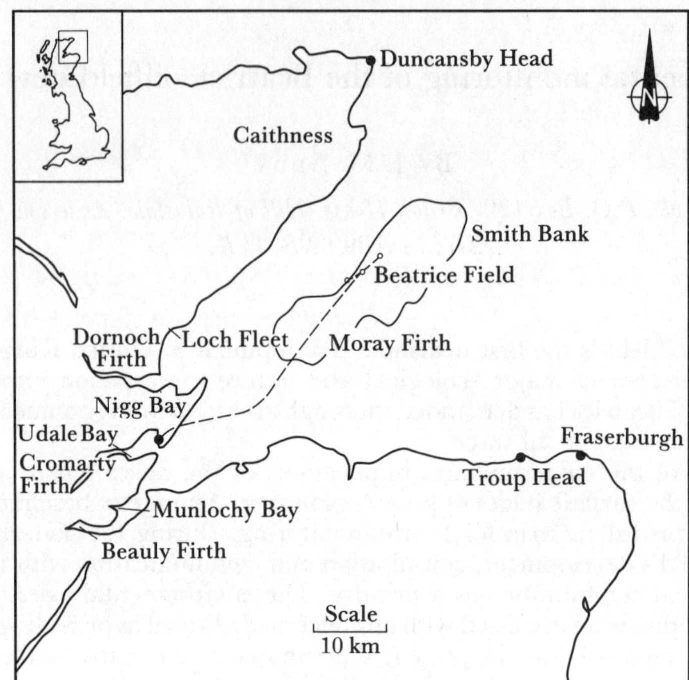

FIGURE 1. Location of the Beatrice oilfield, 12 miles off the coast of Caithness, in the Moray Firth, and the Nigg Oil Terminal in the Cromarty Firth.

representing biological and chemical conditions before the start up of oil production from the field; to use these reference points as a basis for comparison with subsequent data gathered through the continuing monitoring programme.

A large quantity of information on the marine environment of the Moray Firth has been gathered resulting in a number of publications. Much of the information gathered up to 1985 was presented at a symposium at Aberdeen University in March 1985 (ed. R. Ralph 1986). This present paper outlines the range of investigations, carried out in the environmental monitoring programme sponsored by the Beatrice Consortium, and provides examples of results from each part of the programme with the aim of illustrating both the relevance for monitoring of oilfield operations and the biological interest.

ORNITHOLOGICAL STUDIES

The Moray Firth is an area of international importance for a variety of bird species associated with the coastal and marine environment. Throughout the year major concentrations of birds occur both at sea and at several sites along the coastline. It contains some of the best estuary sites in Great Britain and is a main arrival point for immigrant birds which have come from Iceland and northern Scandinavia. The Cromarty Firth itself qualifies under the criteria set down by the International Waterfowl Research Bureau, as an area of international importance.

The ornithological studies therefore constitute some of the more important aspects of the environmental monitoring programme. The main objectives have been to provide a database to allow long-term monitoring of bird populations and to assist the development of appropriate contingency measures for use in the event of any pollution incident.

Work on seabirds and sea ducks has been carried out by the Royal Society for the Protection of Birds (R.S.P.B.) and on shorebirds by the Nature Conservancy Council (up to 1985 after which routine long-term shorebird monitoring has been incorporated in the R.S.P.B. programme).

Breeding seabirds

Available information on the sizes and locations of seabird breeding colonies in the Moray Firth is reviewed by Mudge (1986). The most abundant species being guillemot, with a population of approximately 149000 individuals spread over 13 main colonies. The bulk of the populations of breeding species occur on the coast of east Caithness. Data are presented by Mudge from census study plots which were set up in 1980 to monitor any long term and large annual changes in population size. A comparison with counts made in 1969 (table 1) permits assessment of longer term changes, whereas new data from the period 1980–84 allow shorter term comparisons from year to year.

TABLE 1. WHOLE COLONY COUNTS OF SEABIRDS AT PRINCIPAL SITES IN THE MORAY FIRTH. A COMPARISON BETWEEN 'OPERATION SEAFARER' AND MORE RECENT SURVEYS

(From Mudge (1986).)

	east Caithness		North Sutor of Cromarty		Troup, Lion's and Pennan heads	
	1969	1977	1969	1984	1969	1979
fulmar	16161	21679	779	1393	1180	1612
cormorant	823	284	87	203	0	0
shag	1114–1292	1863	24	30	83	81
kittiwake	32282–34322	53025	400	329	11425	14267
razorbill	12274–13173	14196	60	41	629	1959
guillemot	49483	126251	750	933	8869	21827
tystie	197	408	0	2	+	1
puffin	4722–28203	915	0	0	397	523

Count units: fulmar, apparently occupied sites; cormorant and shag, nests; kittiwake, apparently occupied nests; auks, individual birds.

In summary, it appears that a substantial increase in population size occurred in the 1970s for most species although this trend of increase has now ceased for guillemots, razorbills and kittiwakes. The trend of change over the 1980–84 period for these species may have been downwards (see, for example, figure 2) but there are major difficulties of extrapolation from limited counts so any recent trends cannot be statistically substantiated. It can, however, be concluded that no major decline in population has occurred up to 1984 after the general increases over the 1970s.

In addition to counts in the breeding season, out-of-season attendance has been monitored using automatic surveillance cameras. (Mudge *et al.* 1985). Two species, fulmar and shag, were found to frequent the colonies over a large part of each non-breeding season. There was, however, only a low level of winter colony occupancy by auks, which contrasted with reports from colonies elsewhere. Mudge *et al.* (1986) suggested that the main current constraint is a paucity of local-breeding adults wintering within easy reach of the colonies which in turn

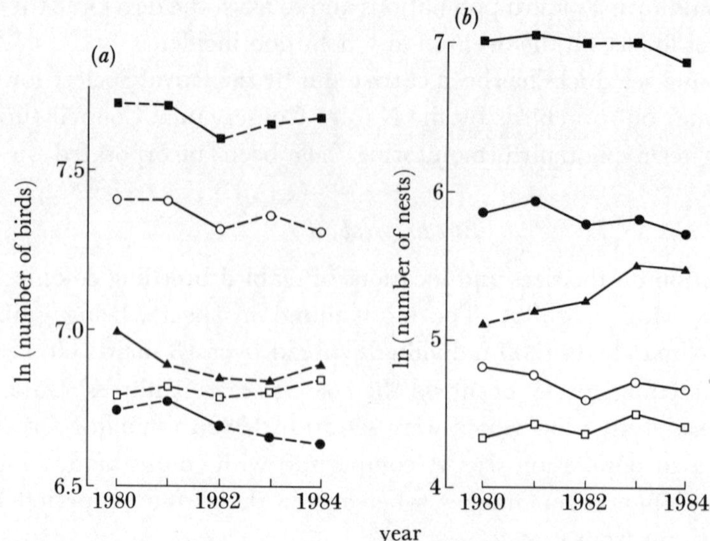

FIGURE 2. Numbers of (*a*) guillemots (individual birds) and (*b*) kittiwake nests in east Caithness plots 1980–84. (From Mudge 1986.)

may be because of recent declines in winter stocks of fish prey in the Moray Firth. Standard ringing techniques have been used to build up an understanding of migration movement and mortality. Coordinated counts offshore at the Beatrice platforms and at the cliff sites have provided information on feeding patterns and daily movements to and from the colonies.

Seabirds at sea

Until recently, very little was known of seabird distribution and movement within the Moray Firth as a whole. However, a large study was carried out in 1982 and 1983 to determine the feeding areas of key seabird species and their pattern of distribution throughout the year (Mudge & Crooke 1986). This has highlighted areas that may be particularly threatened by any oil spillage. A sea area of approximately 8000 km² was systematically studied by using a survey grid of 5′ latitude by 10′ longitude. Using a 'dedicated vessel', observations were made from a sheltered cabin above the wheelhouse by using methods that have evolved for seabird at sea studies in other countries and since developed and tested in British waters (Tasker *et al.* 1984). From monthly surveys over the two-year period, a very large quantity of data has been acquired and is stored and analysed on a computer database by using systems compatible with other systematic ship-based observations elsewhere in the North Sea. The information can be summarized for individual species or groups of species or for different times of year.

Mudge & Crooke (1986) describe the pattern of overall seabird abundance (figure 3) which showed fewest numbers in late winter, but rapidly increasing up to April, before the commencement of breeding. A slight drop occurred in most species during the breeding season. (May to July in most species.) At the end of the breeding season large flocks of auk chicks and moulting adults (all flightless) were present, peaking in August. It is at this time when seabirds in the Moray Firth are probably most vulnerable to any oil pollution incident. In September and October, as flocks disperse, the numbers of birds at sea drop off rapidly.

From the wealth of information gathered on seabird species distribution and variations in

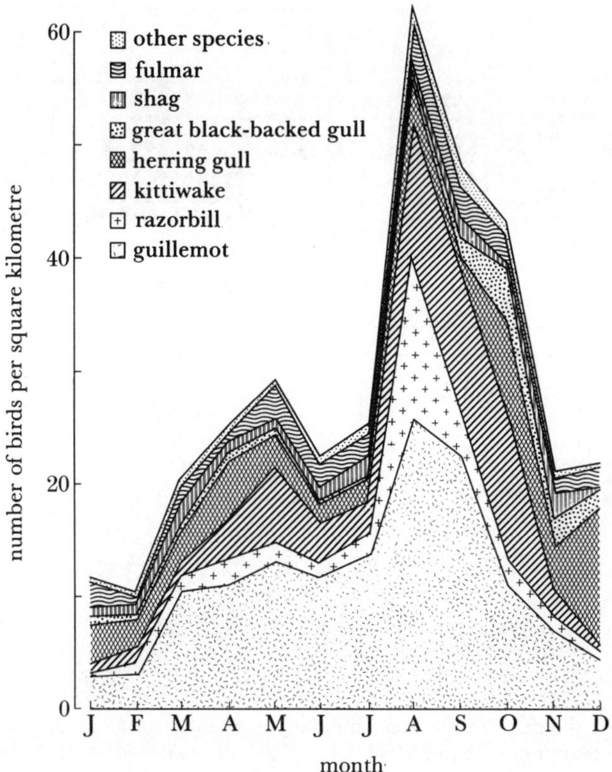

FIGURE 3. Average numbers of birds, and species composition at sea from month to month based upon data from 1982 and 1983 (From Mudge & Crooke 1986.)

this from month to month, perhaps the most useful summaries in relation to oil spill contingency planning are maps of selected species or groups of species at different times of the year. For example, figure 4 contrasts the distributions of razorbills which are found throughout the Moray Firth in summer, with shags which have a much more confined distribution close to their breeding site, throughout the year.

Mudge & Crooke (1986) note several important features which emerge from maps such as those in figure 4. Two areas in particular are held by large numbers of seabirds at all times of the year. One was an inshore area near the border of Caithness and Sutherland, and the other is the Troup Head and Fraserburgh area. The importance of the northeast corner of the Smith Bank was also noted as the location of the main daytime concentrations of seabirds for much of the year, and particularly in the pre and post breeding period.

Seaduck

Surveys of seaducks have been carried out as part of the Beatrice environmental programme since 1981. Campbell *et al.* (1986) reviewed the current status and distribution of seaducks in the Moray Firth and also compared recent data with earlier information to determine any trends and also present an overall conservation assessment of Moray Firth seaduck population. The main species and their estimated populations are shown in table 2, from which the importance of the Moray Firth seaducks in terms of the populations of the British populations is clear. For common and velvet scoter and long-tailed duck the Moray Firth holds more than

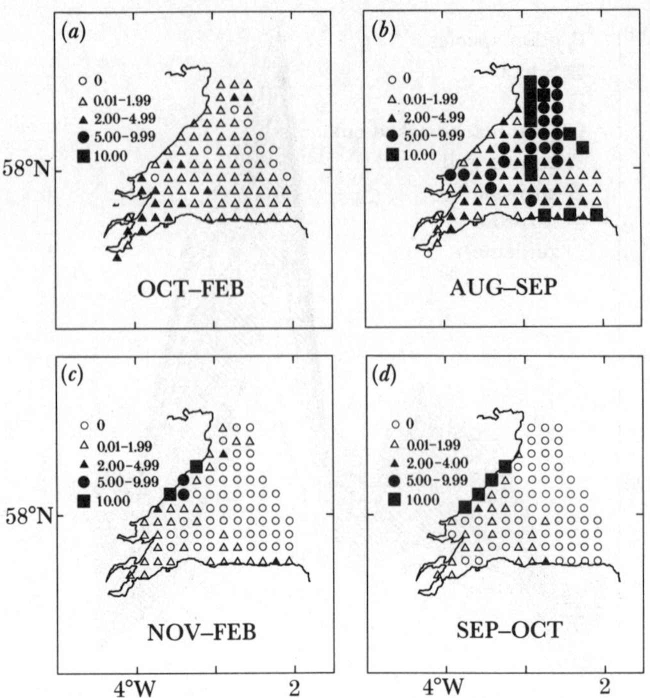

FIGURE 4. Seasonal changes in the distribution and abundance at sea of (*a*, *b*) razorbills and (*c*, *d*) shags. An amalgamation of transect data from 1982 and 1983, Units: birds per square kilometre. (From Mudge & Crooke 1986.)

TABLE 2. SUMMARY OF CURRENT WINTER SEADUCK POPULATION ESTIMATES IN THE MORAY FIRTH

(From Campbell *et al.* (1986).)

	Population estimates		
	Moray Firth	Great Britain	Western Europe
eider	3000	50000	2000000
common scoter	10000	35000	500000
velvet scoter	5000	3000	200000
long-tailed duck	15000	20000	500000
goldeneye	1000	15000	200000

1 % of the estimated European population which on those criteria makes it an internationally important site for those species.

Considering the patterns of numbers and distribution over a period of time, Campbell *et al.* conclude that although individual sites (figure 5) may appear outstanding for one or more species, for conservation purposes the Moray Firth should be considered to be a single seaduck site, within which local flock distribution patterns may vary considerably from year to year.

In the context of oil pollution, the sensitivities of wintering seaduck may be different from the generally more widely dispersed auks. The relative confinement of seaduck flocks in small areas or loose concentrations where they may be present locally at very high density may render them vulnerable not only to widespread or large volume incidents, but also to much more localized or small scale events (Campbell *et al.* 1978).

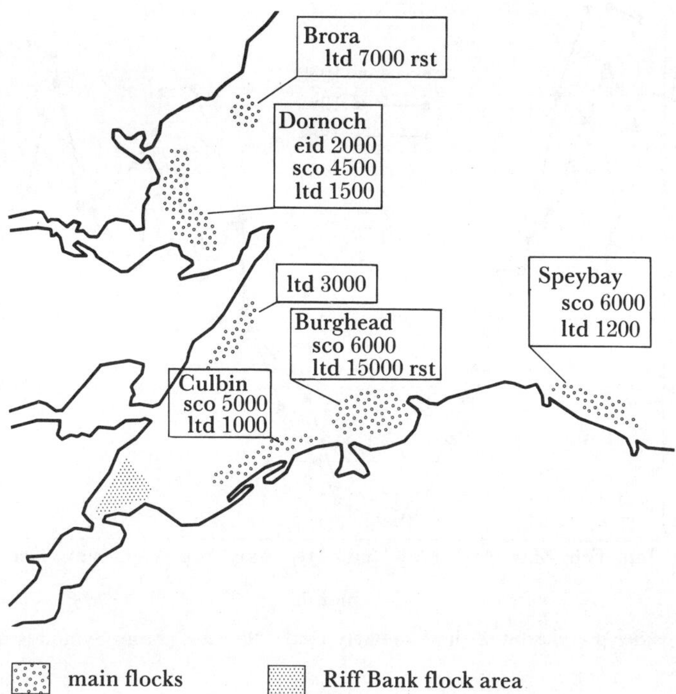

FIGURE 5. The location and approximate size of flocks of wintering eider (eid), scoters (sco) and long-tailed duck (ltd) in the Moray Firth in 1981–82 and 1982–83 (From Campbell *et al.* 1986.) rst, represents roosting flocks.

Shorebirds

The Moray Firth is of international importance for waders and wildfowl, supporting up to 36000 waders on mudflats in the area and up to 45000 dabbling ducks. A research programme between 1981 and 1985 was carried out by the Nature Conservancy Council as part of the Beatrice environmental monitoring programme. The aims of this work were to provide up to date counts of total populations, identify major feeding and roosting sites for each species, estimate movement between sites and assess the relationship between the Moray Firth and other intertidal areas. In all of these the underlying objective was to provide a basis for long-term monitoring of shorebird populations. The procedures used and main results have been summarized by Symonds & Langslow (1986).

Counts over the winter between 1981 and 1985 recorded peak oystercatcher and redshank numbers in excess of 1% of the total northwest European population. Bar-tailed godwit and knot exceeded this in 1984–85. Seasonal and annual trends in population levels were described both for the individual species and the total wader population (see, for example, figure 6). Summer numbers are low, comprising mainly immature oystercatchers. The rapid rise between July and October is made up first of oystercatcher and godwit, followed by redshank. Largely juvenile knot and dunlin arrive in October, followed by a further influx of bar-tailed godwit and adult knot and dunlin in November and December.

Wildfowl arrive in the Moray Firth from their breeding grounds from August to October before dispersing to wintering sites elsewhere in Britain. These autumn concentrations of wildfowl are the largest in Britain, and some areas continue to support large numbers throughout winter.

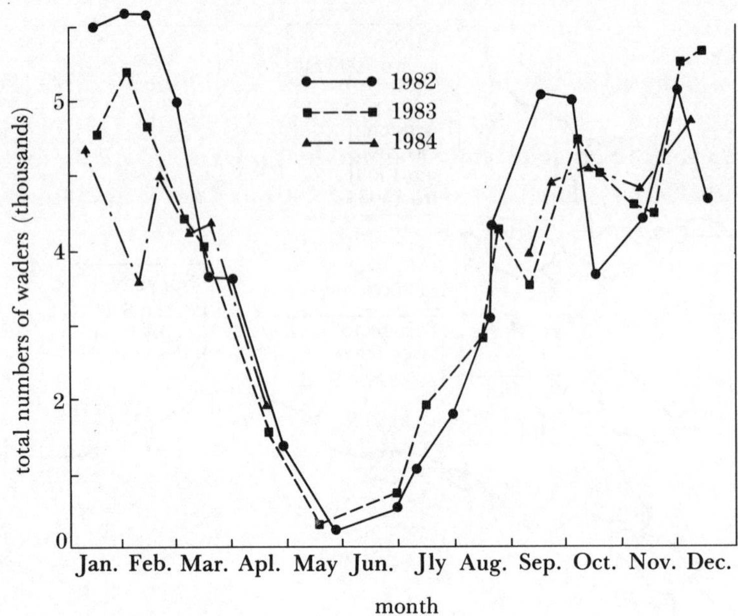

FIGURE 6. Total wader population in the Cromarty Firth 1982–84. (From Symonds & Langslow 1986.)

Loyalty of different species to feeding areas and movement between sites was studied by colour marking birds caught in cannon-netting exercises. Various patterns of movement emerged. Curlew, turnstone, oystercatcher, redshank and ringed plover remained attached to their selected areas, whereas mobile species such as dunlin, knot and bar-tailed godwit moved frequently between feeding areas.

Discussing the results in relation to nature conservation interests, Symonds & Langslow (1986) stress the major importance of the Moray Firth as a migration staging post and wintering area for both waders and wildfowl. Loss or damage to intertidal sites through reclamation pose threats the effect of which would be difficult to quantify. However, it is clear that future conservation planning policies must consider factors such as winter movements of birds and the local distribution of some species, as well as simply the total population sizes. The continuing long-term monitoring of shorebirds in the Moray Firth should greatly help understanding of these aspects.

INTERTIDAL STUDIES

A very wide range of shore habitats lies between Duncansby Head and Fraserburgh and into the Cromarty and Beauly Firths. Shingle ridges and sand dunes are present, and silt and mud in the major estuaries have resulted in the extensive flats at Nigg, Udale and Munlochy Bays, Loch Fleet and the upper Beauly Firth. The rocky shores occur mainly along the outer, more exposed areas of the Moray Firth.

The aims of the intertidal studies carried out by Aberdeen University Marine Studies Ltd have been to monitor shoreline conditions and community structures and to detect any long-term changes. The methods used enable comparison with results from elsewhere. The information, particularly from soft-shore monitoring is highly relevant to other investigations such as shorebirds and wildfowl.

Rocky shores

An intertidal rocky shore monitoring programme was established in the Moray Firth in 1981, covering 34 different locations between Duncansby Head and Fraserburgh (figure 7). The aim was to describe the range of shore communities present, relate these to the physical characteristics of the different shore types and also provide a basis for detecting any long-term biological changes on the shores of the Moray Firth.

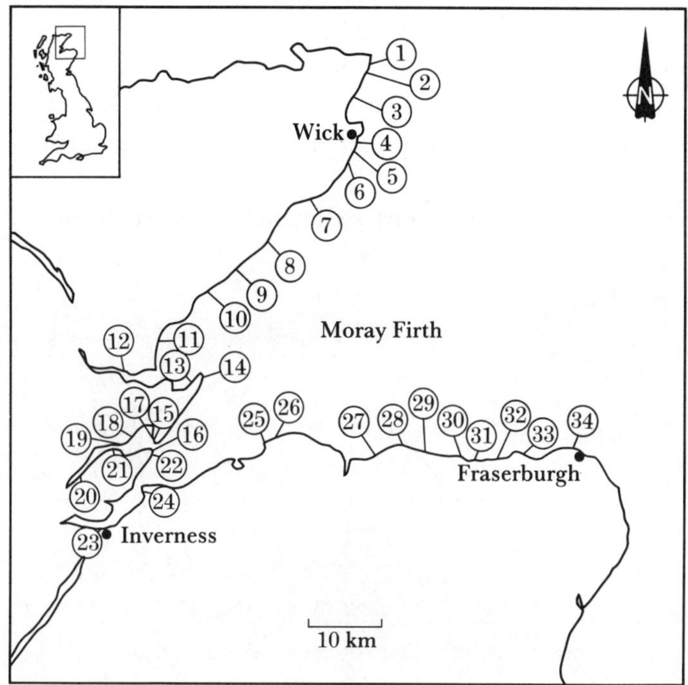

FIGURE 7. Location of rocky shore monitoring transects. (From Terry & Sell 1986.)

The survey methods used and the main results are described by Terry & Sell (1986). The methods used in this monitoring programme have been designed for the variety of shore topographies and biological communities encountered in this region. It is based on analysis of data from shoreline appraisal and transect studies, using a fixed-point quadrat method. Detailed species lists have been compiled for each permanent quadrat at each transect visit. Species composition has been related to exposure and substrate type. The dominant communities and the principal factors affecting species abundance and distribution at each transect location have been described (Terry & Sell 1986). One difficulty in rocky-shore monitoring programmes is interpretation of changes detected, whether natural, or pollution induced. One of the aims of this work has been to ease possible future interpretation by building up a history of the particular communities and identifying the community processes operating.

Nigg Bay sediments

A series of monitoring surveys in Nigg Bay by Aberdeen University Department of Zoology has provided an assessment of the major biological resources of the intertidal sand and mudflats, and the physical factors associated with the distribution and abundance of these resources (Raffaelli & Boyle 1986). Changes in species composition and abundance have been described in relation to natural fluctuations in physical parameters such as sediment silt content.

Information gathered from these studies is relevant to a number of interests. Firstly the aim was to detect any possible effects of the Nigg Oil Terminal, which is situated on reclaimed land at the mouth of the bay. In this context, measurement of physical sediment characteristics are particularly important, and it has been concluded (Raffaelli & Boyle 1986) that fluctuations in sediment silt content probably reflect natural changes from year to year in the sediments of Nigg Bay. Similarly, the biological data permits analysis of change, and the conclusion is drawn that the biological variations have been entirely natural and unrelated to the terminal.

Beyond the immediate monitoring value, the Nigg Bay biological data has considerable interest for conservation planning purposes. Mapping and quantification of shorebird food species (see, for example, figure 8) should greatly facilitate assessment of conservation interests in Nigg Bay.

In addition to the ecological investigations, sediment hydrocarbon concentrations have been

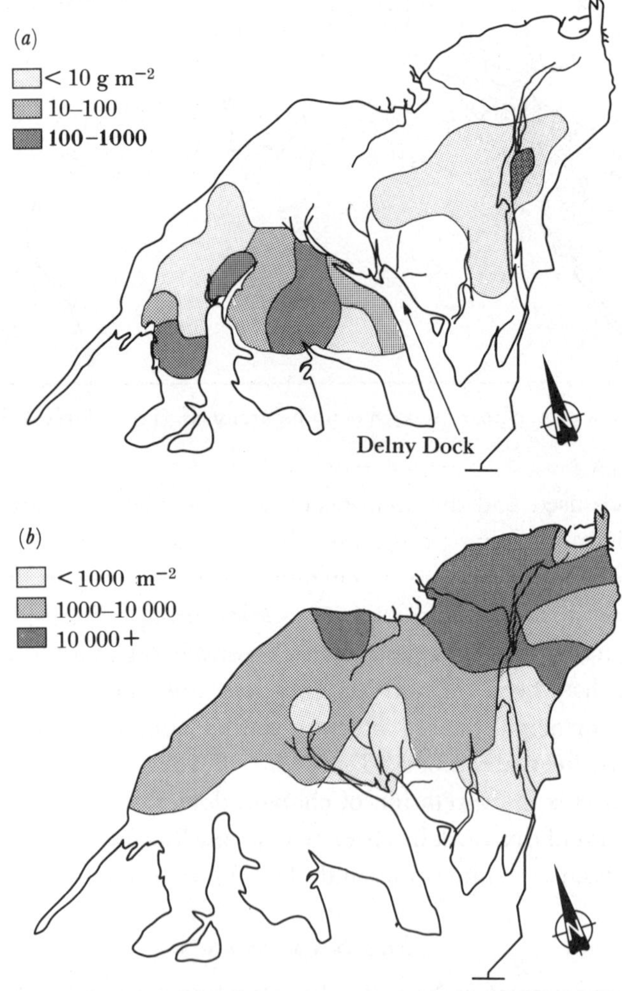

FIGURE 8. (*a*) Biomass of *Mytilus edulis*; (*b*) distribution of *Hydrobia ulvae* in Nigg Bay. (From Raffaelli & Boyle 1986.)

monitored from time to time (by M-Scan Ltd) since before the terminal became operational. As figure 9 shows, the background levels of hydrocarbon in Nigg Bay are low, with some patchiness, which in part can be accounted for by rainwater run-off from the road on the western shore. There has been no build up of hydrocarbons in Nigg Bay sediments since the terminal became operational.

FIGURE 9. Baseline aliphatic hydrocarbon concentrations (parts per million by dry mass of sediment) in Nigg Bay sediments, before commissioning of the Nigg Oil Terminal.

SUBLITTORAL STUDIES

A series of biological, physical and chemical investigations have been carried out since 1977 at the Beatrice Field by the Field Studies Council, Oil Pollution Research Unit and M-Scan Ltd. These are of particular relevance to the fishery interest on the Smith bank, and the surveys are designed to monitor the seabed community and the effect of discharges from the Beatrice platforms. Pre-production seabed ecological surveys in 1977, 1980 and 1981 used extensive replicate sampling from grids of stations over a wide area. This permitted a detailed description of the natural seabed community (Hartley & Bishop 1986). No trends or anomalies were detected that could be attributed to industrial activity. However, the grid stations were all more than 500 m from the platform so it was clear that a study of the effects of drilling discharges would need to involve a different sampling strategy which included stations very close to the platforms. These more recent studies which began in 1982 have described the effects of discharges of cuttings contaminated with oil-based drilling mud (Addy *et al.* 1984). The gradients of sediment hydrocarbon concentration, and the species succession away from the platform are similar to those found in the other North Sea fields (see, for example, Davies *et al.* (1984); Kingston this symposium). These effects, which are consistent with organic enrichment are illustrated in figures 10 and 11 which show the temporal and spatial changes in seabed chemistry and macrobenthic diversity between 1982 and 1983. The effects of Beatrice drilling discharges are very localized and are expected to be largely temporary.

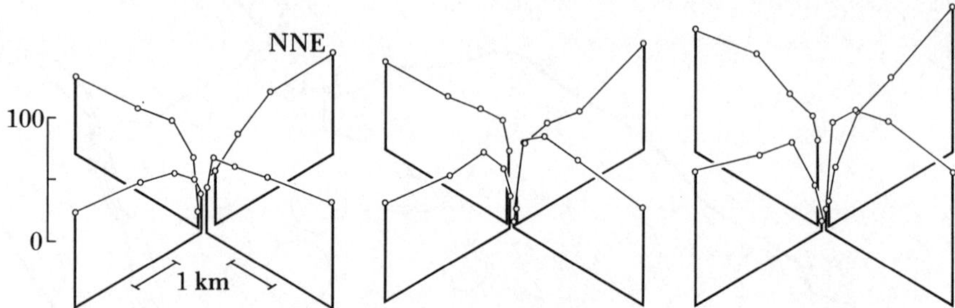

FIGURE 10. Total number of taxa recorded from two grab samples (each 0.1 m^2) at each station in February 1982 (left), May 1982 (centre) and September 1983 (right). (Addy *et al.* 1984.)

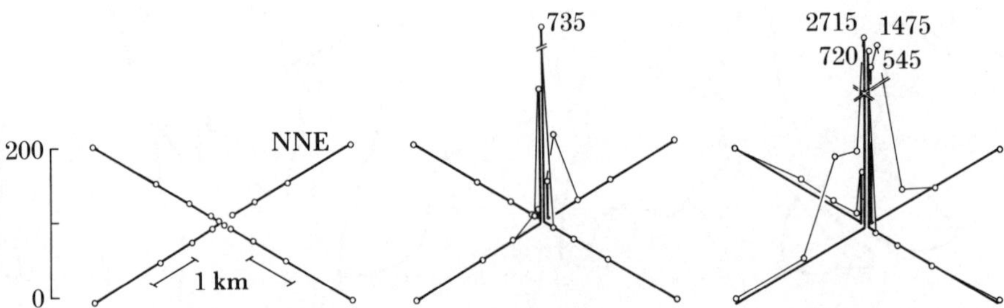

FIGURE 11. Estimated number of *Capitella capitata* per square metre at each station in February 1982 (left), May 1982 (centre) and September 1983 (right). (Addy *et al.* 1984.)

The development drilling phase at Beatrice is now completed and future monitoring, using the same sampling strategy, will follow successional changes and recovery in seabed biology now that the discharge of organically-rich cuttings has ceased.

OTHER STUDIES

Various supplementary studies are done from time to time, usually in relation to specific aspects of operations. For example, air-quality and noise surveys have been performed at Nigg. Sediment hydrocarbon levels are regularly monitored around the terminal effluent discharge point. This provides a check on the receiving environment, complementing the routine chemical analyses done by Britoil plc and the Highland River Purification Board on the discharge itself. The plant and animal communities living on the platform structures in the Beatrice field are surveyed routinely for structural monitoring purposes (Picken 1986). These communities may also give an early indication of any adverse environmental conditions at the centre of operations.

The future

The monitoring programme is not fixed in its scope or methodology. The studies outlined above represent the main parts of the programme to date, but it is emphasized that sufficient flexibility has been built in to enable modifications or special studies in the future. To a large extent the programme has evolved in parallel with the Beatrice development. In the early days the emphasis was on acquisition of descriptive baseline data from extensive biological surveys. Having provided the necessary basic descriptive information, the work in recent years has concentrated on monitoring either selected communities or habitats, or specific operational discharges. This change in emphasis can be illustrated with reference to the offshore seabed investigations, where there has been a shift from extensive wide ranging grids of stations to radiating transects with a concentration of stations very close to the platform to define the effects of drilling discharges.

Similarly, in the case of seabirds, the two-year programme investigating seabirds at sea has provided a wealth of useful descriptive and distributional data but the main emphasis of the ornithological studies is now on the colonies themselves and the populations of wintering sea duck and shorebirds.

It is recognized that the information gathered is frequently of a wider interest from a conservation planning point of view or for its intrinsic academic interest, and publication of results is encouraged. Britoil plc also recognizes the importance of presenting the work in a form readily understandable to a wider audience. To this end, a public relations summary of the programme has been produced and is freely available to schools and any other interested parties, particularly on the Moray Firth coastline.

Conclusion

The various component studies briefly outlined above which make up the Beatrice environmental monitoring programme have concluded that the presence of the platforms, pipeline and terminal have had only very localized effects on the surrounding environment. Long-term monitoring continues as an integral part of the Beatrice development. The studies to date have shown that it is possible to operate an oilfield in a very sensitive coastal area without damaging the marine or coastal resources. For detailed information on any part of the programme, those interested are referred to the published information, or alternatively invited to contact Britoil plc direct.

References

Addy, J. M., Hartley, J. P. & Tibbetts, P. J. C. 1984 Ecological effects of low toxicity oil-based mud drilling in the Beatrice oilfield. *Mar. Pollut. Bull.* **15**, (12), 429–436.

Campbell, L. H., Barrett, J. & Barrett, C. F. 1986 Seaducks in the Moray Firth: a review of their current status and distribution. *Proc. R. Soc. Edinb.* B. **91**, 105–112.

Campbell, L. H., Standring, K. T. & Cadbury, C. J. 1978 Firth of Forth oil pollution incident February 1978. *Mar. Pollut. Bull.* **9**, 335–339.

Davies, J. M., Addy, J. M., Blackman, R., Blanchard, J. R., Ferbrache, J. E., Moore, D. C., Somerville, H. J., Whitehead, A. & Wilkinson, T. 1984 Environmental effects of oil based mud cuttings. *Mar. Pollut. Bull.* **15**, (10), 363–370.

Hartley, J. P. & Bishop, J. D. D. 1986 The macrobenthos of the Beatrice Oilfield, Moray Firth, Scotland. *Proc. R. Soc. Edinb.* B. **91**, 221–245.

Mudge, G. P. 1986 Trends of population change at colonies of cliff-nesting seabirds in the Moray Firth. *Proc. R. Soc. Edinb.* B **91**, 73–80.

Mudge, G. P., Aspinall, S. J. & Crooke, C. H. 1985 A photographic study of seabird attendance at Moray Firth colonies outside the breeding season. Report to Britoil plc by the Royal Society for the Protection of Birds.

Mudge, G. P. & Crooke, C. H. 1986 Seasonal changes in the numbers and distributions of seabirds at sea in the Moray Firth, northeast Scotland. *Proc. R. Soc. Edinb.* B **91**, 81–104.

Picken, G. B. 1986 Moray Firth marine fouling communities. *Proc. R. Soc. Edinb.* B **91**, 213–220.

Rafaelli, D. & Boyle, P. R. 1986 The intertidal macrofauna of Nigg Bay. *Proc. R. Soc. Edinb.* B **91**, 113–141.

Ralph, R. (ed.) 1986 The marine environment of the Moray Firth. *Proc. R. Soc. Edinb.* B **91**, i–vii, 1–358.

Symonds, F. L. & Langslow, D. R. 1986 The distribution and local movements of shorebirds within the Moray Firth. *Proc. R. Soc. Edinb.* B **91**, 143–167.

Tasker, M. L., Jones, P. H., Dixon, T., & Blake, B. F. 1984 Counting seabirds at sea from ships: a review of methods employed and a suggestion for a standardized approach. *Auk* **101**, 567–577.

Terry, L. A. & Sell, D. 1986 Rocky Shores in the Moray Firth. *Proc. R. Soc. Edinb.* B **91**, 169–191.

Phil. Trans. R. Soc. Lond. B **316**, 669–677 (1987)
Printed in Great Britain

Summary and conclusions: environmental effects of North Sea oil and gas developments

BY R. B. CLARK

Department of Zoology, The University of Newcastle upon Tyne NE1 7RU, U.K.

This Royal Society Discussion Meeting has examined the total environmental impact of a whole industry in a single geographical area. Land-based developments related to the exploitation of the North Sea oilfields and their social consequences have been substantial, although neither the worst fears nor the best hopes have been realized. An accommodation has been reached with the fishing industry in the affected area.

Offshore platforms are a source of chronic pollution from production water, but in recent years there has been a marked increase in the use of oil-based drilling muds and it is estimated that 20 Mt per year of petroleum hydrocarbons are added to the sea in oil-contaminated drill cuttings. The effect of these additions has been studied in the laboratory, in mesocosms and in field surveys which, together, yield a consistent picture. Within a radius of a few hundred metres of a platform there is impoverishment of the benthic fauna. Close to the platform the production of anoxic conditions through smothering and the activity of sulphide-producing bacteria is probably more significant than the toxic effect of the oil-based muds. Outside this immediate zone of impact, the oil results in organic enrichment and enhanced populations of some of the fauna. The total area affected is, in the context of the North Sea, minuscule. There is no evidence that plankton is materially affected and the success of commercial fisheries dependent upon the plankton crop is more influenced by fishery practices than by any other factor.

Seabird populations, about which there was formerly much concern, have not so far been affected by oil pollution in the North Sea. There is wide fluctuation in recruitment success, but populations of species thought most vulnerable to oil pollution are generally increasing.

Although marine pollution research has yielded valuable insights into the responses of individuals, populations and communities to perturbation, natural as well as man-made, it is not likely that future problems associated with oil extraction from the sea will be as stimulating to fundamental research. Different problems relating to environmental pollution should now be addressed by marine scientists.

INTRODUCTION

Marine pollution did not suddenly break upon the world with the wreck of the *Torrey Canyon* in 1967, but it stimulated widespread concern about oil pollution and resulted in a large research effort into the consequences of all discharges to the marine environment, whether accidental or deliberate.

Nineteen years later, marine pollution studies have come of age and during that time there have been innumerable conferences on almost all conceivable aspects of marine or any other kind of pollution. This Discussion Meeting has been a little different from its predecessors because it was concerned with the total environmental impact of a whole industry in a particular geographical area – the North Sea.

The development of oil and gas extraction in the North Sea, described by Larminie, has been extensive and surprisingly rapid. Gas extraction started in 1965 and the first oil production platforms were installed in 1969. A chain of oilfields down the centre of the northern half of the North Sea has been exploited and now there is the prospect of other oilfields closer inshore and to the west of Scotland coming into production. All of this has taken place in a most hostile environment and unprecedented water depths, and has demanded new technology for it to have been possible. It is a testimony to the skill and science of the operators that it has been accomplished with so few accidents.

These developments have taken place during a period when there was alarm and a good deal of ignorance about the consequences of marine pollution, and the oil industry was under pressure to severely limit, if it could not completely avoid, environmental damage. Meeting the public demand for 'clean' technology has resulted in the development of a range of monitoring techniques of increasing sensitivity and subtlety to measure environmental change as well as techniques to contain and deal with spilled oil that minimize environmental damage.

The development of sophisticated measures of environmental change has been the work of those hostile to industrial activity in the sea as much as those involved in it, and as much in relation to other pollutants as to oil. Apart from its immediate and utilitarian purpose, this development has been of general benefit to marine ecological research. It has led to a study and better understanding of natural change in marine ecosystems, which proves to be a very noisy background against which to detect a faint pollution-induced signal. Whether or not it is necessary to have such an acute early warning system for pollution effects is a matter I shall not discuss here, but the striving for greater sensitivity in monitoring techniques has resulted in a better understanding of the responses of marine organisms and marine ecosystems to stresses of all kinds, natural as well as man-made. It is notable, too, that natural change can be as great as, and of a similar nature to, pollution-induced change and this has helped set marine pollution into perspective.

Interactions with human affairs

The North Sea is a multipurpose resource. In addition to oil- and gasfields, it has exploitable sand and gravel beds, supports one of the most productive fisheries in the world, and in the south includes some of the most heavily trafficked sea-lanes in the world. It receives the wastes of the largely urbanized and industrialized populations of 31 million people living around it, together with the large influx of summer visitors to the coast, particularly on its eastern side, and a substantial input from the major rivers entering it (Clark 1986). As Eisma showed, it has complex water movements and is far from being a homogeneous body of water. There are equally complicated seasonal movements of fish and of seabirds for breeding and winter feeding. The various human activities, commercial, domestic and recreational, interact with each other and also with these biological events.

It is in this web of interacting forces that the development of North Sea gas and oilfields has taken place. But as Johnston reminded the meeting, its inevitable environmental impact has not been simply at sea and around offshore platforms. A small fleet is required to supply the platforms; the oil has to be brought ashore mainly by pipelines which have been laid on or trenched into the seabed; and ashore there is the need for oil terminals, refineries, harbour facilities, construction yards for the platforms, housing and the rest of a widespread infrastructure.

It is, of course, impossible for a major industrial development to take place without an impact on the human and natural environment, and the impact may take unexpected forms. One of the major preoccupations of the Zetland County Council is the economic boom in the Shetland Islands resulting from the construction of the Sullom Voe oil terminal and the damage this may do to the traditional way of life in the islands, particularly when the boom passes. On the whole, though, it can be claimed that the land-based operations connected with the offshore developments of the industry have been accomplished without undue friction and without the worst fears or some of the hopes being realized.

Offshore, the chief interaction has been with the fishing industry, and here too, it can be claimed that an accommodation between the two has been reached. The scope for interaction between the oil industry and other users of the sea will be much increased if or when oil extraction moves into shallower inshore water, and a very sensitive approach will be needed if such developments are to proceed harmoniously.

Inputs

The concern of the Discussion Meeting has been much more with the impact on the natural environment of oil and petroleum hydrocarbons. Estimates in 1982–83 put the annual oil input to the North Sea between 100 and 170 Mt (Bedborough *et al.*) and the sources are the atmosphere, rivers, land run-off and general shipping as well as activities of the oil industry itself. By far the greatest input of petroleum hydrocarbons in the sea is from rivers, land run-off and the atmosphere, but these are diffuse inputs, whereas those from oil industry operations are for the most part at fixed sites where the input is chronic and therefore more likely to cause local environmental deterioration.

At one time, the chief concern was about oil pollution from tanker operations, but recently attention has focused on offshore drilling activities, and these now represent a major input of petroleum hydrocarbons to the North Sea. With the increased use of oil-based drilling muds, the input from this source has risen to about 20 Mt per year and until low toxicity oils were phased in after 1982, this was largely in the form of diesel with its greater potential to cause environmental damage.

Sublethal signals

It is, of course, an advantage to have early warning of potentially damaging situations, and for this an array of toxicity tests and studies of the sublethal effects of xenobiotics on a range of animals has been developed.

In the early days of marine pollution research, the approach was naïve and undue faith was placed in the LD_{50} test and the existence of pollution indicator species. Both were crudely misapplied and had little predictive value for the natural environment outside the laboratory. But these studies have now reached a new level of sophistication, not least through the work of M. N. Moore and his colleagues at the Institute for Marine Environmental Research.

Their research on molluscs has shown that changes in response to exposure to petroleum hydrocarbons can be detected at the molecular, subcellular and tissue levels. There is induction of the detoxication mechanisms involving the cytochrome P-450 monooxygenase system, destabilization of the lysosome system and, at the tissue level, atrophy of the epithelium of the digestive tubules. These changes have been confirmed in realistic laboratory experiments, in mesocosms, and in the field following actual oil spills. The relevance of these sublethal phenomena to pollution (in the strict sense of that term that the effects are damaging) is firmly established because the metabolic demand of these responses is reflected in a reduced scope for growth and in reduced fecundity with possible implications for recruitment and ultimate population size.

Studies of the sublethal effects of xenobiotics have now matured to the point that they can give early warning of pollution damage and a measure of the recovery of individual organisms from it. Beyond this utilitarian value, the whole field of investigation has proved to be of exceptional scientific interest in giving an insight into the defensive physiological machinery of marine invertebrates to environmental perturbations. It is one of the fairly rare instances in which applied research has had a tremendous scientific spin-off.

Experimental ecology

As a bridge between laboratory and field studies, mesocosms are attractive because they offer greater realism than can be achieved in the laboratory and greater control than is possible in field investigations. The earliest use of mesocosms was in large enclosures for the study of plankton. In their application to pollution research it appeared that while these experiments were very informative about what happened in 'big bags', they were not very relevant to the real environment. Mesocosm experiments involving benthic communities appear to be somewhat more informative, particularly when interpreted in conjunction with the results of field investigations.

Two such studies were reported to the Discussion Meeting. One by Leaver *et al.* was concerned with the impact of drill cuttings and oil-based drilling muds on meiofauna, the other by Gray on the effect of water-accommodated fractions of diesel on a hard substratum intertidal community and a soft bottom subtidal community. Both investigations are long-term; Leaver's experiments have continued for 15 months and Gray's for over 2 years. Both include impact and recovery phases of the ecosystem response.

In the Leaver *et al.* mesocosms, there was a steady decline in meiofaunal abundance even in the control mesocosm. It is not known if this was a consequence of the experimental conditions inherent in the use of enclosures or to natural seasonal fluctuations in the abundance of elements in the meiofaunal community. If the latter, it presents a serious problem in the design of any long-term experiments. At the very least, it represents an important uncontrollable variable which needs to be understood, but seasonal studies of the meiofauna on the Grangemouth intertidal mudflats by C. G. Moore *et al.* begin to give some insight into this.

Next, it is not clear, even when congenial physical and chemical conditions are restored after disturbance, if recovery of the meiofauna is dependent upon recolonization from outside the affected area. This may be a feature of the natural environment but is clearly impossible in mesocosms. Furthermore, in these experiments it is not certain if some of the erratic peaks in recovery of the nematode populations are due to a comprehensive recovery of the community or to the increase in numbers of a single species at a peak time of reproduction.

Gray also had some difficulty in providing satisfactory controls for his rocky intertidal mesocosms, but that is because no rocky shore community is homogeneous and exact replication in each experiment is impossible. Because his mesocosms are not entirely closed, recruitment from unaffected areas is possible for at least some species, and he was able to quantify this for *Littorina*. On the whole, these experiments with macrofauna and flora appear not to have suffered unduly from the uncertainties usually associated with mesocosms, and this may be because attention is focused on a group of species whose biology and interactions are much better known than those of meiofaunal and planktonic communities.

Having noted these problems and uncertainties, the mesocosm experiments do allow some conclusions to be drawn. Deposition of drill cuttings an oil-based drilling muds, whether diesel-based or of the low-toxicity oils that have replaced diesel, have a number of effects which can be separated.

The deposition of drill cuttings, whether oil-contaminated or not, smother the seabed, cut off the substratum from water exchange, reduce the redox potential and increase the soluble sulphide level. The important role of micro-organisms in this sequence of events is explained in the report of Sanders & Tibbetts. This is associated with a substantial fall in the nematode population and the elimination of burrowing copepods, although epibenthic copepods appear able to survive this smothering. If the cuttings are contaminated with diesel or high concentrations of low-toxicity oil, the nematode population continues to decline over the year following deposition of the cuttings, but at low concentrations of low-toxicity oil, shows an erratic recovery. In these circumstances, epibenthic copepods show an enhanced population, perhaps because of organic enrichment by the low toxicity oil. There is also confirmatory evidence of this from field studies in the Beryl field (C. G. Moore *et al.*).

Gray's mesocosms also reveal varied responses of the communities exposed, this time, to high or low dosages of diesel. Attached algae showed little response except, ominously a reduced growth rate of *Laminaria* and *Ascophyllum* during their second year of exposure. This has obvious implications for the alginate industry which harvests these seaweeds. The green algae *Chondrus* and *Ulva*, on the other hand, show enhanced growth.

Of the animals on the rocky substratum, *Littorina* suffered increased mortality at the higher concentration of diesel and *Mytilus* failed to secrete byssus threads, became detached from the rocks and were eaten by crabs. This leads to the intriguing suggestion that a steady dribble of oil from offshore platforms might be beneficial in inhibiting marine growth on the platform legs.

In the soft-bottom mesocosms, mobile species such as amphipods and ophiuroids become less abundant, as do filter-feeders and carnivores. Surface and subsurface deposit feeders on the other hand appear to be less affected, although Leaver *et al.* observed a heavy immediate mortality of the bivalve *Tellina* when drill cuttings contaminated with diesel or a high concentration of low-toxicity oil were added to their mesocosms.

Proof of the pudding

In many ways, it has always seemed preferable to study pollution effects in the real environment than to extrapolate from the apparently more precise results of toxicity tests and microcosm or mesocosm experiments. They can never reflect the compensatory mechanisms that exist in populations and communities in the natural environment. But although there has been no shortage of monitoring programmes of oil pollution effects, they have revealed the familiar problems that beset measuring change in marine ecosystems and relating it to man-made influence. If the pollution is major and from a known source, such as the wreck of an oil tanker on the coast, it is hardly necessary to mount an expensive monitoring exercise to measure damage and assign the cause, even though that is not as certain as might appear at first sight. When the pollution is small, the effects it may have are hard to identify with certainty because of the complexity of the natural environment. Hence the attraction of controlled experiments.

Most, if not all, marine environments show temporal and spatial inhomogeneity and the sampling protocol must be suitably designed to take account of it. This is expensive and time-consuming and in the past, few monitoring programmes have been on sufficient geographical or time scale to distinguish natural from man-made events. In the 4–5 years after the opening of the five refineries around the Milford Haven oil terminal, there was no successful recruitment there of the shore gastropods *Gibbula* and *Monodonta*. As clear a case of cause and effect as one might wish for, except that there was poor or no recruitment of these species in those years on the entire west coast of Britain to the northern limit of their geographical range (Lewis 1982). Few monitoring programmes look so far afield.

Over the past 15 years, considerable effort has been devoted to studying the effects of oil pollution in the natural environment in most situations and forms in which it occurs. At this meeting, attention has been directed chiefly to the offshore platforms in the North Sea.

C. G. Moore *et al.* have examined the meiofauna around the Beryl A platform and also in intertidal mudflats near the Grangemouth refinery, and their results need to be read in conjunction with the mesocosm experiments of Leaver *et al.* Sanders & Tibbetts have studied the effects of drill cuttings on microbial populations at two unnamed offshore installations, and Kingston has taken a comprehensive view of the impact of drill cuttings and the oil-based drilling muds with which they are contaminated on the benthic macrofauna at a variety of North Sea oilfields.

It is difficult from these observations to separate the toxic effects of the oil-based drilling muds from those of the activity of sulphur-reducing bacteria, leading to the production of sulphides, in areas where the seabed is blanketed by drill cuttings and organically enriched by the presence of oil-based muds whether these contain diesel or various formulations of low-toxicity oils.

Close to the platform where the greatest depth of cuttings accumulates, blanketing of the seabed is severe, but a short distance away (50–250 m) where cuttings are 10 cm deep, there is the greatest activity of sulphur-reducing bacteria and production of sulphides. Around the Beryl A platform, meiofauna is impoverished for a distance of 800 m (C. G. Moore *et al.*) and, viewing North Sea oilfields generally, the distance from the platform over which there is an impact on the benthic macrofauna is about 500 m (Kingston). It is anticipated that the impacted area would be smaller around platforms where there are strong bottom currents and dispersion of the cuttings contaminated with oil-based muds would be greater, as in the southern North Sea.

Degradation of the oil in drilling muds results in organic enrichment of the substratum. Moore & Murison reported an enhanced population of epibenthic copepods around the Beryl A platform in 1985, although not in 1984, and they attributed this to the degradation of diesel in the oil-based muds used there until 1982 after which low toxicity muds were employed. No such enhancement of nematode populations has occurred and biodegradation and release is presumably faster at the surface of the substratum than in the less oxygenated layers below, so that surface dwelling animals (epibenthic copepods) might be expected to benefit earlier than animals living deeper in the substratum (nematodes).

The rate at which diesel and low-toxicity muds degrade is uncertain. From the microbiological studies of Sanders & Tibbetts, it appears that diesel is degraded faster than low-toxicity oils with a high boiling point, but slower than low-toxicity oils with a low boiling point. From Kingston's study of a large number of North Sea oilfields it appears, surprisingly, that Statfjord A and Brent which have used diesel-based drilling muds show more enhancement of the opportunistic benthic macrofauna than at Beatrice which has used low-toxicity muds.

An important finding of Kingston's survey was the extreme patchiness of drill cuttings around platforms. This has obvious implications for any sampling programme and accounts for extraordinarily high and anomalous levels of contamination found in some surveys around platforms. If routine monitoring of contamination levels is required by law, the objectives of the monitoring programme should be explicit and the monitoring schedule designed appropriate to satisfy those objectives (Segar & Stamman 1986). Current legal requirements do not meet this need.

Oil discharges do not affect only animals living in the seabed, but are particularly likely to affect the plankton living in surface waters. Variations in the standing crop of plankton in the North Sea are exceptionally well known thanks to the continuous plankton survey carried out over many years by the Institute for Marine Environmental Research. Attempts have been made to correlate these with climatic and natural water quality changes. As Reid showed, there is nothing to implicate petroleum hydrocarbons with these fluctuations. Although, of course, there remain uncertainties and it is possible to conceive of circumstances in which oil might have some consequence for plankton production, this does not seem a serious risk.

Interest in plankton is primarily because the success of commercial fisheries depends critically on the abundance of plankton. Fisheries have not been seriously discussed, apart from the exclusion of fishermen around platforms and the damage trawls may cause to exposed pipelines, or vice versa (Johnston). It has long been known that the success of a fishery depends more on the practices of the fishermen themselves than on any other human activity.

For many years, the chief concern about oil pollution was its impact on seabirds. It has a very visual impact through the press and television and it produces a strong public emotional reaction. By now, as Dunnet showed, we know that the recruitment of seabirds, like most other maritime organisms, is erratic. Many seabird populations have increased in recent years for unknown reasons and despite losses from oil pollution. Ornithologists are more concerned about the impact of intensive fishing on the food supply of seabirds than the losses from oil pollution, which distressing though they may be in the public eye, are biologically insignificant.

The development of the Beatrice oilfield in the Moray Firth generated considerable alarm because it was the closest to shore of any of the North Sea oilfields. It is close to internationally important seabird colonies and winter feeding grounds, and within the area of economic fisheries. Because of the sensitivity of the area, comprehensive monitoring programmes have been established there (Addy). This survey has revealed the seasonal movements of seabirds in the Moray Firth and may show the interaction of economic fisheries with seabird populations, but happily has not shown any impact of the development of this inshore oilfield except in the immediate vicinity of the platforms. It has, however, done much to reassure local public opinion about the oil industry operations in the Moray Firth, which appears to have been the main purpose of the exercise.

It is scarcely an exaggeration to suggest that seabirds have caused more problems for the oil industry than any other factor. It was the picture of oiled guillemots coming ashore on the

Cornish coast after the wreck of the *Torrey Canyon* that did most to arouse public opinion to the effects of oil pollution and this impression has been reinforced by television and the press at every oil spillage since then.

Tanis & Mörzer-Bruijns (1962) estimated that between 250000 and 750000 birds per year, mainly guillemots, were lost in the northeast Atlantic and North Sea from oil pollution. That may be an order of magnitude too high, but is still an impressive figure. It was thought that these seabirds with their very low reproductive rate could not sustain such continuing losses, and the decline of the breeding colonies in Brittany, France, and southwest England appeared to confirm these fears.

With the development of the North Sea oilfields and the increased traffic of oil by sea, northern seabird colonies were expected to be under greater threat. There was also concern that migratory landbirds crossing the North Sea would be attracted to gas flares and killed in them. Happily, neither of these fears proved justified. Censuses of breeding colonies of guillemot and razorbill around British coasts in 1969 and again in 1974 showed that while fringe colonies at the southern end of their geographical range in Brittany, Cornwall and southwest Wales were certainly declining, those in Scotland were, in many cases, increasing in size (N.E.R.C. 1977).

The underlying reason for this change is probably climatic, but as Dunnet has shown, the extent of seabird population fluctuations and their immediate causes are certainly complex. Public concern about seabirds has led the oil industry to be particularly wary of inflicting damage on them, but development of the North Sea oilfields has had no discernible impact; the development of industrial fishery practices has probably had more by depleting the food resources of the birds.

Conclusions

The past 20 years have seen a large number of meetings, symposia and conferences on matters relating to oil pollution. There have been many alarms and excursions, but they have faded from sight or fallen into perspective. To a great extent we now have the measure of oil pollution. Refinery waste discharged across the foreshore or into shallow water causes progressive environmental deterioration in that area. A tanker wrecked on the coast causes immense local damage, but restoration is more a logistical one of disposing of oily waste than a biological one. The operation of offshore platforms results in a disturbance of the benthic fauna, but only within a few hundred metres and in the context of the North Sea the total area affected is minuscule. Oil pollution has had no significant impact on fisheries or seabird populations.

In this case, it is easy to be complacent and suggest that it is hardly worth investigating oil pollution any further. But technology does not stand still and new developments in the industry will always need fresh evaluation if their environmental impact is to be contained. Dispersants in use 20 years ago were more toxic than the oil they were meant to disperse; once this fact had been explored, it proved possible to introduce equally effective low-toxicity dispersants. In the same way, the rapid increase in the use of oil-based drilling muds in North Sea oilfields led to an evaluation of their impact on the environment, and this has been reduced by the introduction of low-toxicity oil-based muds. No doubt as oil extraction moves closer inshore, further environmental problems will be revealed and it will be necessary to investigate them and reduce them.

The sort of investigation required to satisfy this need will generally be *ad hoc*, target-oriented and narrow in scope. What place has fundamental science in this? The final session of the United States National Academy of Sciences review of oil in the sea, in 1981, was devoted to the identification of future research needs. It is fairly easy to list areas of incomplete understanding and gaps in our knowledge if the objective is to account for the transport, degradation and fate of all the components of a crude oil, and of all the degradation products; to account for the passage and metabolism of these substances in marine food webs, their sublethal effects on plants and animals, and the responses of marine populations and ecosystems to these sublethal as well as lethal effects of petroleum hydrocarbons. The list is unending. Much of the work would be of trivial scientific interest and for the most part would have no practical significance for the activities of the oil industry.

Papers presented at this meeting have contributed much new information but few surprises. If anything, they have confirmed the earlier view that oil pollution has a detectable but very localized impact, and that the impact can be contained within acceptable limits; or if not acceptable, that the technology exists to reduce the impact – at a price. Perhaps we should conclude that oil pollution research is no longer justified and that it can no longer give useful direction to fundamental marine research. There are more important challenges for fundamental marine science to address, and if it is felt necessary to respond to pressures to do 'useful' rather than 'pure' science, there are more urgent pollution problems which cannot be solved without considerable advances in fundamental scientific understanding, such as acid precipitation or carbon dioxide and the 'greenhouse effect', both of which may well have an important but unquantified marine dimension.

One argument that has not been seriously addressed at this meeting is the awkward one advanced in some quarters that 'no detectable effect' of an input to the sea cannot be equated with 'no effect'. The dispersion and dilution of pollutants may avoid local damage or reduce it to a trivial level, but the waste has nevertheless been added to the North Sea, causing, it is claimed, a widespread if less conspicuous debilitation of marine ecosystems. It is difficult to answer this claim on scientific grounds, but it reflects an attitude which underlies proposed European Community directives and requries consideration if not an answer. In practical terms, that may prove a more intractable problem for marine scientists than any we have discussed at this meeting.

References

Clark, R. B. 1986 *Marine pollution*. Oxford University Press.

Lewis, J. R. 1982 The composition and functioning of marine ecosystems in relation to the assessment of long-term effect of oil pollution. *Phil. Trans. R. Soc. Lond.* B **297**, 257–267

N.E.R.C. 1977 *Ecological research on seabirds*. N.E.R.C. Publ. Ser. C, no. 18. London: Natural Environment Research Council.

Segar, D. A. & Stamman, E. 1986 Fundamentals of marine pollution monitoring programme design. *Mar. Pollut. Bull.* **17**, 194–200.

Tanis, J. J. C. & Mörzer-Bruijns, M. F. 1962 Het onderzoek naar Stookolievogels van 1958–1962. *Levende Nat.* **65**, 133–140.